Springer
Proceedings in Physics 48

Springer Proceedings in Physics

Managing Editor: H. K. V. Lotsch

30 *Short-Wavelength Lasers and Their Applications* Editor: C. Yamanaka

31 *Quantum String Theory*
Editors: N. Kawamoto and T. Kugo

32 *Universalities in Condensed Matter*
Editors: R. Jullien, L. Peliti, R. Rammal, and N. Boccara

33 *Computer Simulation Studies in Condensed Matter Physics: Recent Developments*
Editors: D. P. Landau, K. K. Mon, and H.-B. Schüttler

34 *Amorphous and Crystalline Silicon Carbide and Related Materials*
Editors: G. L. Harris and C. Y.-W. Yang

35 *Polycrystalline Semiconductors: Grain Boundaries and Interfaces*
Editors: H. J. Möller, H. P. Strunk, and J. H. Werner

36 *Nonlinear Optics of Organics and Semiconductors*
Editor: T. Kobayashi

37 *Dynamics of Disordered Materials*
Editors: D. Richter, A. J. Dianoux, W. Petry, and J. Teixeira

38 *Electroluminescence*
Editors: S. Shionoya and H. Kobayashi

39 *Disorder and Nonlinearity*
Editors: A. R. Bishop, D. K. Campbell, and S. Pnevmatikos

40 *Static and Dynamic Properties of Liquids*
Editors: M. Davidović and A. K. Soper

41 *Quantum Optics V*
Editors: J. D. Harvey and D. F. Walls

42 *Molecular Basis of Polymer Networks*
Editors: A. Baumgärtner and C. E. Picot

43 *Amorphous and Crystalline Silicon Carbide II: Recent Developments*
Editors: M. M. Rahman, C. Y.-W. Yang, and G. L. Harris

44 *Optical Fiber Sensors*
Editors: H. J. Arditty, J. P. Dakin, and R. Th. Kersten

45 *Computer Simulation Studies in Condensed Matter Physics II: New Directions*
Editors: D. P. Landau, K. K. Mon, and H.-B. Schüttler

46 *Cellular Automata and Modeling of Complex Physical Systems*
Editors: P. Manneville, N. Boccara, G. Y. Vichniac, and R. Bidaux

47 *Number Theory and Physics*
Editors: J.-M. Luck, P. Moussa, and M. Waldschmidt

48 *Many-Atom Interactions in Solids*
Editors: R. M. Nieminen, M. J. Puska, and M. J. Manninen

49 *Ultrafast Phenomena in Spectroscopy*
Editors: E. Klose and B. Wilhelmi

50 *Magnetic Properties of Low-Dimensional Systems II: New Developments*
Editors: L. M. Falicov, F. Mejía-Lira, and J. L. Morán-López

51 *The Physics and Chemistry of Organic Superconductors*
Editor: G. Saito and S. Kagoshima

Volumes 1 – 29 are listed on the back inside cover

Many-Atom Interactions in Solids

Proceedings of the International Workshop, Pajulahti, Finland, June 5–9, 1989

Editors: R. M. Nieminen, M. J. Puska, and M. J. Manninen

With 117 Figures

Springer-Verlag
Berlin Heidelberg New York London
Paris Tokyo Hong Kong Barcelona

Professor Risto M. Nieminen
Dr. Martti J. Puska
Laboratory of Physics, Helsinki University of Technology, SF-02150 Espoo, Finland

Professor Matti J. Manninen
Department of Physics, University of Jyväskylä, SF-40100 Jyväskylä, Finland

ISBN-13: 978-3-642-75755-6 e-ISBN-13:978-3-642-75753-2
DOI:10.1007/978-3-642-75753-2

Softcover reprint of the hardcover 1st edition 1990

2154/3150(3011)-543210 – Printed on acid-free paper

Preface

This book contains the proceedings of the international workshop on *Many-Atom Interactions in Solids*, which was held June 5–9, 1989, in Pajulahti, Finland. The purpose of the workshop was to bring together physicists, chemists and materials scientists working in the field of interatomic interactions and their applications in computer simulations of condensed matter. The workshop attracted a good fraction of the active groups in the field, and created lively discussion and interchange of ideas.

The contributions in this volume have been grouped by the editors into review-type articles and more specific applications to different topics. The order of the articles does not follow the order of the presentations in the workshop.

The editors wish to express their gratitude first of all to all the workshop participants for creating an enjoyable and fruitful workshop and to the contributors for their efforts in putting together these proceedings. We hope that this volume will be a useful resource for practitioners in and newcomers to this exciting field. We would like to thank Jens Nørskov for his help in planning the scientific programme and Eija Järvinen for taking care of most of the practical arrangements of the workshop. The workshop was made possible by financial support from the Finnish Ministry of Education, the Research Institute for Theoretical Physics (Helsinki), NORDITA (Copenhagen), and Helsinki University of Technology.

Espoo, November 1989

Risto M. Nieminen
Martti Puska
Matti Manninen

Contents

Part I General Reviews

Introduction to Many-Atom Interactions in Solids
By M. Manninen, R.M. Nieminen, and M.J. Puska 2

Volume and Pair Forces in Solids and Liquids
By V. Heine and J. Hafner (With 4 Figures) 12

The Effective-Medium Theory
By K.W. Jacobsen (With 6 Figures) 34

The Embedded Atom Method: A Review
By M.S. Daw (With 3 Figures) 48

From Exact to Approximate Theory: The Tight Binding Bond Model and Many-Body Potentials
By D.G. Pettifor (With 5 Figures) 64

Implications of the Embedded-Atom Method Format
By R.A. Johnson (With 3 Figures) 85

Molecular Dynamics Simulations of Materials Phenomena
By U. Landman, W.D. Luedtke, and R.N. Barnett (With 13 Figures) . . . 103

Atoms Embedded in Electron Gas
By M.J. Puska (With 5 Figures) 134

The Structural and Electronic Properties of Molten Metals, Semimetals, and Semiconductors
By J. Hafner (With 7 Figures) 144

Part II Metals

First-Principles Interatomic Potentials in Transition Metals: Multi-ion Interactions and Their Analytic Representation
By J.A. Moriarty (With 4 Figures) 158

The Effective-Medium Theory Beyond the Nearest Neighbour Interaction
By H. Häkkinen (With 1 Figure) 168

Defect Migration in Perfect Lattices
By R. Thetford (With 7 Figures) 176

Dynamic Simulations of Dislocation Core Structures in Gold Using Many-Atom Interactions
By J. von Boehm and R.M. Nieminen (With 4 Figures) 191

One-Electron Energy Corrections in the Effective-Medium Theory: Application to Copper Clusters
By S. Myllyrinne, M. Manninen, and J.K. Nørskov (With 2 Figures) ... 199

Dynamic Methods for Electronic Problems
By R. Virkkunen and R.M. Nieminen (With 3 Figures) 204

Part III Semiconductors and Covalent Materials

Simulated Annealing Studies of Sulphur and Selenium Clusters
By R.O. Jones and D. Hohl (With 6 Figures) 214

Effective-Medium Theory: Considering Corrections and Delimitations
By A. Zwartkruis and J.E. van Himbergen (With 5 Figures) 232

Improved Tight Binding Parameters for Silicon
By L. Goodwin and A.J. Skinner (With 4 Figures) 242

A Pair Potential Model of Covalent Bonding in Silicon
By G. Ackland (With 4 Figures) 249

Derivation of Angular Forces for Semiconductors and Transition Metals
By A.E. Carlsson (With 4 Figures) 257

Molecular Dynamics and Tight-Binding
By K. Laasonen and R.M. Nieminen (With 2 Figures) 264

Part IV Surface Phenomena

Linear Surface Model and the Environment-Dependent Effective Pair Potential
By G. Tichy (With 1 Figure) 274

Simulation of Surface Premelting
By P. Stoltze (With 6 Figures) 281

Hydrogen Diffusion on Metal Surfaces – Nonadiabatic Effects
By G. Wahnström (With 4 Figures) 291

Molecular Dynamics Simulations for Molecular Beam Epitaxy
By K. Terakura, K. Hara, M. Ikeda, M. Nomura, M. Mikami, and T. Oguchi (With 11 Figures) 303

Initial Stages of Oxidation of Al(111)
By U. Yxklinten, M. Limbäck, and B.I. Lundqvist (With 3 Figures) 314

Index of Contributors 319

Part I

General Reviews

Introduction to Many-Atom Interactions in Solids

M. Manninen[1]*, R.M. Nieminen*[2]*, and M.J. Puska*[2]

[1]Department of Physics, University of Jyväskylä,
SF-40100 Jyväskylä, Finland

[2]Laboratory of Physics, Helsinki University of Technology,
SF-02150 Espoo, Finland

Abstract. A short introduction is given to many-atom interactions in solids. The basic ideas of the new methods that relate the energy of an atom to the collective interaction with the neighbouring atoms are presented and the different approaches are compared. The limitations of the methods as well as some open questions are discussed.

1. Introduction

The dynamics of atoms in condensed matter is the key to understanding many physical properties and processes involving structural rearrangements. Especially at higher temperatures the classical picture of atoms moving according to the Newton equations of motion is an adequate description of the physics involved. The central problem is then to find the interactions between the atoms and the forces acting on the atoms at any instant of time.

The total energy of condensed matter (solid or liquid) at any atomic arrangement can be calculated, in principle, by solving the many-body Schrödinger equation with the known electrostatic interactions between the electrons and the atomic nuclei. This gives a multidimensional potential energy surface which determines the time evolution of the atoms (coordinates of the nuclei) via the classical Newton equations of motion. The seemingly overwhelming problem of the electron-electron interactions can be approximatively overcome by using the local density approximation[1] of the density functional theory. However, the *first-principles* computation of the electronic structure and total energy still remains monstrous, say, for a sample of 10000 atoms of a liquid: for large systems and complicated processes requiring long simulations a simpler picture of the interactions between the atoms is necessary. Nevertheless, impressive advances in (super)computer capacity and algorithm design (*e.g.* using the Car-Parrinello method[2]) have resulted in increasingly larger ($\approx$ 100 atoms) systems being attacked using *ab initio* electronic structure methods.

Traditionally, the interactions between the atoms have been described with pairwise interactions, the so-called pair potentials. A practical example taken from computer simulations are the Lennard-Jones-type potentials valid for inert gases. For metals, pair potentials can be determined from the pseudopotential theory and its generalizations. Recently, several new methods to determine the interatomic interactions in metals and semiconductors have been presented. Common to all these is that they implicitly include some many-atom interactions, but in a way which makes their use in computer simulations as easy as the use of conventional pair potentials. In this introduction

Springer Proceedings in Physics, Vol. 48 Many-Atom Interactions in Solids
Editors: R.M. Nieminen · M.J. Puska · M.J. Manninen

we give a brief overview of the ideas of the many-atom interactions and the general qualitatively new results they imply.

2. The pair-potential picture

There are two different approaches which describe the interatomic interactions using pair potentials. The first approach is to write the total energy of the solid as a sum over pair potentials:

$$E_{tot} = -\frac{1}{2}\sum_{i \neq j} \phi(\mathbf{R}_i - \mathbf{R}_j), \tag{1}$$

where ϕ defines the pair potential and $\mathbf{R}_i$ the atom coordinates. Theoretically this expression can be argued to be valid for rare gases where the long range attractive interaction is a result of weak polarization forces (van der Waals) and the short range repulsive interaction is a result of the overlapping of the closed electron shells (orthogonalization). The most commonly used potential for rare gases is the Lennard-Jones potential

$$\phi(r) = 4\epsilon\left(\left(\frac{\sigma}{r}\right)^{12} - \left(\frac{\sigma}{r}\right)^{6}\right). \tag{2}$$

The form of the potential and the dependence on only two parameters ϵ and σ imply that all the results of the Lennard-Jones systems can be scaled from one to another. Thus in simulations using this potential one can talk about a Lennard-Jones system rather than any specific rare gas. Besides being able to describe well many of the properties of rare gases, the Lennard-Jones systems, being extensively studied[3], offer an excellent testing ground for many complex processes and their simulations.

The simplicity of the total energy expression of Eq. (1) has inspired its use also for materials other than rare gases. However, the Lennard-Jones potential does not describe metals well, for example. A variety of other potentials have been suggested. A widely used example is the Morse potential

$$\phi(r) = \epsilon\left(e^{-2\alpha(r-r_0)} - 2e^{-\alpha(r-r_0)}\right), \tag{3}$$

which has three parameters to be fixed to the experimental properties of the material in question (*e.g.* cohesive energy (ϵ), lattice constant (r_0), and bulk modulus (α)). However, there are many properties of metals, *e.g.* the vacancy formation energy, the inward surface relaxation, and the relations between the elastic constants (departures from the Cauchy relations) that the Morse potential and also more complicated pair potentials[4] fail to reproduce.

It is *not* possible to show theoretically that the total energy of a metal can be written in the form of Eq. (1). Nevertheless, by making an ansatz that the energy is expressed in that way, theoretical methods can be used to derive the potential as done for example by Carlsson *et al.*[5] using electronic band structure total energy techniques. Theoretically derived pair potentials have anyway the same difficulties as those fitted to experimental results.

The second approach to pair potential description of the total energy is based on the linear response theory and is mainly used in (simple) metals. In nearly free electron metals (nontransition metals) the valence electrons form an almost uniform electron gas and the ions can be described by weak pseudopotentials. The second order perturbation theory (linear screening of the pseudopotentials) can then be used to calculate the total energy of the metal as[6]

$$E_{tot} = \frac{1}{2} \sum_{i \neq j} \phi(\mathbf{R}_i - \mathbf{R}_j) + E_{vol}, \tag{4}$$

where E_{vol} is an energy term which does not depend on the structure of the metal but depends on the total volume of the system. The pair potential ϕ is calculable from the bare pseudopotential and the screening properties of the electron gas. In this approach the pair interaction has a very different meaning than in expression (1) since only part of the total energy comes from the sum over the pair potentials (in Al for example the pair potential sum gives only a few percents of the total energy[7]). The pair potential derived from the pseudopotential theory has always a long range, the asymptotic behaviour being governed by the Friedel oscillations of the form

$$\phi(r) \propto \frac{\cos(2k_F r + \eta)}{r^3}, \tag{5}$$

where k_F is the Fermi wave vector of the electron gas and η an average phase shift. Moreover, the pair potential is also dependent on the volume of the system.

Several different ways to derive the pseudopotential and the pair potential have been presented. The simplest are based on model pseudopotentials[8] whereas in more advanced treatments also nonlinear and/or nonlocal effects of the screening are built in[7,9]. Generally, the interatomic potentials based on the pseudopotential theory work fairly well for those properties of the bulk metals and liquids where the disturbances of the electron density are so small that linear screening can be used. In the cases of surfaces and open volume defects, such as vacancies, these pair potentials should not be used, since the density changes are so large that they cannot be described by the linear response theory.

In semiconductors the mere pair potential cannot describe the directed covalent bonding. The pair potential picture is then extended to include three-body forces, the most commonly used model being that of Stillinger and Weber[10]. In metals the pseudopotential theory can be extended systematically to higher orders in perturbation theory to derive three-body forces and higher many-body interactions. However, these potentials are also long-range and have not been used widely in molecular dynamics simulations. The perturbation theory approach, even in higher orders, is not expected to be adequate for studying the structure of surfaces where the electron density goes to zero.

In conclusion, for materials other than the rare gases the pair potential picture cannot give good quantitative results. However, one should not completely abandon the simple picture of Eq. (1), since the widely used Lennard-Jones model can be kept as a reference system for any computer simulations. For complicated materials, say high-T_c superconductors, even if the new methods can be used[11] a simple pair potential picture can still be a useful first tool, since only a few parameters are needed to construct an empirical pair potential[12].

3. Many-atom interactions

During this decade several new approaches have been suggested for describing the interatomic interactions in metals. In most of these the total energy of the metal can be written in the generic form

$$E_{tot} = \sum_i F(n_i) + \frac{1}{2}\sum_{i \neq j} \phi(\mathbf{R}_i - \mathbf{R}_j), \tag{6}$$

where n_i is an effective electron density at the site of the atom i (or the coordination number of the atom), F is a nonlinear function describing the interaction of the atom with its electronic surroundings, and ϕ a pair potential (*not* the same as that in Eq. (1) or in Eq. (4)). The electron density n_i in Eq. (6) is the properly averaged electron density coming from the neighbouring atoms. It is approximated as a superposition of individual atom densities as

$$n_i = \sum_{j \neq i} n_a(\mathbf{R}_i - \mathbf{R}_j). \tag{7}$$

The electron density n_a is not necessarily a free atom density but a (pseudoatom) density associated with each atom in a solid. The function F has usually one minimum corresponding to the electron density at a vacancy in the bulk solid. The pair potential is taken to be of short range and can also be purely repulsive.

The functions F, n_a, and ϕ can be derived theoretically or determined empirically by choosing proper parametrized functions and fitting the parameters to some experimental data. The first empirical method was the so-called embedded atom method (EAM) introduced by Daw and Baskes[13]. In their formulation the purely repulsive pair potential term is interpreted as the screened Coulomb repulsion between the positive ion cores. The cohesion comes from the energy function F, which is the "embedding energy" of an atom in a homogeneous electron gas of density n_i. Finnis and Sinclair[14] used the tight binding model to show that the energy function F in Eq. (6) is negative and can be thought to be proportional to the square root of the n. Both Daw and Baskes and Finnis and Sinclair then proceeded by using parametrized functions F, n_i, and ϕ, where the parameters were fitted to some experimental data.

Ercolessi *et al.*[15] also used an empirical model based on Eq. (6), but interpreted n_i as a generalized coordination number. Their model, known as the glue model, is essentially the same as the embedded atom approach.

The underlying ideas in all the models are formulated in the effective medium theory of Nørskov and Lang[16] and the pseudoatom model of Stott and Zaremba[17], which were developed to describe an impurity in a metal. They are both based on the notion that the impurity energy is related to the *local* electron density of the host metal at the site of the impurity. The many-atom potentials of the type of Eq. (7) try to extend the impurity model to the bulk metal. Jacobsen *et al.*[18] and Manninen[19] have derived approximate formulae of the type of Eq. (6) using the basic idea of the effective medium theory for a single impurity, *i.e.* the screening of the atom is calculated in the homogeneous electron gas corresponding to the local electron density at the site of the atom in the real metal. Jacobsen *et al.* made an assumption that locally the system differs only slightly from an fcc structure (coordination is about the same as in fcc) and used the ideas of the atomic sphere approximation. Their resulting formula is usually written as

$$E_{tot} = \sum_i E_{C,i}(\bar{n}_i) + E_{AS} + E_{1el}, \tag{8}$$

where the $E_{C,i}$ depends on the embedding energy of the atom i in homogeneous electron gas of density $\bar{n}_i$. E_{AS} is the so-called atomic sphere correction, and E_{1el} is a correction term which depends on the change in the local density of states. The atomic sphere correction is assumed to be zero in a perfect fcc structure, and the one-electron energy term E_{1el} is approximated to be zero for nontransition metals (in transition metals it takes into account the effect of the d-electrons).

Equation (8) is a slightly different expression than Eq. (6) of the embedded atom method. However, the practical analytic expressions used in calculating functions $\bar{n}_i$, $E_{C,i}$, and E_{AS} are such that for nontransition metals (like Al and Cu) the final formulae used in the computations can be written as Eq. (6), but now the functions F, n_i, and ϕ are determined from *ab initio* calculations for the atom in a homogeneous electron gas [20].

The interatomic interactions of the form of Eq. (6) can be derived using other approximate ways. Daw[21] has used the extended Thomas-Fermi method and ideas similar to the Gordon and Kim[22] model to show the existence of the formula of the type of Eq. (6). The Thomas-Fermi-Dirac-von Weizsäcker approximation does not, however, give as good *ab initio* results as the Kohn-Sham method which is the basis of the theory of Jacobsen *et al.*. Kress and DePristo[23] have also studied the derivation of the many-atom potentials of the form of Eq. (6) and suggested some corrections to the formulas of Jacobsen *et al.*[18].

4. Some basic results

The advantage of the total energy expression of the form of Eqs. (6) and (7) is that they are essentially as efficient to use in computer simulations as the simple pair potential expression, Eq. (1). So far the most widely used of these methods is the semiempirical EAM. It has been used to simulate the properties of the bulk transition metals (phonons[24], defects and impurities[13], liquid metals[25], alloys[26]) and transition metal surfaces (surface energy and structure[27], adsorbate layers[27], surface segregation[28], and surface phonons[30]). All these properties are described fairly well by the model. Similar good results have been obtained using the tight-binding model of Finnis and Sinclair[14], the glue model[15], or the (*ab initio*) effective medium theory[31].

The inward surface relaxation of metals is one basic property that all the new many-atom interactions describe correctly. The physical explanation is that the coordination number of the atoms in the surface layer is smaller than that of the bulk atoms. Then by reducing the interlayer distance the effective coordination number (electron density) increases and becomes more bulk-like. The simple pair potential picture of Eq. (1) leads practically always to an outward relaxation.

The advantage of the *ab initio* effective medium theory has been that it can be used to derive systematics of the cohesive properties of metals[31]. For example it explains the systematics of the cohesion in transition metals and the increasing tendency to surface reconstruction when going from Cu to Ag and to Au[32]. The derivation of the EMT has many inherent approximations the effects of which are sometimes difficult to estimate. Nevertheless, it also gives ideas on how one could systematically improve the derived many-atom interactions. For example, it shows how the one-electron energy electron correction E_{1el} appears. This correction is assumed to be small in nontransition metals but becomes important for example in studying finite clusters[33].

5. Open questions

The new many-atom interactions are certainly a breakthrough in simulating metals and semiconductors by computers. However, there are still some open questions which have been overshadowed by the apparent success of the methods. In the semiempirical methods some experimental results are used as inputs in the model. This can give a good tool to derive other properties, but does not guarantee that the physical picture is correct. In the *ab initio* EMT on the other hand the quantitative effect of all the approximations is not carefully estimated (and in many cases it is extremely difficult). In the following we mention some open questions which should be studied and which also elucidate the difficulties in determining reliable many-atom interactions.

Nearly-free electron metals.

In simple nontransition metals, such as the alkalis and their alloys, the conventional *ab initio* pseudopotential picture of Eq. (4) gives quite accurate results for the equilibrium properties of the bulk (including vacancy formation energies and the phonon spectra[34]), and for the properties of the liquids, *i.e.* the pseudopotential theory works well if the density changes are small. In fact, these properties are described *better* in the pseudopotential theory than in the effective medium theory, which gives quite poor results for the equilibrium lattice constant and the bulk modulus for sodium[18]. Even in the case of aluminium the pseudopotential theory[34] gives at least as good phonon spectra as the effective medium theory[18]. The failure of EMT in the case of sodium calls for studying the derivation of the EMT in the limit of a weak pseudopotential, since one would expect that approximations made in EMT are smallest in the cases of alkali metals.

Long-range effects.

So far all the applications of the many-atom interactions have been made with short-range potential functions: either only the nearest neighbor interactions are included or the interactions fall off exponentially. However, interactions beyond the third neighbors are needed to obtain any energy difference between the fcc and hcp phases and to get a nonzero stacking fault energy. Semiempirically it is easy to construct potentials which result in the correct stacking fault energy. However, theoretically the long-range interactions are difficult to deal with. In EMT a method is proposed how the interactions can be extended heuristically as far as needed[18,35] (the resulting bulk fcc lattice constant and the cohesive energy being independent of the range of the interactions). However, Häkkinen[36] has shown that both in Al and in Cu the stacking fault energy is very sensitive to the range of the interactions. A similar tendency is usually obtained by using empirical pair potentials[37]. In the pseudopotential theory the pair interaction extends to infinity, but when summed up it gives reasonable results (correct sign and order of magnitude) for almost any pseudopotential[38].

Vacancy formation energy.

The vacancy formation energy of metals has been a topic of extensive studies for decades. One of the latest studies is due to Gillan [39] who calculated the formation energy for a relaxed vacancy in Al using the pseudopotential method within the local density approximation. His result 0.56 eV agrees quite well with the measured value of 0.66 eV [40,41]. In the case of Al the effective medium theory or semiempirical theories cannot correctly describe the ratios between the cohesive energy, the surface energy and the vacancy formation energy. In the many-atom models the vacancy formation energy can be derived easily and accurately once the potential parameters are known. The failure shows that something essential is still missing from the models used and that this has to be kept in mind in studying open volume defects and surfaces. On the other hand, it is somewhat disconcerting that (unlike for surfaces) earlier direct density-functional calculations for vacancy formation energies leave a lot to be hoped for and do not agree with experiment, for example in the case of Al [42].

One-electron energy corrections

The derivation of the effective medium theory includes a term, E_{1el}, which depends on the density of one-electron energy levels. It has been argued[18] that for bulk nontransition metals the effect of this term is small (this argument is supported by the good results obtained without this term). There are cases, however, where this term must become important. An extreme case is a small cluster where the density of states is discrete[33]. However, also in infinite metals many properties can be argued to be partly results of the one-electron energy spectrum: the lattice structure[43], the stacking fault energy[37,44], and Peierls transitions[45]. The fcc-structure is the bulk ground state in the EMT-construction of Ref. [18] if the one-electron energy correction is omitted. The modelling of other lattice structures, like bcc and diamond, requires that the (large) deviation of the actual one-electron energy spectrum from that for a free electron gas is taken into account as E_{1el}. The existence of the one-electron energy term should be kept in mind and one should study whether this kind of correction could easily be transformed to a form usable in computer simulations. One example is its incorporation into molecular dynamics in terms of distance-dependent tight-binding parameters [46].

Other derivations of many-atom potentials

Although many derivations of the many-atom interactions already exist there is still room for more theoretical work. The effective medium theory makes use of the screening properties of the atom in a homogeneous electron gas[20]. Another possibility would be to take as a reference the homogeneous electron gas with a vacancy, as is done in some of the pseudopotential approaches[7,34]. A band structure calculation for the perfect metal could perhaps also serve as a reference system instead of the homogeneous electron gas.

It would also be extremely useful to have some theoretical benchmark results calculated, say, using the local density approximation and *e.g.* supercell band structure techniques. Such benchmarks should include the vacancy formation energy, the surface energy, the stacking fault energy, elastic constants, and the phonon spectra. Accurate theoretical results would make it easier to find the cause of the inaccuracies of the model used and show the limits of their applicability.

Many-atom interactions in semiconductors.

In semiconductors the pair-potential picture fails to describe the directional covalent bonds and at least three-body interactions are needed. The most commonly used are the Stillinger-Weber potentials[10]. Their use is, however, already more difficult than the use of the many-atom interactions of Eq. (6) used in metals. Baskes[47] has suggested an extension of the semiempirical embedded atom method to semiconductors. The idea was to include nonspherical terms to the electron density. Ackland[48] has suggested a model for silicon where the correct coordination number plays a crucial role. Both these models still need careful scrutiny. The effective medium theory gives a clear guideline that the opening of the energy gap in the electron band structure stabilizes the diamond structure with respect to fcc. The effect should be incorporated via the one-electron energy term, but this is difficult in practice and has not been done thus far in a useful form. In conclusion, more theoretical work is needed before it can be seen if the ideas which have been shown to be useful in metals can be utilized also in semiconductors.

6. Conclusions

The new many-atom interactions of the type of Eq. (6) have shown to be useful and effective in computer simulations in metals. They do not have the problems which make the simple pair potentials, Eq. (1), virtually useless in metals, and contrary to the pseudopotential picture they can attack problems with low symmetries and large electron density changes. Due to the many approximations needed for obtaining applicable formulations the results from different calculations can vary markedly. More theoretical work is still needed for understanding the successes and weaknesses of the suggested many-atom interactions. These proceedings offer a good collection of reviews of the work done so far.

References

[1] *Theory of Inhomogeneous Electron Gas*, edited by S. Lundqvist and N. H. March (Plenum, New York 1983).

[2] R. Car and M. Parrinello, Phys. Rev. Lett. 55, 2471 (1985) and R. Car, in these proceedings.

[3] F. F. Abraham, J. Vac. Sci. Technol. B **2**, 534 (1984) and references therein.

[4] M. I. Baskes and C. F. Melius, Phys. Rev. B **20**, 3197 (1979).

[5] A. E. Carlsson, M. L. Klein, and H. Ehrenreich, Phil. Mag. A **41**, 241 (1980).

[6] N. W. Ashcroft, in *Interatomic Potentials and Simulation of Lattice Defects*, edited by P. C. Gehlen, J. R. Beeler, Jr., and R. I. Jaffee (Plenum, New York 1972), W. A. Harrison, *Pseudopotentials in the Theory of Metals* (Benjamin, New York 1966).

[7] M. Manninen, P. Jena, R. M. Nieminen, and J. K. Lee, Phys. Rev. B **24**, 7057 (1981).

[8] W. M. Shyu and G. D. Gaspari, Phys. Rev. **163**, 667 (1967).

[9] M. Rasolt and R. Taylor, Phys. Rev. B **11**, 2717 (1975).

[10] F. H. Stillinger and T. A. Weber, Phys. Rev. B **31**, 5262 (1985).

[11] A. Zwartkruis and J. E. van Himbergen, in these proceedings.

[12] S. Valkealahti and D. O. Welch, Physica C (in press).

[13] M. S. Daw and M. I. Baskes, Phys. Rev. Lett. **50**, 1285 (1983); see also the review in these proceedings.

[14] M. W. Finnis and J. R. Sinclair, Phil. Mag. **A50**, 45 (1984).

[15] F. Ercolessi, E. Tosatti, and M. Parrinello, Phys. Rev. Lett. **57**, 719 (1986).

[16] J. K. Nørskov and N. D. Lang, Phys. Rev. B **21**, 2131 (1980).

[17] M. J. Stott and E. Zaremba, Phys. Rev. B **22**, 1564 (1980).

[18] K. W. Jacobsen, J. K. Nørskov, and M. J. Puska, Phys. Rev. B **35**, 7423 (1987).

[19] M. Manninen, Phys. Rev. B **34**, 8486 (1986).

[20] M. J. Puska, in these proceedings; M. J. Puska, R. M. Nieminen, and M. Manninen, Phys. Rev. B **24**, 3037 (1981).

[21] M. S. Daw, Phys. Rev. B **39**, 7441 (1989).

[22] R. G. Gordon and Y. S. Kim, J. Chem. Phys. **56**, 3122 (1972).

[23] J. D. Kress and A. E. DePristo, J. Chem. Phys. **87**, 4700 (1987) and **88**, 2596 (1988).

[24] M. S. Daw and R. L. Hatcher, Solid State Commun. **56**, 697 (1985).

[25] S. M. Foiles, Phys. Rev. B **32**, 3409 (1986).

[26] S. M. Foiles, M. I. Baskes, and M. S. Daw, Phys. Rev. B **33**,7983 (1986).

[27] M. S. Daw, Surf. Sci **166**, L161 (1986), S. M. Foiles, Surf. Sci. **191**, L779 (1987), T. E. Felter, S. M. Foiles, M. S. Daw, and R. H. Stulen, Surf. Sci. **171**, L379 (1986).

[28] S. M. Foiles, Phys. Rev. B **32**, 7685 (1985).

[29] J. S. Nelson, E. C. Sowa, and M. S. Daw, Phys. Rev. Lett. **61**, 1977 (1988).

[30] M. W. Finnis, A. T. Paxton, D. G. Pettifor, A. P. Sutton, and Y. Ohta, Phil. Mag. A **58**, 143 (1988).

[31] K. W. Jacobsen, Comm. Cond. Matter Phys. **14**, 1679 (1988), and in these proceedings.

[32] K. W. Jacobsen and J. K. Nørskov, in *The Structure of Surfaces II*, edited by J. F. van der Veen and M. A. Van Hove,Springer Ser. Surface Sci., Vol. 11 (Springer, Berlin, Heidelberg 1988) p. 118.

[33] S. Myllyrinne, M. Manninen, and J. K. Nørskov, in these proceedings.

[34] L. Dagens, M. Rasolt, and R. Taylor, Phys. Rev. B **11**, 2726 (1975), R. Taylor and A. H. MacDonald, J. Phys. F **10**, 2387 (1980), G. Jacucci and R. Taylor, J. Phys. F **9**, 1489 (1979).

[35] H. Häkkinen and M. Manninen, J. Phys. Condens. Matter (in press).

[36] H. Häkkinen, in these proceedings.

[37] G. J. Cotterill and M. Doyama, Phys. Rev. **145**,465 (1966).

[38] J. P. Simon, J. Phys. F **9** 425 (1979).

[39] M. Gillan, J. Phys. Condens. Matter **1**, 689 (1989).

[40] W. Triftshäuser, Phys. Rev. B, **12**, 4634 (1975).

[41] M. J. Fluss, L. C. Smedskjaer, M. K. Chason, D. G. Legnini, and R. W. Siegel, Phys. Rev. B **17**, 3444 (1978).

[42] B. Chakraborty, R. W. Siegel, and W. E. Picket, Phys. Rev. B **24**, 5445 (1981).

[43] H. L. Skriver, *The LMTO Method*, Springer Ser. Solid-State Sci., Vol. 41 (Springer, Berlin, Heidelberg 1984).

[44] G. J. Ackland, G. Tichy, V. Vitek, M. W. Finnis, Phil. Mag. A **56**, 735 (1987).

[45] C. Kittel, *Introduction to Solid State Physics* (John Wiley and Sons, New York 1986).

[46] K. Laasonen and R. M. Nieminen, in these proceedings.

[47] M. I. Baskes, Phys. Rev. Lett. **59**, 2666 (1978).

[48] G. Ackland, in these proceedings.

Volume and Pair Forces in Solids and Liquids

V. Heine [1] *and J. Hafner* [1;2]

[1]Cavendish Laboratory, Madingley Road, Cambridge CB3 0HE, UK
[2]Institut für Theoretische Physik, Technische Universität, A-1040 Wien, Austria

Abstract. The main purpose of this article is to discuss *many-atom* interactions in solids and liquids: their origin in the theory of electronic structure, some experimental evidence for their importance, and the manner of their incorporation in various approximate 'glue' schemes for calculating the total energy of a system. The use of pseudopotentials with second order perturbation theory is discussed in relation to understanding many aspects of the structure of solid and molten elements and alloys. Progress in the *ab initio* total energy calculations is commented on, and their usefulness underlined in creating a database for testing or fitting more approximate 'glue' models.

1. Introduction

The main purpose of this article is to survey from a contemporary point of view the existence of *many-atom* interactions in solids and liquids. In section 2 we shall trace their origin in tight-binding (TB) theory and in the nearly-free-electron (NFE) theory of electronic structure. We shall show that there is good evidence for the reality of their existence. The ideas apply not only to metals but also to covalent semiconductors like Si.

In order to study computationally the many processes that occur in solids, including defects and events at surfaces, one must be able to calculate the total energy. The calculations have to be efficient for large enough systems to model complex processes, and must have sufficient relative accuracy of the order of 10^{-3} eV per atom so that one can determine equilibrium positions of the atoms. We shall make some remarks in section 5 about the ab initio quantum mechanical total energy calculations, which are making good progress. However one would of course like to have much more rapid methods for many purposes. One of these is the use of second order perturbation theory in the context of the NFE approach with pseudopotentials. Developed in the 1960's [1], does it still have any use? We shall survey briefly in section 3 its many successes over the intervening years. At best, in appropriate situations, it continues to give results which are still good by modern standards. At

Springer Proceedings in Physics, Vol. 48 Many-Atom Interactions in Solids
Editors: R.M. Nieminen · M.J. Puska · M.J. Manninen

worst it can give a semiquantitative interpretation of many observed phenomena in the structures and phase diagrams of elements and alloys which (presumably) would also emerge from the more accurate total energy calculations. (We shall reserve the expression "total energy calculation" for the complete solution of the Schroedinger equation in density functional theory (DFT), although second order perturbation theory is of course also a calculation of the total energy from the same functional and pseudopotentials. The latter we shall abbreviate to "perturbation theory".) In other situations such as liquids and amorphous structures the total energy methods are as yet not a real match to the perturbation theory if the aim is to explore systematic trends in the structures of melts and glasses. But even in the context of the total energy calculations the perturbation approach might be helpful by setting up useful starting configurations for refinement by total energy methods. This would avoid spending most of the computer time in exploring irrelevant regions of phase space.

However the perturbation theory fails whenever the material is not approximately of uniform density, such as at a vacancy or a surface. In recent years several schemes have arisen to embody the ideas of NFE or TB theory, with various approximations and empirical elements [2-6].We shall refer to all these as "glue models", and in section 4 consider how they model the important many-body forces.

There is one set of schemes which falls between the total energy methods and the glue models. In chemical calculations on large molecules some quite good results can be obtained by TB type of calculations using a rather minimal basis set of local orbitals, e.g. a single s-orbital for hydrogen etc..Quite a large effort has gone into making such an approach work [7-13]. At one extreme the local orbitals have been defined mathematically from linearised muffin tin orbitals [10], in which case the method is essentially an ab initio total energy calculation. On the other hand one can choose the orbitals more empirically. The recursion method is good for solving the one-electron equation to a controlled accuracy in a manner that includes properly the local environment around every atom [14-16]. In so far as the method solves the Schroedinger equation it includes exactly the many-atom effects. Even with empirical orbitals the method can be built upon density functional theory to define in ab initio terms the TB parameters and the additional short-range repulsions between atoms [11]. What seems to be a workable formulation can be given for the Hellmann-Feynman forces [7,13]. However there appear to be problems about defining orbitals that work well without change over a wide range of interatomic distances, e.g. to give correctly the bulk modulus of a material [12]. The problems are related to the question of neglecting three-center terms in the TB theory, and how to treat the overlap of the orbitals.

2. The Origin of Many-Atom Forces as Seen in Tight-Binding and Nearly-Free-Electron Theory

The matter was well put by Pauling in 1938 [17]: " the metallic bond is very closely related to the ordinary covalent or electron-pair bond,.....with the bonds resonating among the available positions in the usual case that the number of positions exceeds the number of bonds." In aluminium the three electrons per atom "resonate" among the twelve bonds to the nearest neighbours. Such resonance lowers the energy as chemists study for their archetype, namely benzene. We mix together in the total wave function a large number of Slater determinants each corresponding to a different configuration of bonds, and the variational principle tells us that the energy is lowered by such a mixing below the energy of any one typical arrangement. In terms of the uncertainty principle, the energy is lowered by delocalizing the bond to let it run around all possible positions. Thus in aluminium we expect significant 13-atom forces (one atom and its 12 neighbours): any expansion of the cohesive energy in n-atom terms with n=2,3,.... will not converge until n is of order 12 . It is not even clear that it makes much sense.

The point is well illustrated in TB theory. Consider a system with one s-orbital and one electron per atom. We take the atomic ("diagonal") energy as our zero, and the hopping integral between neighbouring atoms as h (negative) , the same between all atoms. The bonding orbital in the dimer molecule has energy $-|h|$ which we take as the energy of a bond (per electron). However in the solid the energy is of order

$$-\sqrt{C}\ |h| \tag{1}$$

per atom, where C is the coordination number of nearest neighbours. The extra factor $\sqrt{C}$ in (1) follows very simply from TB theory (p.82 of [16]). It is the extra lowering of the energy from Pauling's resonance effect, and for $C\sim12$ it is a very substantial factor. With $\sqrt{12} = 3.46$, we can say that the pairwise bonding energy (corresponding to the dimer molecule) is $-|h|$ and the extra resonance energy is $-2.46\ |h|$. The factor $\sqrt{C}$ in (1) does not hold just for s-bonds: see e.g. p.81 of [16] for the generalization to d-bands.

There is direct experimental evidence for the resonance effect. Consider a vacancy in a close-packed metal and let us stay with the simple s-band model. The bulk cohesive energy is

$$U_{bulk} = \sqrt{12}\ |h| = -3.46\ |h|\ . \tag{2}$$

The vacancy energy is defined by taking one atom out and adding it to the bulk: it is

$$U_{vac} = (13\sqrt{12} - 2\sqrt{11} - \sqrt{12})\ |h| = 1.77\ |h| \tag{3}$$

$$= 0.51\ U_{bulk}\ .$$

However a model of purely pairwise bonding gives $U_{vac}/U_{bulk}=1$. The observed value is typically even lower than (3). Similar considerations apply to the adhesion of atoms at surfaces of different roughness and the binding energy of a dimer on a surface (p.83 of [16]). We believe this all represents powerful evidence for the resonance effect being the dominant contribution to metallic cohesion. In no way can metals be described even approximately in terms of pairwise forces, which is why we are here, still struggling to make quantitative sense of them, 50 years after Paulings article [17] in 1938.

The ultimate delocalization is seen in the NFE theory of sp-bonded elements. We have for the total energy per atom of an element to second order in perturbation theory (p.268 of [1])

$$U = U_o(V) + U_E + U_{bs} \tag{4a}$$

$$U_{bs} = \Sigma'_q\ |\ S(q)\, v^{ion}(q) / \varepsilon(q)\, |^2\, \chi(q)\, \varepsilon(q)\ . \tag{4b}$$

Here $U_o(V)$ is the energy of a neutral atom in the sense of an ionic pseudopotential in a sphere of atomic radius R_a containing a uniform electron gas. The U_E is the small deviation of the Ewald electrostatic energy from the part not in U_o, the $S(q)$ is the structure factor, $v^{ion}(q)$ the ionic pseudopotential, $\varepsilon(q)$ the screening function and $\chi(q)$ the perturbation characteristic (negative), all defined in the usual way [1]. The perturbation expression (4b) works well for any configuration which is more or less uniformly dense, e.g. a typical metallic structure, a metal with a phonon or a dislocation, a liquid metal, but not for a case like a vacancy. In the latter the potential perturbation $S\, v^{ion} / \varepsilon$ is not particularly large, but the phase factors in $S(q)$ combine in the higher order terms in a coherent way to give large corrections. The expression (4) can be rewritten

$$U = U_1(V) + N^{-1}\, \Sigma_{ij}\, \Phi(R_{ij};\ V)\ . \tag{5}$$

Here U_1 is the energy of a screened pseudo atom and $F(R_{ij})$ the pairwise screened interaction between the atoms i and j. Note that U_1 and U_o are not equal because one has to add and subtract various bits to carry out the Fourier transform in going from (4) to (5) [18]. They both represent the energy of a neutral atom, but in U_o it is a sphere of radius R_a carved out of the material, whereas in U_1 it is a more extended object defined by linear screening of v^{ion}.

The important point about (4), (5) is that the one-atom energy U_o or U_1 is the dominant term not only for the binding energy of the elements, but also for the heat and volume of formation of binary alloys. It already contains the resonance and most of the bonding effect because it represents the atom in an electron gas of bonds. In either U_o or U_1 it depends on volume through the kinetic delocalisation (Fermi) energy of the electron gas and other contributions. The picture is only valid if the electron gas, i.e. the atomic density, is more or less uniform. The F is a small *residual* pairwise interaction, e.g. for small deviations from uniformity. U_o rather than U_1 turns out to be a better first order approximation for considering volume, pressure and compressibility effects.

We can see evidence for the multi-atom forces also from this NFE picture. Here we identify many-atom forces with the volume-dependence discussed above. Of course there is no direct analogue in NFE theory for the $\sqrt{C}$ effect in (1) because the NFE formulation is only applicable to the case of uniform bulk density. Rather we see the mobility of the electrons in the NFE gas as equivalent to the resonance among highly unsaturated bonds in the TB picture. The traditional pointer to volume dependent forces is the Cauchy "anomaly" between certain elastic constants, e.g. the difference between c_{44} and $c'=1/2(c_{11}-c_{12})$ for cubic metals. This has to vanish identically for a structure in equilibrium under central pairwise forces only. In an alloy U_1depends on the <u>average</u> electron density in which the atom finds itself, which will differ in general from that of the pure element. To a first approximation the volume of the alloy may be calculated by minimizing the elastic energy associated with the compression and expansion of the individual pseudoatoms. This calculation predicts systematic deviations from both Zen's and Vegard's laws which assume that either the volume or the mean atomic radius of the alloy is the weighted mean of the constituent atomic volumes or radii. These deviations have quite a characteristic signature. In general a large difference in the electron density leads to a large negative volume of formation, but when the difference in electron density has the opposite sign to the valence difference because of an overwhelming difference in atomic volume, a volume expansion on alloying can occur instead [19].

The change in the effective atomic radius is also seen in the crystal structure of the alloys, depending as it does on the radius ratio of the constituents (Section 3). The volume energy also dominates the heat of formation of the alloy. Pettifor [20] has pointed out that the changes in U_O on alloying give a repulsive contribution to the heat of formation from the loss in kinetic energy, and an attractive contribution from the gain in exchange and correlation energy. At higher electron densities of the alloy the latter contribution dominates and this sets the main trend in the heat of formation of sp-bonded alloys.

In the textbooks of the 1930's the TB and NFE approaches were presented as opposite poles, a tendency which can even be seen sometimes today. In reality they give qualitatively the same picture when the orbital overlap is substantial as in sp-bonded materials. One should always interpret a NFE band structure chemically by ascribing p-bonding, s-antibonding etc. character to the states. Already in the 1930's Mott and Jones [21] knew that even diamond is a reasonably good example of a NFE gas with a small band gap compared to the measured free-electron band width. Of course the approximation (4), (5) is not sufficient to stabilize the TA phonon modes, though the NFE picture is evident in a slightly modified form [22,23]. But the more basic point we are trying to make here is that we must expect multi-atom forces even in materials such as the diamond-type semiconductors which we usually think of in the very localized terms of saturated covalent bonds. There is evidence for this. For example Payne et al have done total energy calculations of the Σ=5 [001] twist and other grain boundaries in germanium, and they find 3- and 5-fold coordinated atoms in what may be the lowest energy structures even though with some strain one could couple the two grains with purely tetrahedral bonds [24,25]. Evidence of NFE behaviour of a different kind comes from SiC polytypes. They have perfect tetrahedral bonding but each atomic double layers can be placed in two orientations as in the difference between the wurtzite and zinc blende structures. The point is that total energy calculations on different stacking structures show significant interactions between atoms which are up to six bonds apart [26]. These relatively long ranged interactions are interpreted as the remnant in the semiconductor of Friedel oscillations in a metal [27]. The same applies to rather distant interatomic forces seen in phonons in Si [28] .

At first sight such NFE effects in a diamond-type semiconductor conflicts with the very successful picture of tetrahedral bonding, which we must therefore examine in more detail. In the latter we start with atomic s and p orbitals, form the hybridised sp^3 tetrahedrally directed valence orbitals, and take the bonding combination between neighbours. If we expand the total wave function Ψ in terms of these bonding orbitals as a basis set, then the form of Ψ in terms of Bloch bands and that with local orbitals are mathematically identical because the valence band is full. Moreover the bond order is always two and the energy is the same, independent of stacking structure. The major error in this picture is that we should really expand in terms of the full sp basis set, i.e. both bonding and antibonding orbitals. When we do that we find that the occupancy of the valence band is not nearly sp^3: even for silicon in the diamond structure Paxton finds a configuration $s^x p^{4-x}$ with x= 1.44 [8,9]. When he calculates for hexagonal silicon (wurtzite structure) he finds different bond orders for the two inequivalent types of bond, and the one with slightly larger bond order is found to have the shorter bond length in the SiC total energy calculations [8,9]. Such variation of bond order through different amount of mixing across the band gap is analogous to the behaviour of a NFE band structure.

3. Is Nearly-Free-Electron Pseudopotential Perturbation Theory Still Useful?

About twenty years ago pseudopotential perturbation theories led the way to a quantum-mechanical understanding of cohesion and structure in sp-bonded metals and alloys [1]. Over the last ten years the results obtained by perturbation theory have gradually been superseded and extended by the application of non-perturbative total energy techniques. Does this mean that NFE perturbation theories are now outdated?

We do not think that this is the case. Even if all computational problems were solved, it is still essential to relate any numerical result to some simple underlying physical model for it to be useful. In this section we want to illustrate how immensely helpful NFE perturbation theories continue to be for a real understanding of the main physical effects. Firstly, if we say that calculations can now be done more accurately with total energy methods, this does not mean that the results of the perturbation theories become invalid. The new techniques merely yield more accurate results, especially in cases were higher order corrections become important. Secondly, even with present-day computer facilities, calculations for very complex crystal structures and for liquid and glassy materials remain cumbersome. Certainly the new algorithms based on simulated annealing and related techniques [24,25,29] open new and fascinating possibilities, but for the time being perturbation theories continue to be very valuable in exploring the structures of complex intermetallic compounds, of liquid and glassy materials.

To begin with, let us compare the results obtained for the binding energies and the structural energy differences for a difficult case such as a

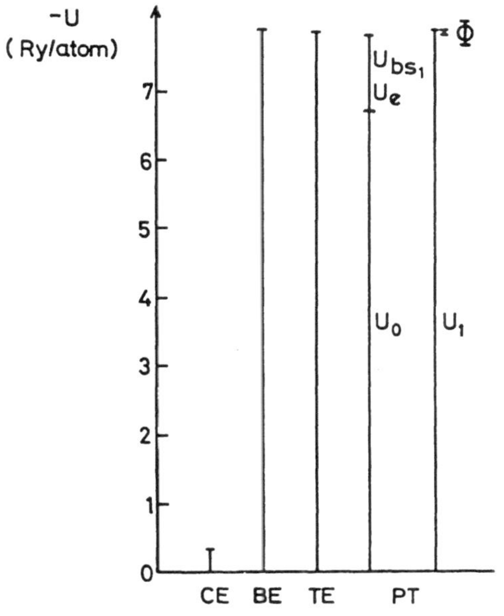

Fig. 1 Contributions to the binding energy in silicon: BE - binding energy, CE - cohesive energy (experimental), TE - result of total energy calculation (Ref. 31), PT - result of second - order perturbation theory (Ref. 30), decomposed according to either equ. (4) $U = U_0 + U_\Xi + U_{bs}$ or equ. (5) $U = U_1 + \Phi$. (Note that in this case the contribution of the pair potential to the binding energy is actually repulsive.)

tetrahedrally coordinated semiconductor. Fig.1 shows the binding energies of Si in the diamond structure, calculated using first- and second order perturbation theory [30] and using a non-perturbative total-energy method [31]. The main point is that the largest part of the binding energy is contained already in the first-order result and that the truly covalent effects not describable in terms of volume- and pair forces (or equivalently in terms of overlapping spherical pseudoatoms) are small corrections at the level of the binding energy. This remains true for the structural energy differences: even second order NFE perturbation theory predicts the diamond structure to be more stable than any of the close-packed metallic structures for Si, Ge, and Sn [30, 32], and the correct sequence of pressure-induced phase transitions (diamond → ß-tin → simple hexagonal → close-packed: see Fig.2 for a comparison of perturbation and total energy calculations for Ge). Admittedly the quantitative differences are now somewhat larger, but this does not affect the main lesson: the trend in the crystal structures is set already by the interplay of the volume- and pair-forces, higher order corrections are small and important

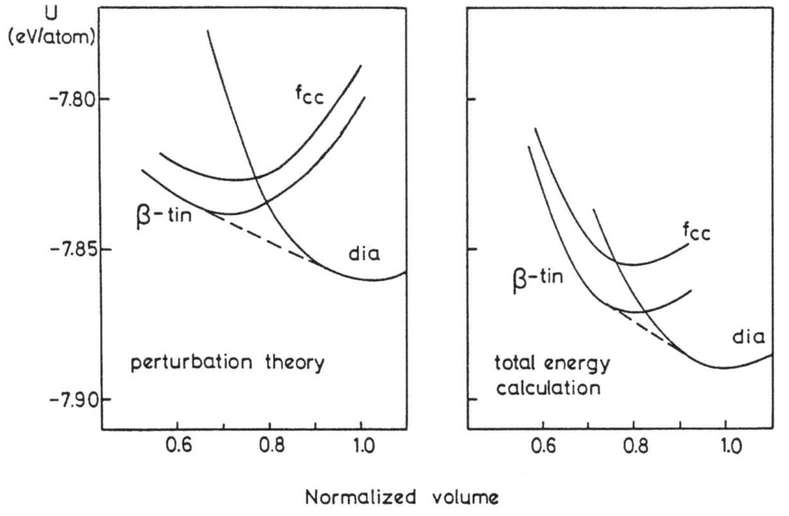

Fig. 2 Binding energy of Ge in the diamond, β-tin and face-centred cubic structures as a function of volume, as given by a total-energy calculation (Ref. 31) and by second-order perturbation theory (Refs. 30, 32). The dashed line indicates the common tangent determining the pressure induced phase transition.

mainly for dynamical properties. The comparison is somewhat hampered by the fact that perturbation- and total-energy-calculations are based on quite different pseudopotentials. The question 'how useful is a good norm-conserving pseudopotential within the framework of perturbation theory' remains to be answered. For a recent compilation of structural energy differences from NFE perturbation theory and total energy calculations, see Ref. [33] .

The explanation of the trends in the crystalline and liquid structures across the periodic table is perhaps one of the most important achievements of NFE perturbation theory. These trends include (i) the change from close-packed metallic structures in groups I to III to open covalent structures with decreasing coordination numbers in groups IV to VI, (ii) in group IIb the increasing distortion of the hexagonal lattice in Mg to Cd as the atomic number increases, culminating in a unique structure in Hg, (iii) in group III a similar phenomenon with a close-packed structure in Al and a unique low-symmetry lattice in Ga, but then a return to a less distorted one in In and a close-packed one in Tl, and (iv) in groups IV to VI the tendency to a more metallic behaviour and more close-packed structures in the heavy elements, most pronounced in the series Si → Ge

$\rightarrow$ Sn $\rightarrow$ Pb. Similar trends are observed in the structure of the liquid elements where they appear as anomalies in the first few peaks of the pair correlation function and static structure factor [34,35]. The main trends have been explained nearly twenty years ago by Heine and Weaire [1] using a **k**-space argument, but there was always some difficulty in understanding the return to metallicity for the very heavy polyvalent elements. Later, this argument has been reformulated in terms of real-space interactions [36], which allows one to link the trend in the atomic structures to the systematic variations of the interatomic potentials with electron density and pseudopotentials. The effective pair potential $\Phi(R)$ consists of a short-range repulsive part which is essentially of screened Coulomb form with a medium- and long-range oscillatory tail. The diameter of the repulsive core D_{rep} is given to a good approximation by [1,36]

$$D_{rep} = 2\,R_c + 2\,\lambda_{TF} = 2\,R_c + 2\,a\sqrt{R_s} \tag{6}$$

where R_c is the diameter of the ionic core, λ_{TF} is a Thomas-Fermi screening length and a is a constant. Note that λ_{TF} varies as the square-root of the electron-density parameter R_s (R_s is the radius of a sphere which contains on the average one electron). The wavelength of the oscillatory part of the potential is set by the Friedel wavelength λ_F

$$\lambda_F = 2\pi\,/\,2k_F = b\,R_s \tag{7}$$

where k_F is the Fermi wavevector. The amplitude of the oscillations is set by the screened on-Fermi-sphere matrix element $v^{ion}(2k_F)/\varepsilon(2k_F)$ of the pseudopotential. It is found that the trends in the structures are driven by two essential mechanisms: (i) The "core-effect" - at higher electron densities (smaller R_s) the repulsive core (6) of the pair potential moves over the first minimum as the wavelength of the Friedel oscillations (7) decreases more rapidly than the diameter of the repulsive core. This tends to destabilize close-packed structures and explains the trends from metallicity to covalency and to decreasing coordination numbers. (ii) The "amplitude effect" - as the pseudopotential matrix element for $|\mathbf{k}| = 2\,k_F$ decreases, the Friedel oscillations become weaker. In the extreme case the interatomic potential is turned into a screened Coulomb form . This explains the return to close-packed structures in the heavy elements.

These arguments are illustrated most convincingly by the successful calculation of the complex liquid structures of the molten semiconducting and semimetallic elements. In the crystalline phase the infinite number of

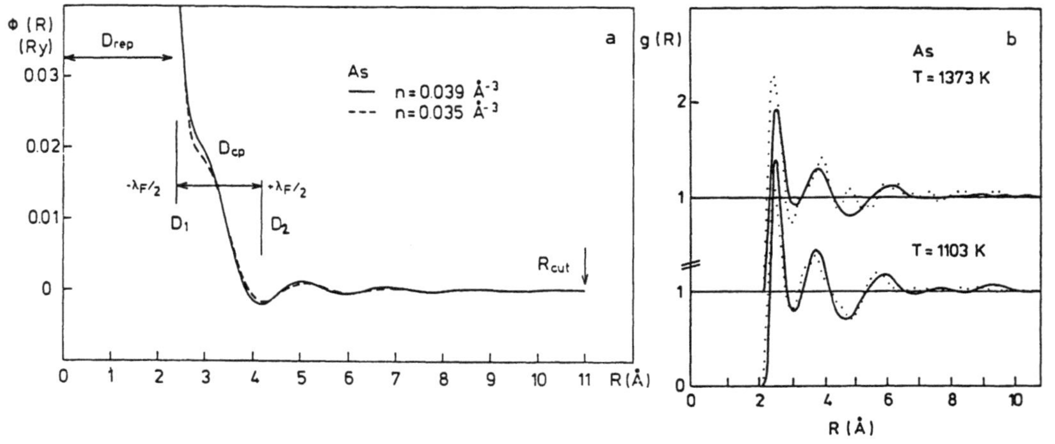

Fig. 3 (a) Interatomic potential $\Phi(R)$ for liquid As (see text) .

(b) Pair correlation function g (R) for liquid As. Full-line molecular dynamics calculation, dotted line - neutron diffraction (Ref. 37).

structural alternatives can never be explored in an exhaustive manner: we are always restricted to a calculation of the relative energies of a selected set of structural alternatives. In the liquid phase computer simulation techniques allow one to locate the minimum-energy configuration in phase space in a way that is not biased by any a-priori choices. It is therefore an important argument for the validity of NFE perturbation theory that these simulations yield a quantitative description of complex metallic , semimetallic, and semiconducting fluids such as Ga, Si, Ge, P, As, Te,.... [32,37-39]. The characteristic variation of the coordination numbers, bond distances and bond-angles is correctly described, even in cases where they are similar to those of a crystalline lattice whose structure is conventionally explained in terms of tight-binding arguments.

This is illustrated in Fig.3 for the example of liquid arsenic. Here the nearest neighbour distance D_{cp} in a close-packed structure falls on a repulsive hump of the Friedel oscillations in $\Phi(R)$, and this causes a splitting of the nearest neighbour shell into two subshells centred roughly at

$$D_{1,2} = D_{cp} \pm \lambda_F/2 \tag{8}$$

such as to put the atoms close to the minima of the Friedel oscillations. The coordination number C in the nearest neighbour shell may be estimated from the simple relation (see p.424 in [1])

$$\{(C+1)/13\}^{1/3} = D_1/D_{cp} \tag{9}$$

with the result C=3-4. An actual molecular dynamics simulation of liquid arsenic [37] shows that the coordination number in the liquid is nearly exactly three, as in rhombohedral arsenic. This is well confirmed by diffraction studies (see Fig.3(b)). Note that this is just a real-space argument for the occurrence of a Peierls distortion (which is conventionally described in a TB-language [40,41]), with the additional bonus that it allows us to predict the magnitude of the distortion.

Two points should be noted. The first is the important role of the volume forces. It is the volume-pressure which places the nearest neighbours in a close-packed structure at a distance where the force constants are negative and force the structure to distort in the way actually observed. The second point concerns the nature of the pseudopotential. Even the simplest local pseudopotential (an Ashcroft empty-core potential) is sufficient to model the trend from metallic to covalent structures. The return to metallic structures in the heavy elements is a more subtle effect. It is intimately related to relativistic effects which tend to bind s-electrons more strongly than p-electrons, hence a quantitative description requires nonlocal pseudopotentials even though qualitatively the effect may be modelled by stretching the trends in the parameters of the local pseudopotentials [42]. The relativistic effects are most evident in the electronic structure where they lead to a splitting of the s- and p-bands, both in the crystalline and in the liquid metals.

NFE perturbation theory continues to make important contributions to the theory of binary alloys. We have already mentioned in Sect.2 that even first-order perturbation theory predicts the volume of formation of binary sp-bonded alloys better than any of the empirical laws [19], and the trend in the heat of formation [20,35]. The volume effects are intimately related to the effective atomic size. It is well known that even in intermetallic compounds whose structures seem to be well explained in terms of the geometrical requirements of sphere-packing, quite large deviations from the ideal size-ratio can occur. A striking example is afforded by the AB_2 Laves phases: the ratio R_A/R_B varies between $R_A/R_B \sim 1.07$ and $R_A/R_B \sim 1.55$ (calculated in terms of the nominal atomic radii), compared with an ideal radius ratio of $R_A/R_B = 1.225$. However, this does not mean that the Laves phases are not properly considered as the packing of different sized spheres. Pseudopotential calculations [33,43] show that the effective radius of the atom varies with the average electron density via the variation of the electronic screening length, of the Friedel wavelength, and of the effective pseudopotential in the alloy. Incidentally the physical mechanism underlying the variation of the effective atomic size in metallic alloys was correctly described fifty years ago by Pauling [44] in terms of competing electronic rearrangements (it is better to avoid the term charge transfer in this context) associated with electronegativity and electroneutrality. Quite generally this leads to a compression of the elements with the lower valence and explains an effect known for a long time under the name of "chemical compression". In $CaAl_2$ for example we find $R_{Ca}/R_{Al} \sim 1.18$ at a nominal value of $R_{Ca}/R_{Al} \sim 1.40$ [43]. These calculations show that, like in a close-packed element, the interatomic distances in topologically close-packed structure such as a Laves phase fall into the attractive minima of the interatomic potentials calculated for the proper composition of the alloy. Again these considerations may be extended to the disordered phases. Here they contribute to our understanding of the ubiquitous occurrence of short-range icosahedral order in liquid and glassy alloys [45,46], but also of the formation of different patterns of topological short-range order if the effective size ratio is too large to be compatible with icosahedral ordering [47].

Chemical ordering is another important topic. Perturbation calculations show that in binary alloys with large differences in the electron density of the components the interatomic potentials are incompletely screened at the nearest neighbour distance (note the role of the volume forces in determining this distance), and this leads to a preferred interaction between pairs of unlike atoms and chemical ordering [48]. A notorious example is given by the structures of both crystalline and liquid alloys of lithium with some polyvalent metals whose crystalline and liquid structures show salt-like ordering which is strongest at the stoichiometric composition (examples are Li-In, Li-Pb, Li-Sn,...). Note however that the bonding is still predominantly metallic, as shown by the very short-range nature of the ordering potential. Indeed selfconsistent calculations of the electronic structure show that although the conduction band of e.g. Li_7Pb_2 or $Na_{15}Pb_4$ is entirely dominated by the orbitals of the polyvalent metal, ionicities defined in terms of net ionic charges are close to zero, due to the delocalized nature of these orbitals [49].

At even larger differences in the electron densities of the components, the changes in the screening becomes so large that the first attractive minimum in the pair interactions between the polyvalent atoms (which in the pure metal is entirely covered by the repulsive core) is exposed [50], see Fig.4(a). In that case the structure of the alloy is dominated by the strong interactions among the polyvalent ions, leading to the formation of polyanionic cluster compounds. Examples are the alloys of lead and tin with the heavy alkali metals K, Rb, and Cs, and the germanides and silicides of the lighter alkali metals. In all these compounds tetrahedral clusters of the group-IV atoms are isolated from each other by alkali atoms [51,52]. In most of these alloys the forces stabilizing the clusters are so strong that the tetrahedral units are found even in the liquid alloys

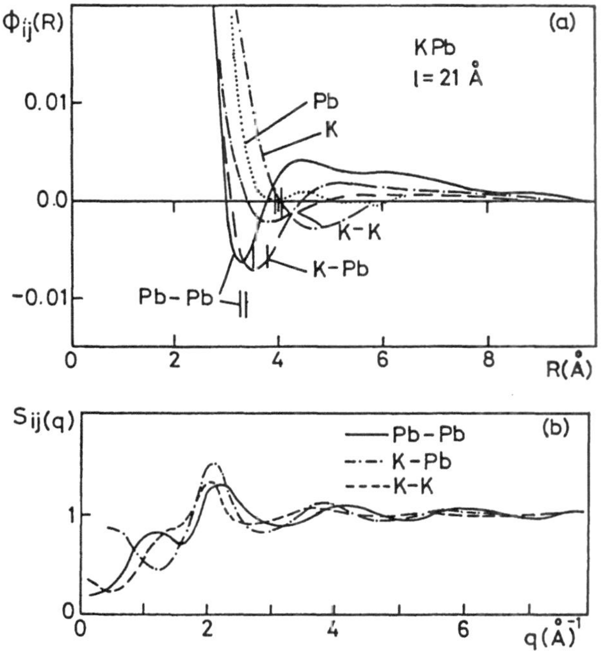

Fig. 4 (a) Interatomic potentials Φ_{ij} (R) in pure K and Pb metals, and in equiatomic alloy KRb. In the calculation of the potentials for the alloy, a finite mean-free path of the electrons of l = 21Å (consistent with the electronic transport properties) has been assumed. The vertical bars mark the nearest neighbour distances in the crystalline compound KPb.

(b) Partial static structure factors $S_{ij}(q)$ for a liquid KPb alloy. The peak in $S_{Pb\text{-}Pb}(q)$ at q ~ 1Å^{-1} indicates the formation of Pb-clusters (after Ref. 50).

[53]. It is quite surprising that NFE perturbation theory (albeit with slight corrections to account for the extremely short electronic mean-free-path in these strong-scattering systems [50]) describes the tendency to form polyanionic clusters so well, see Fig.4(b). Of course higher order corrections are important to get all the details right. With volume and pair-forces only we have bonds with no directional properties and no limit to the number of bonds an atom can form beyond the limit set by packing requirements. Thus the strong preferrential interactions between the anions will lead to the formation of anion tetrahedra, but occasionally also to the formation of polytetrahedral clusters or incomplete tetrahedra (triangles) with more or less than four atoms in a cluster. True covalent

bonding effects prefer perfect tetrahedra. However, selfconsistent electronic structure calculations [49] have shown that the bonding in these clusters is *very* different from a classical two-center two-electron bond, even if valence rules are satisfied (note that a Ge_4^{4-} cluster is isoelectronic to an As_4 molecule).

Another case where the interplay of chemical and topological ordering is important consists of the molten alloys of III-V and II-VI compounds. Recent work [54] has shown that NFE perturbation calculations allow us to understand the rather different behaviour of III-V and II-VI alloys on melting: III-V alloys undergo a semiconductor-to-metal transition, accompanied by a substantial increase in the coordination number (i.e. they behave similar to the group IV elements), whereas the II-VI alloys show only a modest increase in the conductivity and no change in the coordination number. Again this is well understood in terms of the interplay of pair- and volume-forces.

A remarkable application of perturbation methods beyond the limited scope of sp-bonded materials is the generalised pseudopotential perturbation theory for transition metals [55]. Again the volume energy makes a substantial contribution to the binding energy. Due to the directional properties of the d-orbitals, angular interactions are found to be more important than in sp-bonded metals and essential for the description of the dynamical properties.

4. The Implication of Many-Atom Forces for Glue Models

The main message of this section is obvious from section 2: we must expect a valid glue model to include the $\sqrt{C}$ effect of (1) in some form.

The first point is that there is no satisfactory way of representing the function $\sqrt{C}$ in the form

$$\sqrt{C} \approx \alpha C + \beta C\,(C-1)/2 + \gamma\, C\,(C-1)\,(C-2)/6 + . \qquad (10)$$

where successive terms represent the number of pairwise bonds, bond pairs i.e. involving three atoms, etc. We challenge the reader to try it. The trouble is that $\sqrt{C}$ increases most rapidly at small C whereas the higher terms in the power series have maximum gradient at high C. This is what is behind our statement that we believe there is no way of writing the energy as a sum of 2-atom, 3-atom etc terms. (But see also Postscript, section 7.)

A more sophisticated version of this problem arises in trying to develop the total band structure energy

$$U_{bs}\,(E_F) = \int_{E_{min}}^{E_F} E\, n\,(E)\, dE \tag{11}$$

in TB theory. This involves the density of states distribution n(E) which is a function of its moments μ_r:

$$\mu_r \; = \int_{E}^{E_{max}} E^r \; n\,(E)\, dE\,. \tag{12}$$

This function can be written down more or less using continued fractions [56]. The μ_r in turn are sums over chains of r atoms [57] and hence related to C. Thus (12) as a function of E_F is determined by the chains of atoms, but it is a very complicated dependence and certainly not a linear one as we would wish in order to write something in the spirit of (10).

Of course, for a small variation only around some value C_o one can write

$$\begin{aligned}\sqrt{C} &\approx \sqrt{C_o} + (C - C_o)\;/2\;\sqrt{C_o} + \;.....\\ &= \left(\tfrac{1}{2}\sqrt{C_o}\right) + \left(1/2\sqrt{C_o}\right) C + \;.....\end{aligned} \tag{13}$$

We could interpret the first (constant) term in (13) as part of the self-energy of the atom, and the second term as a pairwise bonding term proportional to C. But such an interpretation would be entirely spurious as is obvious from our derivation, and the numbers would depend on what one chose for one's C_o.

The scheme of Ercolessi and Tosatti [6] seems to offer useful possibilities. The total energy is written

$$U \;=\; \sum_i G\,(C_{eff,i}) \;+\; \tfrac{1}{2} \sum_{ij} \Phi\,(R_{ij}) \tag{14}$$

where G is the resonance energy and Φ a remaining pairwise repulsion at short distances. One has to define the effective coordination number C_{eff} for each atom i in terms of its neighbours j at varying distances R_{ij}

$$C_{eff,i} \;=\; \sum_j \; f\;(R_{ij}) \quad . \tag{15}$$

The function f is chosen as equal to unity at the nearest neighbour distance of the bulk equilibrium structure, in which case (15) simply reduces to C. Then f is somewhat greater or less than unity for closer or more distant neighbours. The method of Finnis and Sinclair [2] is equivalent to (14) with G taken proportional to $(C_{eff})^{1/2}$ in accordance with (1). However, it may be better to take the square root form for G as a first approximation and fit empirically around that with a function b(C):

$$G(C) = a\sqrt{C} = b(C) \quad . \tag{16}$$

A small problem is that the energy (14) is invariant under the transformation

$$\begin{aligned} &G \rightarrow G' = G(C) - d\,C \\ &\Phi \rightarrow \Phi' \rightarrow = \Phi(R) = d\,f(R), \end{aligned} \tag{17}$$

a linear term in G is equivalent to an extra pairwise force in Φ. For definiteness one should therefore define b(C) such that it is zero for the bulk value of C, say.

In contrast to (14), the 'effective medium' method of Norskov [4] is a generalisation of eq. (5). The effective electron density at the site of atom i is defined in some way in terms of its neighbours, and the selfenergy U_1 is the dominant term in (5): excellent as far as it goes. However, it is not clear that it really does justice to the $\sqrt{C}$ effect. Of course the method involves resonance and electron gas effects in its use of an appropriately defined free electron medium, but it is not clear to us that it incorporates specifically the dependence (1) on the coordination number, say at a transition metal surface or defect. In conclusion, the method certainly reduces to (5) for a NFE bulk metal, but there is doubt about whether it also reproduces the effect of (1) for its dependence on coordination number. (But see also Postscript, section 7.)

A little aside: there are of course situations of a restricted range within which pairwise forces are a good and legitimate representation of the energy. Consider the ordering of two transition metals A and B on the sites of a fixed bulk crystal structure. All the band and $\sqrt{C}$ effects are then constant, and one can define a pairwise interaction as a perturbation on a solid of average A/B atoms [58].

Finally, with any empirical or semi-empirical scheme, there is always the question of what data to fit the parameters to. This is a point to which we will return at the end of the next section.

5. Some Points About *ab Initio* Calculations

At the present time, the method of pseudopotentials with plane wave basis functions is the most developed technique for routine calculations for complex processes modelled in supercells often containing 50 to 100 atoms or even more [59]. We will confine our remarks to that.

The present audience will know that in 1985 a seminal paper by Car and Parrinello (CP) [29] gave the subject a big boost. We want to make a few remarks about that, without properly reviewing the whole subject, because in a sense CP introduced not just one idea but three new ideas simultaneously. No wonder that everyone was confused. People are still pursuing these ideas, separately and in various combinations, studying how they can be exploited further. The first idea is that it is wasteful to do an excellent job on one part of the calculation while still doing a lousy job (e.g. being far from convergence) on another part. CP approached convergence simultaneously in the solution of the one-electron wave equation, self-consistency in the electron density, and moving the atomic positions to find their equilibrium positions. To this one now adds convergence in sampling k-space, the size of the basis set, and perhaps different convergence characteristics of the upper and lower occupied bands. The second idea was that one can express the solution of a differential equation (the Schrodinger equation) in various mathematical forms. CP chose a time-dependent equation of motion, even for a static problem. The vogue has now shifted to energy minimisation by a conjugate gradient method with preconditioners. Both approaches appear to cope reasonably well with the electron self-consistency that was such an unstable problem in the traditional method. The third idea may ultimately turn out to be the most important one because it concerns how the size of the computation scales up with the size (ie number of atoms n) of the supercell. Consider a basis set of N functions, where N is proportional to n. The basic step is multiplying some approximate wave function ψ by a Hamiltonian $H = T + V$. Normally such a matrix multiplication takes N^2 steps. However, multiplying ψ by a diagonal matrix only takes N steps, and this applies to T when ψ is expanded in plane waves and applies to a local potential V when ψ is represented on a grid of points in real space. The rate-limiting process is the fast Fourier transform between the two representations which takes N log N steps. The challenge is to achieve this also for non-local pseudopotentials, including the calculation of the Hellmann-Feynman forces on the atoms.

We reiterate here a couple of points from Sections 1 and 3. Even in the context of total energy calculations, one wants an intuitive understanding of the results obtained, and here the perturbation method can be useful in some circumstances as shown by Section 3. Secondly a dynamic calculation on, say, a melt is computationally very expensive. In such a case a preliminary perturbation calculation can give a useful starting configuration for the atoms.

Finally we consider - very much in the spirit of this workshop - the contribution of total energy calculations to the development of the more empirical glue models of Section 4. As mentioned at the end of that section, what data does one fit empirical parameters to? The trouble is that fitting to experiment is not as satisfactory as it sounds. One wants to fit configurations with a range of coordination numbers, e.g. a surface, a vacancy, an interstitial etc. One may know the experimental energy, but unfortunately not usually the exact relaxed atomic positions. The difficulty is that the energy is just one number whereas a relaxed defect or surface requires much more information to specify it. This is where total energy calculations can fulfill an important role in providing both the total energy and the relaxed atomic positions. Or one can fit to the energy calculated with the atoms in their unrelaxed ideal locations. The situation is particularly bad if one wants to include something like the hot solid or a melt in the data base. In that case it would require a complete run of molecular dynamic simulation for every attempted adjustment of the model, which becomes prohibitive in computer time. How much better to use some artificial configuration to represent a snapshot of the thermal agitation in the hot solid or melt, for which the total energy has been calculated. It does not have to correspond exactly to the real system at any particular temperature as long as it incorporates more or less the right range of interatomic spacings and coordination numbers. Another point: fitting to experiment means that one is fitting only to low energy configurations that are observed easily. But if, for instance, the model is to give correctly an activation barrier, then it has to describe correctly the high energy configurations that are not easily observed. This is also necessary if one is to be confident that the model will not yield some quite spurious energy minima.

In conclusion, therefore, we suggest that total energy calculations could provide a most useful service in building up and testing the glue models. What is needed is a database of appropriate atomic configurations and their energies, all calculated from the same pseudopotential. These would constitute a testbed for any proposed glue model.

6. Acknowledgement

One of us (J.H.) gratefully acknowledges the hospitality at Clare College and the Cavendish Laboratory, Cambridge, where this paper was written, with the financial assistance of the Royal Society and the Science and Engineering Research Council.

7. Postscripts after the Workshop

1. Re the definition of many-atom interactions. In order to define many-atom forces precisely, one first has to define two-atom forces: and it is here that there is an ambiguity. We first note that the result (1) from μ_2 is exact, involves only two-atom coupling in the Hamiltonian, and does not depend on the directions in which the C neighbours lie even in the case of a d-band. Do we therefore describe the first approximation (1) for the cohesive energy as representing *pair-wise* forces? From one point of view we could say 'yes' because it is derived from purely pairwise effects, as just noted. On the other hand one could define pairwise forces as meaning an energy proportional to the number of pairwise bonds, ie proportional to C. It is all a matter of definition. Moriarty (this workshop) seems to be using the former definition when discussing his 3-atom and 4-atom interactions. This differs from the point made in Sect. 4 that $\sqrt{C}$ *already* implies many-atom effects in the second sense above. Thus Moriarty's 3- and 4-atom effects are in addition to those implicit in the $\sqrt{C}$ formula.

2. Re the $\sqrt{C}$ effect in effective medium theory (EMT). We remarked in Sect. 4 that we were not clear how the $\sqrt{C}$ effect is contained in EMT when applied to a NFE material. It is interesting that this has been clarified by the workshop. The upward curvature of the embedding energy as a function of electron density has a similar effect on the total energy as the upward curvature of (1) as a function of C. Moreover this curvature of the embedding energy with electron density arises from the kinetic energy of the embedding electron gas which seems to correspond to the resonance/delocalisation effect mentioned in Sect. 2

References

[1] V. Heine and D. Weaire, Solid State Phys. 24, 250 (1970).
[2] M.W.Finnis and J.E.Sinclair, Phil. Mag. A50, 45 (1984).
[3] M.J.Stott and E. Zaremba, Phys.Rev. B22, 1564 (1980).
[4] K.W.Jacobsen, J.K.Norskov and M.J. Puska, Phys.Rev. B35, 7423 (1987).
[5] S.M. Foiles, M.I.Baskes and M.S.Daw, Phys.Rev. B33, 7983 (1986).
[6] F.Ercolessi, M.Parrinello and E.Tosatti, Phil.Mag.A58, 213 (1988).

[7] A.P.Sutton, M.W.Finnis, D.G.Pettifor and Y.Ohta, J.Phys.C : Solid State Phys. 21, 35 (1988).
[8] A.T.Paxton, Phil. Mag. B58, 603 (1988).
[9] A.T. Paxton in *Atomistic Modelling of Matterials Beyond Pair Potentials* (Plenum Press, New York).
[10] O.K. Andersen and O.Jepson, Phys.Rev. Lett. 35, 2571 (1984).
[11] W.M.C.Foulkes and R. Haydock, Phys. Rev. B, 39, 12520 (1989).
[12] W.M.C.Foulkes and R. Haydock, to be published.
[13] B.M. Deb, ed. *The Force Concept in Chemistry* (Van Nostrand Reinhold, New York).
[14] R. Haydock, Solid State Phys. 35, 216 (1980).
[15] M.J. Kelly, Solid State Phys.35, 296 (1980).
[16] V. Heine, Solid State Phys. 35, 1 (1980).
[17} L. Pauling, Phys. Rev. 54, 899 (1938).
[18] M.W. Finnis, J.Phys.F: Metal Phys. 4, 1645 (1974).
[19} J.Hafner, J.Phys.F: Metal Phys. 15, L43 (1985).
[20] D.G. Pettifor, Solid State Phys. 40, 43 (1987).
[21] N.F.Mott and H.Jones, *The Theory of the Properties of Metals and Alloys* (Clarendon Press, Oxford, 1936) pp.125-128.
[22] V.Heine and R.O.Jones, J.Phys.C: Solid State Phys. 2, 719 (1969).
[23] Ref. [1] pp.354-359.
[24] M.C.Payne, P.D.Bristowe and J.D.Joannopoulos, Phys.Rev. Lett. 58, 1348 (1987).
[25] E.Tarrow, PO.D.Bristowe, J.D.Joannopoulos and M.C.Payne, J.Phys. Condensed Matter 1, 327 (1989).
[26] C.M.Cheng, R.J.Needs and V.Heine, J.Phys.C: Solid State Phys.21, 1049 (1988).
[27] J.J.A.Shaw and V. Heine, J.Phys. Condensed Matter (1989) to appear.
[28] A.Fleszar and R. Resta, Phys.Rev. B 34, 7140 (1986).
[29] R.Car and M. Parrinello, Phys.Rev.Lett. 55, 2471 (1985).
[30] A.Morita and T.Soma, Solid State Comm. 11, 927 (1972).
[31] K.J.Chang and M.L.Cohen, Phys.Rev. B31, 7819 (1985); B34, 8581 (1986).
[32] J.Hafner, Phys.Rev. B10, 4151 (1974); J.Hafner, A.Arnold, and N.Mauser, J.Phys.- Condensed Matter 1, 1033 (1989).
[33] J.Hafner, in *Cohesion and Structure*, vol.2, ed. by D.G.Pettifor and F.R.de Boer (North Holland, in print).
[34] J.Hafner and G.Kahl, J.Phys.F: Metal Physics 14, 2259 (1984).
[35] J.Hafner, *From Hamiltonians to Phase Diagrams,* Solid State Sciences vol. 70 (Springer, Berlin 1987).
[36] J.Hafner and V.Heine, J.Phys.F: Metal Physics 13, 2479 (1983).
[37] J.Hafner, Phys.Rev.Lett. 62, 784 (1989).

[38] W.Jank and J.Hafner, Europhys.Lett. 7, 283 (1988).
[39] J.Hafner, this volume.
[40] P.B.Littlewood, CRC Critical Reviews in Solid State and Materials Science 11, 235 (1984).
[41] J.P.Gaspard, I.Marinelli, and A.Pellegati, Europhys.Lett. 3, 195 (1987).
[42] J.Hafner and W.Jank, Phys.Rev. B (submitted).
[43] J.Hafner, Phys.Rev. B21, 406 (1980).
[44] L.Pauling, *The Nature of the Chemical Bond*, 1st. ed., Sect.11 (Cornell University Press, Ithaca 1939).
[45] J.Hafner, J.Phys.F: Metal Physics 18, 153 (1988); J.Hafner, S.S.Jaswal, M.Tegze, A.Pflugi, J.Krieg, P.Oelhafen and H.J.Guentherodt, ibid. p.2583.
[46] D.R.Nelson, Phys.Rev. B28, 5515 (1983).
[47] J.Hafner and M.Tegze, J.Phys.- Condensed Matter in print.
[48] A.Pasturel, J.Hafner, and P.Hicter, Phys.Rev. B32, 5009 (1985).
[49] M.Tegze and J.Hafner, Phys.Rev. B39, 8263 (1989).
[50] J.Hafner, J.Phys.- Condensed Matter 1, 1133 (1989).
[51] H.Schaefer and B.Eisenmann, Rev.Inorg.Chem. 3, 29 (1981).
[52] J.Witte and H.G. von Schnering, Z.Anorg.Chem. 327, 260 (1964).
[53] H.T.J.Reijers, W. van der Lugt, and C. van Dijk, Physica B+C 144, 404 (1987).
[54] J.Hafner and W.Jank, J.Phys. Condensed Matter in print and to be published.
[55] J.A.Moriarty, Phys.Rev. B38, 3199 (1988).
[56] Pages 288-289 of Ref [14].
[57] Page 81 of Ref [16].
[58] F. Ducastelle and F Gautier, J Phys. F: Metal Phys. 6, 2039 (1976).
[59] For a review see J. Ihm, Rep. Prog. Phys. 51, 105 (1988).

The Effective-Medium Theory

K.W. Jacobsen

Laboratory of Applied Physics, Technical University of Denmark, DK-2800 Lyngby, Denmark

The physical picture behind the effective-medium theory for bonding in metallic systems is presented. The theory is illustrated by a number of applications including the oxygen-induced reconstruction of the copper (110)-surface. The importance of the so-called one-electron energy for a correct description of the transition metals is discussed.

Introduction

The description and understanding of bonding in metallic systems has improved considerably in recent years with the development of density-functional theory[1] as one of the milestones. The electronic-structure problem is tremendously simplified with the treatment of the exchange and correlation problem in the local-density approximation (LDA) and this has made possible very accurate large-scale calculations of structural properties of many metallic systems. Alongside with the improvements in these *"ab initio"* methods the development of a more approximate and at the same time simpler effective-medium theory of bonding has taken place. The effective-medium theory is also based on density-functional theory. The idea behind the development of simpler methods than the full local-density approach is at least twofold. First, it is useful to get a simple physical model for the bonding in metallic systems to be able to understand various phenomena by extracting the dominating physical mechanism behind it.Secondly, the study of many temperature- or time-dependent phenomena with Monte Carlo or Molecular Dynamics techniques require a much more efficient way of calculating the total energy than full-blown LDA calculations.

In the present paper we will concentrate on the first of these aspects and the physical picture behind the effective-medium theory will be outlined and illustrated through a few selected applications. It will be shown how simple it is to make analytic estimates of energies like a surface energy or a surface-reconstruction energy in this model. For an overview of the applications of the effective-medium theory the reader is referred to ref. [2]. The calculational details are described in [3, 4, 2].

The paper is organized as follows. In the first section the basic physical picture behind the effective medium theory will be presented together with a short discussion of the derivation from density-functional theory. The second section will illustrate the picture by application to studies of chemisorption, surface relaxations and reconstructions. It will be shown how simple analytical estimates of total energies of simple metals can be made. The third section is devoted to a discussion of the hybridization- or one-electron energy for the transition metals. In this context it is discussed how concepts from the LMTO method are very well suited to be used in the framework of the effective-medium theory. Section 4 contains a discussion of chemisorption with special emphasis on the importance of the one-electron energy in the description of oxygen-induced surface reconstructions.

Springer Proceedings in Physics, Vol. 48 Many-Atom Interactions in Solids
Editors: R.M. Nieminen · M.J. Puska · M.J. Manninen

1 The Effective Medium Picture

At the beginning of this decade a method to describe chemisorption of light gas atoms on metals was put forward by Stott and Zaremba (the "quasiatom" method)[5, 6] and by Nørskov and Lang ("the effective-medium theory")[7, 8]. The basic idea behind the effective-medium theory is the following: an atom in a metallic environment does not feel the detailed structure of the nearby metal atoms due to the effective screening in metallic systems. It may therefore be possible to model the environment of the atom in question by a simpler host than the true solid or surface. In the effective-medium theory the chosen model-host is a homogeneous electron gas, and the binding energy of the atom in the solid is related to the embedding energy of the atom in the homogeneous electron gas, which can be calculated within the local-density approximation once and for all. This seems to be an appropriate choice for metallic systems, because the ability of the atom to attract or repel valence electrons should be well represented in this model-host. The only parameter describing the homogeneous electron gas is its density, and the environment of the atom in the real host is therefore described effectively by just a single parameter.

The effective-medium theory has been used to discuss such chemisorptive properties as binding energy and bond lengths of hydrogen,[9] oxygen,[10] and carbon[11] on transition metal surfaces, and to study hydrogen in metals.[12]

Recently, the effective-medium theory has been modified and extended to be able to describe not only a single gas atom embedded in a metallic host but also a set of interacting atoms forming a metallic system.[3, 2] In the framework of density-functional theory[1, 13] an expression for the binding energy of a system of interacting atoms can be derived from an ansatz for the total electron density of the system. Due to the variational properties of the total energy as a functional of the electron density, a deviation of the ansatz for the density from the true ground state density will give rise to only second and higher order corrections in the total energy. A reasonable ansatz for the ground-state electron-density should therefore provide a good estimate of the total energy of the system.

Before we define the ansatz for the electron density used in the effective-medium theory, it is useful to introduce the following notation: An atom embedded in a homogeneous electron gas induces changes in the electron density and the electrostatic potential. These induced changes will be denoted by the prescript Δ, so $\Delta n(\mathbf{r})$ denotes the difference between the total electron density of the homogeneous electron gas with the atom embedded and the electron density of the homogeneous electron gas alone. The total charge contained in the induced density equals the negative of the nuclear charge of the atom. The shape of the induced density $\Delta n(\mathbf{r})$ depends on the density of the homogeneous electron gas. We shall often refer to the density of the homogeneous electron gas as the embedding density of the atom.

In the effective-medium theory, the ansatz for the electron density $n(\mathbf{r})$ of a system of atoms is constructed as a superposition of the induced densities described above:

$$n(\mathbf{r}) = \sum_i \Delta n_i(\mathbf{r}), \tag{1}$$

where the sum runs over all the atoms in the system. As mentioned, the induced densities $\Delta n_i(\mathbf{r})$ depend on the embedding densities and the electron density $n(\mathbf{r})$ is therefore not well-defined before we specify what embedding density we should use for each atom. The embedding density $\bar{n}_i$ for atom i should depend on its environment, *i.e.*, the position of neighbouring atoms. In the effective-medium theory, $\bar{n}_i$ is defined as the average over a neutral sphere around atom i of the electron density coming from the other atoms in the system:

$$\bar{n}_i = \left\langle \sum_{j \neq i} \Delta n_j \right\rangle_i . \tag{2}$$

The averaging is performed over a spherical region of space for two reasons. First, we have to use a local region around the atom for the averaging. The usual division of space in the so-called Wigner-Seitz cells could be used for most bulk properties. If we consider open defects or surfaces, however, the Wigner-Seitz cells are very large (infinite for a surface atom), and it is physically meaningless to use the average over such a cell. Second, the model host used (*i.e.*, the homogeneous electron gas) exhibits complete spherical symmetry and we cannot expect to describe detailed directional dependencies of the density. We can, therefore, just as well average over a simple spherical region of space. (Directional bonding can be included in a different way, as discussed later in the treatment of the transition metals).

The requirement that the spheres should be neutral gives a relation between the "size" of an atom and its environment. Consider an atom embedded in a homogeneous electron gas. This system is neutral as a whole due to the presence of a compensating positively charged background. In the real metallic system in which the atom is embedded, this background is of course not present. Therefore, a natural way to define the radius of a neutral atomic sphere is to require the total charge, consisting of electrons (*i.e.*, the sum of the electron gas density and the density induced by the atom) and the charge of the nucleus, to vanish inside the sphere. For a given atom this gives a relation between the density of the homogeneous electron gas (the embedding density) $\bar{n}$ and the neutrality radius s. To a good approximation, this relation is exponential in the density range of interest for metallic systems for all the atoms that have been investigated. It can be written

$$\bar{n}(s) = n_0 \exp\left[-\eta(s - s_0)\right], \tag{3}$$

where the exponent η depends only on the atomic number of the embedded atom. The parameters n_0 and s_0 will be defined and discussed later. Eq. 3 establishes a relation between the "size" of the atom (*i.e.*, the neutrality radius s) and the environment (*i.e.*, the embedding density $\bar{n}$) in which the atom is situated.

The reader is referred to Ref. [3] for a detailed derivation of the binding energy of a metallic system based on the ansatz for the density discussed above. Here we want to emphasize the physical picture that follows from this energy-expression through applications to different phenomena like spontaneous and adsorbate-induced surface relaxations and reconstructions.

The effective-medium result for the binding energy E_B (*i.e.*, the total energy of the system minus the energy of the separated atoms) is given as a sum of three terms[3]

$$E_B = \sum_i E_{c,i}(\bar{n}_i) + E_{\mathrm{AS}} + E_{\mathrm{1el}}. \tag{4}$$

The three terms on the right hand side are the cohesive function, the overlap energy, and the one-electron energy, respectively. Their interpretations are as follows:

The cohesive function

The first term in Eq. 4 is related to the energy for embedding each atom in a homogeneous electron gas of the appropriate density. The *cohesive function* $E_{c,i}$ for a given atom i is defined as

$$E_c(\bar{n}) = E^{\mathrm{hom}}(\bar{n}) - \alpha(\bar{n})\bar{n}. \tag{5}$$

In this expression $E^{\mathrm{hom}}(\bar{n})$ denotes the embedding energy of the atom in a homogeneous electron gas of density $\bar{n}$ (that is, E^{hom} is the difference between the energy of the system consisting of the atom embedded in the homogeneous electron gas and the energy of the homogeneous electron gas and the atom, separately). The last term in Eq. 5 describes the electrostatic attraction between the density tails from the neighbouring atoms sticking into the neutral sphere and the charge associated with the atom in the sphere. This electrostatic contribution appears explicitly in Eq. 5 because it is not included in the embedding calculation due to the compensating positively charged background. The quantity α is the integral over the neutral sphere of radius s of the induced electrostatic potential $\Delta\phi$

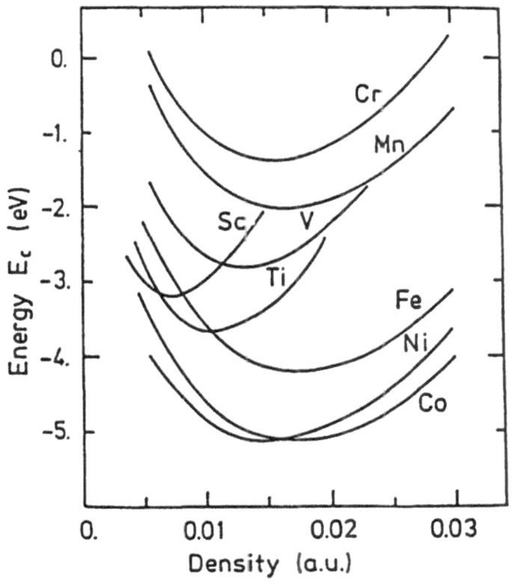

Figure 1: The cohesive functions E_c for the 3d transition metals calculated in the local-density approximation[3].

$$\alpha = \int_s \Delta\phi(\mathbf{r})\, d\mathbf{r}. \tag{6}$$

For a given atom, the cohesive function can be calculated once and for all as a function of embedding density.

The cohesive functions E_c calculated by Puska[3] in the local-density approximation are shown for the $3d$ transition metals in Fig. 1. In the energy range shown on the figure the embedding energies E^{hom} are sharply rising functions of density and the electrostatic contributions $-\alpha\bar{n}$ get more and more attractive as the density is increased. The resulting cohesive functions all exhibit a single minimum for a certain density n_0 that only depends on the kind of atom (*i.e.*, the atomic number) considered. The simple interpretation of these curves is that a given atom in a metallic system will seek an environment with a certain electron density in order to minimize its energy.

The overlap energy

The second term (E_{AS}) in Eq. 4 is the overlap energy or the atomic-sphere-correction energy. This term is electrostatic in nature and comes about because we have chosen to describe the electrostatics by using neutral spheres. It is given by[3]

$$E_{\text{AS}} = -\frac{1}{2}\int d\mathbf{r}\, O(\mathbf{r})n(\mathbf{r})\phi(\mathbf{r}), \tag{7}$$

where the overlap function $O(\mathbf{r})$ is equal to $(m-1)$ if the space-point r lies in m spheres. n and ϕ denote the total electron density and the electrostatic potential, respectively. The overlap energy describes the electrostatic repulsion that occurs when neutral spheres are forced to overlap. For a metal with a perfect close-packed lattice the neutral spheres are the Wigner-Seitz spheres. If the lattice constant is changed, the sphere radii change as well to conserve the total volume. In the so-called atomic-sphere approximation (ASA) which is known to give good results for close-packed systems both the overlap of the spheres and the holes between them are neglected. The overlap energy is therefore assumed to vanish for a perfect fcc-lattice. However, this approximation is not reliable for configurations where the atomic spheres do have overlaps and in these situations the overlap energy comes into play. The explicit expressions for the overlap energy in the cases of pure metals and metals with adsorbates can be found in refs. [3, 2, 4].

The overlap energy is very important in the description of the elastic properties of a metal. Consider for example a homogeneous distortion of an fcc-metal that leaves the local volume in the metal unchanged to lowest order in the distortion parameter. The distortion could for example be the one associated with the elastic constant C_{44} or a frozen zone-boundary phonon. For such a distortion the change in the embedding density of one of the metal atoms would be of second order in the distortion parameter. Because the metal atoms in equilibrium are positioned so that the embedding density minimizes the cohesive function the contribution from the cohesive function to the distortion energy will be of fourth order in the distortion parameter. The elastic constant C_{44} or a zone-boundary phonon frequency are therefore independent of the cohesive function in this approximation. For the simple metals these quantities are therefore determined by the electrostatic overlap energy alone.

The one-electron energy

The third term E_{1el} in the binding energy (Eq. 4) is called the one-electron energy and is important in situations where the electronic environment in which an atom is embedded deviates considerably from a homogeneous electron gas. The one-electron energy is defined in terms of the local density of states. Let $\Delta n_i(\mathbf{r}, \epsilon)$ denote the local density of states induced by embedding the atom in a homogeneous electron gas. Similarly, let $\Delta n_i^{met}(\mathbf{r}, \epsilon)$ be the change in the local density of states induced by embedding the atom in the real metal. The one-electron energy is then defined by[3]

$$E_{1el} = \sum_i \int_s d\mathbf{r} \int_{-\infty}^{\epsilon_F^{met}} d\epsilon\, \epsilon \Delta n_i^{met}(\mathbf{r}, \epsilon) - \int_s d\mathbf{r} \int_{-\infty}^{\epsilon_F} d\epsilon\, \epsilon \Delta n_i(\mathbf{r}, \epsilon), \tag{8}$$

where ϵ_F^{met} and ϵ_F denote the Fermi-energy in the metal and the homogeneous electron gas, respectively. The same potential for the atom should be used in calculating both $\Delta n_i(\mathbf{r}, \epsilon)$ and $\Delta n_i^{met}(\mathbf{r}, \epsilon)$. If the environment of the atom is free-electron-like, as it is for a simple metal, this term can be neglected. In Section 4 we shall discuss the implications of this contribution to the energy in the case of transition metals with and without gas-atoms chemisorbed.

2 Applications to Simple-Metal Systems

As outlined in the previous section the binding energy of a simple metallic system is composed of two different pieces: a) the contributions from the density-dependent cohesive functions for all the atoms in the system and b) the electrostatic repulsion between overlapping atomic spheres. In this section we shall illustrate some of the consequences of this simple expression for the binding energy through a number of applications.

Bulk properties of simple metals

To begin with a simple example consider a simple metal with the atoms positioned on a perfect fcc lattice. The overlap energy is in this case identically zero in the atomic-sphere-approximation as mentioned in the previous section. In the equilibrium configuration all the atoms are positioned at the embedding density n_0 that minimizes the cohesive function. The cohesive energy is therefore simply the minimum-value of the cohesive function and the equilibrium lattice constant is obtained by relating the embedding density to the radius s_0 of a neutral atomic sphere by the exponential relation Eq. 3. Furthermore, the bulk modulus becomes proportional to the curvature of the cohesive function at the minimum-point. An analysis of the LDA-calculation of the atom embedded in a homogeneous electron gas shows that this curvature is mainly determined by the kinetic energy of the electron gas. Typically the value for a lattice constant calculated in this way is within 5% of the one obtained by a full self-consistent LDA-calculation. Bulk moduli usually agree within 20-30% with results of benchmark calculations.

It can be noted already at this stage that because of the non-linearity of the cohesive function the binding-energy-expression Eq. 4 has a many-body (*i.e.*, non-pair-potential) feature included.

To see this consider the energy change when an atom (a) is approaching another atom (b) that is positioned in some metallic environment. Assume first that atom b is already sitting in an environment that supplies the equilibrium embedding density for this atom. The extra embedding density contributed by atom a will then have a rather small effect on the energy of atom b because the cohesive function is at its minimum. If by comparison atom b had been positioned at an embedding density very different from the equilibrium one the extra embedding density would have a greater effect on the energy of this atom.

Simple-metal surface

Let us now consider a simple-metal surface where we first assume that the atoms do not relax away from the bulk lattice positions. The lower coordination number of a surface atom implies that the embedding density for a surface atom is less than the optimum density n_0. To estimate the surface energy we first calculate the energy $E(N)$ of an atom with coordination number N (*i.e.*, with N nearest neighbours positioned a bulk nearest-neighbour-distance away). If the energy-zero is chosen as the energy of a bulk atom this energy is given by[14, 2]

$$E(N) = E_2\left[\left(\frac{N}{12}\right)^{\gamma} - 1\right]^2 + \tilde{\alpha}\left[\left(\frac{N}{12}\right)^{\gamma} - \frac{N}{12}\right], \tag{9}$$

where E_2 is the second derivative of the cohesive function at the minimum point divided by $2n_0^2$ and $\tilde{\alpha}$ is the electrostatic parameter α defined in Eq. 6 multiplied by n_0. The dimensionless exponent γ is given by the fall-off of the electron-density. The parameters are in the case of Aluminium (Copper) given by $E_2 = 1.12\text{eV}(1.33)$, $\tilde{\alpha} = 8.96\text{eV}(17.1)$, and $\gamma = 0.87(0.91)$. In expression 9 the term proportional with $\tilde{\alpha}$ is the electrostatic overlap energy. The parameters E_2 and $\tilde{\alpha}$ can also easily be estimated from the elastic properties of the metal because as already mentioned E_2 is proportional with the bulk modulus B and $\tilde{\alpha}$ is proportional with C_{44}.

The lower embedding density for a surface atom results in two different contributions to the surface energy as can be seen from Eq. 9. First, the cohesive function is no longer at the minimum and, second, there is a larger atomic sphere associated with the surface atom and therefore an increased overlap energy. Using the parameters for aluminium listed above the surface energy E_S per atom for a (110)-surface is found to be $E_S = E(8) - E_0 = 0.43$ eV which is in reasonable agreement with the measured value of 0.6 eV. In the upper part of Fig. 2 the surface energies for Cu, Ag, and Au calculated in this way are compared to the experimental values.

The expression for the energy of an atom as a function of coordination number (Eq. 9) has been used to analyze the (1×2) missing-row reconstruction observed on the (110)-surfaces of Au, Pt, and Ir[14]. The (1×2) reconstructed surface can be obtained by removing every second $[1\bar{1}0]$-row from the topmost layer of the unreconstructed surface. If we look at a 1×2 surface-unit-cell of the unreconstructed surface we have 2 atoms with 7 neighbours and 2 atoms with 11 neighbours. For the missing-row reconstructed unit cell there are, on the other hand, 1 atom with 7 neighbours, 2 atoms with 9 neighbours, and 1 atom with 11 neighbours. The total number of "lost" neighbours or "broken bonds" is seen to be the same for the reconstructed and unreconstructed surfaces but the distribution of nearest neighbours per surface atom is different. The energy change $\delta E^{1\times 2}$ associated with the reconstruction is therefore quite small (of the order 40 meV per 1×2-unit cell for Au). It is given by

$$\delta E^{1\times 2} = -(E(11) + E(7) - 2E(9)) \approx -4\left.\frac{d^2E}{dN^2}\right|_{N=9}. \tag{10}$$

From Eq. 9 it can be seen that the reconstruction is associated with a lowering of the kinetic energy and an increase of the electrostatic energy. In Fig. 2 the reconstruction energy is shown for Cu, Ag, and Au. It is clearly seen that the tendency to reconstruct is much larger in the case of Au. Fig. 3 shows how the elastic constants B and C_{44} can be used to make a "phase diagram"

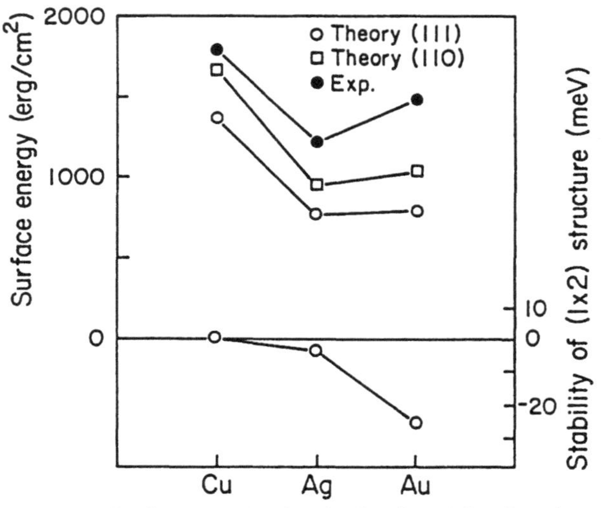

Figure 2: Surface energies for the (110) and (111) surfaces of Cu, Ag, and Au calculated[14] using Eq. 9 compared to the experimental values[32] deduced from measurements of the surface tension at the melting point. The experimental values are thus rather uncertain and represent an average of various crystal planes and defects. The lower part of the figure shows the stability of the (1×2) reconstruction for these metals (reconstruction energy per 1×1 unit cell). The larger tendency towards reconstruction of the Au(110) surface is clearly seen. The points for Ag and Au have been obtained by using the experimental bulk modulus and C_{44} together with the Cu-value for the exponent γ.

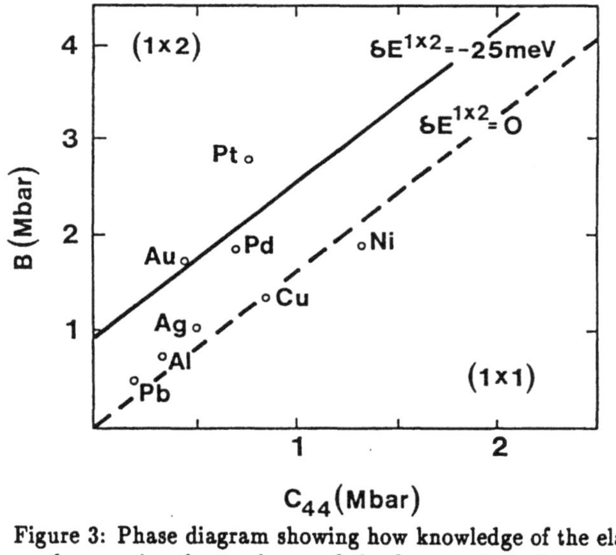

Figure 3: Phase diagram showing how knowledge of the elastic constants B and C_{44} can be used to characterize the tendency of the fcc metals to form the (1×2) missing row structure. The experimental data are from Ref. [33]. The $\delta E^{1\times 2} = 0$ line separates the (1×1) and (1×2) regions according to Eq. 10. Due to the approximate nature of the calculation the separation is not complete. There are a number of metals that do not reconstruct that are close to but on the wrong side of the line. Also shown is the $\delta E^{1\times 2} = 25$meV line which separates the metals that reconstruct from those that do not.

for the missing-row reconstruction of the fcc-metals. The tendency to reconstruct increases with increasing B and decreasing C_{44} in accordance with the statement that the reconstruction is driven by a gain in kinetic energy and opposed by the electrostatics. For a more detailed discussion and references to experimental and theoretical work on the missing-row reconstruction the reader is referred to Ref. [14].

Let us now turn to a discussion of relaxations of metal surfaces. Experimentally it is usually found that the first metal layer relaxes inwards so that the spacing between the first and the second layer is decreased compared to a bulk interlayer spacing. This fact is easily understood in the effective-medium approach. The inwards surface relaxation is a consequence of the change in coordination number of a surface atom compared to a bulk atom. The lower-than-optimum embedding density for a surface atom results in an inwards relaxation because the surface atoms move towards regions with higher electron density. The inwards relaxation will be counteracted by the overlap energy that rises as the atomic spheres for the first layer penetrate into the atomic spheres in the second layer.

The physical mechanism behind this relaxation can be analyzed further by referring to the fact that the minimum in the cohesive function comes about due to a competition between the kinetic repulsion caused by electron gas states being orthogonalized to the atomic states (E^{hom}) and the electrostatic attraction ($-\alpha\overline{n}$) between the atomic charge and the electron tails sticking into the atom (Eq. 5). The lower embedding density of the surface atoms results in a decrease in the kinetic repulsion. The electrostatic attraction can then pull the surface atoms closer to the second layer. This interpretation makes contact with the explanation proposed by Finnis and Heine[15].

It can be noted that the overlap energy for a simple metal is a sum of a density-dependent function and a pair-potential[3]. The binding-energy expression for a simple metal is then of the same functional form as the one used in the more empirically based embedded atom method[16].

Simple metals interacting with adsorbates

The binding energy expression Eq. 4 does also apply to a metal interacting with chemisorbed atoms and in the case of a simple metal the one-electron energy can be neglected. The chemisorption energy, bond length, and vibrational frequency for an atom chemisorbed on a metal surface can roughly be found from the cohesive function characteristic of the atom in question. This has been discussed in detail in Refs. [8, 9, 10, 11]. Also the trends in the heat of adsorption of hydrogen in metals can be understood just from knowledge about the interstitial electron density in the metals and the hydrogen cohesive function[12]. Here we shall concentrate on two effects where the response of the metal is important, namely adsorbate-induced surface relaxations and substrate-mediated adsorbate-adsorbate interactions.

Adsorbate-induced surface relaxation. We have discussed above that clean metal surfaces usually relax inwards compared to the bulk-terminated surface because the surface atoms try to increase the electron density in which they are embedded. If a layer of adsorbate atoms are chemisorbed on the metal surface the adsorbate atoms will also make a contribution to the embedding density of the surface atoms. This additional contribution to the embedding density of the surface atoms will lead to an outwards relaxation of the metal surface. The inwards relaxation of a clean metal surface can thus be removed by adsorption of atoms on the surface. The outwards relaxation is localized to the metal atoms close to the adsorbed atoms and depends on the coverage of the chemisorbed atoms. For example, a copper atom at the (110)-surface will relax 4% outwards if the embedding density is increased due to a single neighbouring hydrogen atom. The relaxation increases to 6% if the surface atom has two neighbouring hydrogen atoms[17]. A hydrogen-coverage-dependent outwards relaxation of a Cu(110)-surface has been observed in LEED experiments[18].

Substrate-mediated adsorbate-adsorbate interactions. The metal surface can mediate a repulsive interaction between adsorbed atoms. This can be seen in the following way. The energy gained by two surface atoms, each having a small increase δn in the embedding density due to the adsorbates, is bigger than the energy gained by adding an embedding density $2\delta n$ to a single surface atom. This follows from the fact that the energy gain per atom must saturate when the optimum density for a surface metal atom has been obtained. It will therefore be energetically favorable for the adsorbed atoms to spread out over the surface to give extra embedding density to as many surface atoms as possible, *i.e.,* there is an effective substrate-mediated repulsion between the chemisorbed atoms. This effect is very small for hydrogen, but can be substantial for adsorbates like oxygen or carbon.

It should be noted that the substrate-mediated adsorbate-adsorbate repulsion can change to an attraction if the ad-atoms go sub-surface. The presence of extra atoms just below the surface induces substantial outwards relaxations in the first metal layer. If only a single atom is positioned below the surface the relaxations will be localized to a limited region around the ad-atom and there will be an energy cost associated with the deformation of the first surface layer. If instead a full layer of ad-atoms is positioned below the surface of the metal, the ad-atoms can help each other in pushing out an undistorted first metal layer. There is therefore an effective *attraction* between subsurface ad-atoms and, as a consequence, ad-atoms tend to cluster in islands below the surface.

3 Transition metals

For the simple metals the one-electron energy is neglected due to the fact that the environment of an atom in a simple metal is well approximated by a homogeneous electron gas. The situation is different if we consider a transition metal where the narrow *d*-bands require a special treatment. The *d*-bands and therefore the one-electron energy is important for the understanding of many aspects of the transition metals. The *d*-band formation gives a considerable contribution to the cohesive energy of these metals especially for the metals with the *d*-bands half filled[19, 20]. Also the relative stability of the different possible crystal structures is to a large extend determined by the one-electron energy. This follows from the so-called force-theorem which has been used directly by Skriver[21] to calculate the structural energy differences for many metals.

Let us first consider a 3*d*-metal-atom embedded in a homogeneous electron gas. The atomic 3*d*-state becomes broadened to a resonance due to the coupling to the free-electron-like states in the electron gas. The resulting resonance is seen in Fig. 4 where the local density of states for the atom in a homogeneous electron gas is shown. One way to view this broadening is to assign a certain lifetime to the level associated with the process that an electron hops from the localized *d*-state to the states in the surrounding electron gas and back. The width of the level is then according to Fermi's golden rule proportional to the square of the matrix element for the hopping from the *d*-state to the gas-states. The d-state of a transition-metal atom positioned in a transition metal will of course also be broadened due to the hopping of electrons to and from the *d*-state. Here the possibility now arises for an electron to hop from one *d*-state to the free-electron-like states and then on to a *d*-state on another atom. This process results in a further broadening of the resonances into a *d*-band. This band-formation is of course not included in the calculation with just one transition-metal atom embedded in a homogeneous electron gas and has to be taken into account in the one-electron energy Eq. 8. However, it should be noted that the modelling of this band-formation energy can be done on the basis of information obtained for the atom in a homogeneous electron gas alone. This is so because the band-width is determined by the matrix element of the process that an electron hops from one *d*-state to the free-electron-like gas-states and then on to another *d*-state. But this matrix element can be estimated directly from knowledge about the matrix element that determines the width of the *d*-resonance for the atom in the homogeneous electron gas.

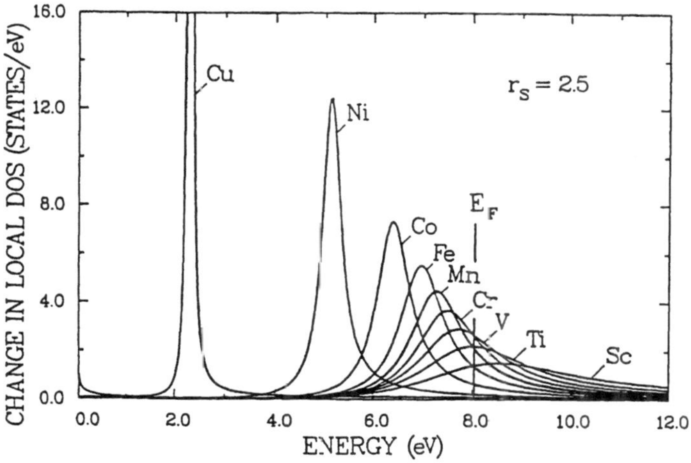

Figure 4: The local density of states for the $3d$-transition-metal atoms in an $r_s = 2.5$ homogeneous electron gas[3].

The result that the *interatomic* hopping matrix element that determines the d-band-width is given by an *intra-atomic* property has been discussed in detail previously by Pettifor[22], by Heine[23], and by Moriarty[24]. Here we shall only make a few comments on one method, namely the LMTO-method, which we find is particularly well suited to estimate the one-electron energy in the effective-medium theory in a systematic and efficient way.

Some of the key features of the LMTO-method can be summarized as follows[25]. The electronic structure problem can be approximated by a tight-binding problem in a certain energy-window. The determination of the tight-binding Hamiltonian to be used separates into two separate problems, namely the determination of the so-called potential parameters and the determination of the so-called structure constants. The potential parameters are determined by the potential in a single atomic sphere alone so these are the intra-atomic properties mentioned above. A comparison of the potential parameter that determines the d-band width for the $3d$-transition-metals calculated self-consistently in the bulk metal with the same parameter determined for a single atom in a homogeneous electron gas of the relevant density shows agreement within approximately 10%. This indicates that the potential for the atom in the homogeneous electron gas is indeed a good approximation to the potential in the metal. The structure constants on the other hand do not depend on the potential at all but only on the atomic positions.

A very simple model for the one-electron energy in the case of the $3d$-transition-metals with an fcc-lattice is obtained by approximating the d-band by a square-shaped band and then estimating the band-width from the potential parameters calculated for the atom in a homogeneous electron gas[3]. Fig. 5 shows the results for the Wigner-Seitz radii and the cohesive energies calculated with this approximation for the one-electron energy. The values calculated by Moruzzi *et al.* [26] are shown for comparison. All the trends are seen to be well described. In general, the radii are on the low side and this leads to an overestimate of the band-formation energy ($\propto 1/s^5$). If the band-formation energy is calculated at the same lattice constants used by Moruzzi *et al.* [26] the values indicated by squares in the figure are obtained. The effect of the one-electron energy is small for the transition metals to the right in the periodic system. It is therefore not unreasonable for these metals to include the one-electron energy approximately in the cohesive function or the overlap energy so that the simple form of the total-energy expression can be maintained. This is effectively what is done in the more empirical schemes like the embedded atom method[16]. However, there are situations where the transition-metal d-band has to be taken explicitly into account. This will

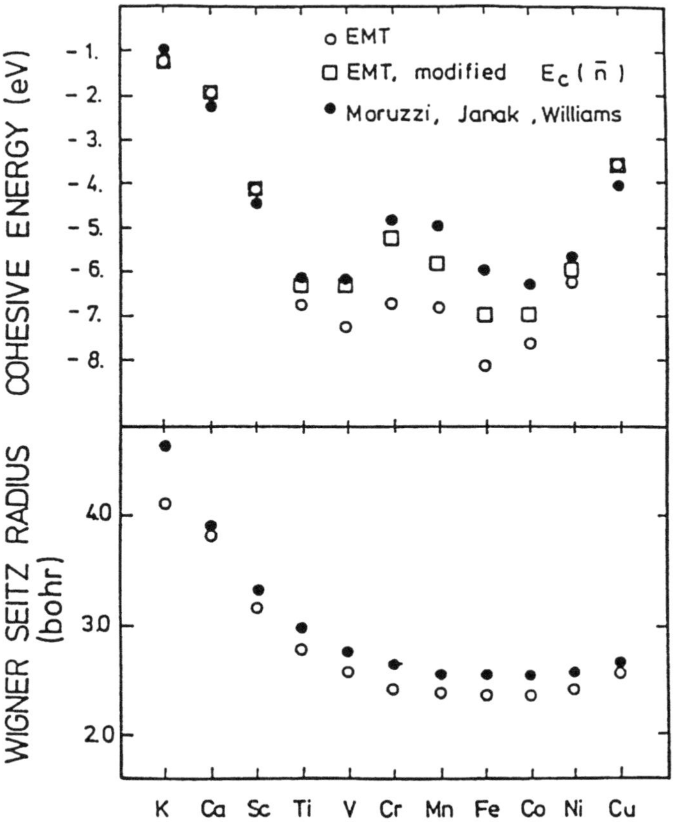

Figure 5: Comparison of calculated[3] cohesive energies and equilibrium Wigner-Seitz radii for the $3d$ transition metals with the results of self-consistent KKR calculations.[26] Each modified EMT point is the cohesive energy, where the cohesive function (E_c) has been rigidly shifted to lower density so that the Wigner-Seitz radius is equal to the one used by Moruzzi *et al.* ([26]).

be illustrated in the following section where a treatment of oxygen chemisorption on the copper (110)-surface will show that even the completely filled d-band of copper can have strong effects on the determination of the surface structure.

4 Oxygen chemisorption on transition metals

The narrow d-bands in the transition metals can also affect the properties of the surface with respect to chemisorption of for example hydrogen or oxygen. This is again included in the effective-medium theory through the one-electron energy. It is well known that the hybridization between, say, the s-level on a chemisorbed hydrogen and the transition-metal d-band leads to an extra binding that increases in magnitude as the filling of the d-band decreases. This effect can be described by using the Newns-Anderson model for a resonant state interacting with a band[27, 9]. The chemisorption of oxygen on transition-metal surfaces can in some cases lead to a complete restructuring of the metal surface. It is one such example for oxygen on a Cu(110)-surface that will be discussed in the following.

Figure 6: The minimum energy configuration for half a monolayer of oxygen on the copper (110) surface as calculated by the effective-medium theory. The oxygen atoms are positioned in the long bridge sites on a missing-row reconstructed surface.

The O/Cu(110)-surface exhibits a (2×1) LEED pattern for an oxygen-coverage of 1/2 which has lead to several suggestions of the underlying atomic structure. A large number of experiments have been performed on this system in order to determine the structure with varying results. There seems now to be a general agreement that the structure is of the missing-row type where every second of the rows in the [001] direction in the outermost layer is removed[28].

The result[29] of a minimization of the energy in the system with respect to the atomic coordinates using the effective-medium expression for the total energy is shown in Fig. 6. The obtained structure is indeed the missing-row structure that is observed experimentally. In the calculation the one-electron energy has been approximated by a Newns-Anderson model for the hybridization between the oxygen *p*-level and the copper electron states. The copper density of states has been approximated by a sum over a broad (*sp*-)band and a narrow *d*-band. The gain in energy of the system compared to the lowest-energy-structure with an unreconstructed surface is -0.4 eV per unit cell. This can be viewed as the result of two competing mechanisms. It costs about 0.3 eV to restructure the clean copper surface but the oxygen chemisorption energy is enhanced by 0.7 eV compared to the chemisorption energy for the unreconstructed surface. The reason for this increase in chemisorption energy is to be found in the one-electron energy. Without this contribution to the energy the oxygen would prefer to be positioned on the unreconstructed surface with a high coordination number. (The sum of the cohesive function and the overlap energy will in general result in a high coordination number of the adsorbate. This is so because the overlap energy is decreasing with increasing bond length, and the bond length is increased if as many atoms as possible contribute to the embedding density of the adsorbate). The outermost copper atoms on the missing-row reconstructed surface have a lower coordination number than on the unreconstructed one. This leads to a rise in the position of the *d*-level of these atoms. As a consequence some of the anti-bonding levels that are a result of the coppper-*d* – oxygen-*p* hybridization are shifted above the Fermi-level and emptied. Therefore the magnitude of the chemisorption energy is increased.

Conclusion

The effective-medium theory has been shown to be able to describe a range of phenomena including surface relaxations and adsorbate-induced surface reconstructions of metallic systems. The ingredients in the theory are the following: the electron density as a superposition of atomic densities, the associated density-dependent cohesive functions, the overlap energy that can be interpreted as the electrostatic repulsion of atomic spheres, and the one-electron energy which must be considered if the system is not free-electron like.

For simple metals the binding-energy expression is a sum of a density-dependent function and a pair-potential. The theory therefore provides a theoretical basis for the more empirical methods with the same functional form for the binding energy: "the embedded atom method"[16], "N-body potentials"[30] and "the glue model"[31].

For the transition metals the one-electron energy can be of great importance even for a closed-*d*-shell metal like copper as illustrated by the oxygen-induced reconstruction of the (110)-surface.

Acknowledgments

It is a great pleasure to thank my co-workers on the effective-medium theory Jens K. Nørskov, Martti Puska, and Per Stoltze for many interesting discussions.

References

[1] H. Hohenberg and W. Kohn, *Phys. Rev. B* **136**, 864 (1964).

[2] K. W. Jacobsen, *Comments Cond. Mat. Phys.* **14**, 129 (1988).

[3] K. W. Jacobsen, J. K. Nørskov, and M. J. Puska, *Phys. Rev. B* **35**, 7423 (1987).

[4] K. W. Jacobsen, *Ph. D. Thesis*, unpublished (Univ. of Copenhagen, 1987)

[5] M. J. Stott and E. Zaremba, *Sol. St. Comm.* **32**, 1297 (1979).

[6] M. J. Stott and E. Zaremba, *Phys. Rev. B* **22**, 1564 (1980).

[7] J. K. Nørskov and N. D. Lang, *Phys. Rev. B* **21**, 2131 (1980).

[8] J. K. Nørskov, *Phys. Rev. B* **26**, 2875 (1982).

[9] P. Nordlander, S. Holloway, and J. K. Nørskov, *Surf. Sci.* **136**, 59 (1984).

[10] B. Chakraborty, S. Holloway, and J. K. Nørskov, *Surf. Sci.* **152/153**, 660 (1985).

[11] K. W. Jacobsen and J. K. Nørskov, *Surf. Sci.* **166**, 539 (1986).

[12] J. K. Nørskov and F. Besenbacher, *J. Less-Common Met.* **130**, 475 (1987).

[13] W. Kohn and L. Sham, *Phys. Rev.* **140**, A1133 (1965).

[14] K. W. Jacobsen and J. K. Nørskov, in *The Structure of Surfaces II*, edited by J. F. van der Veen and M. A. van Hove, pg. 118 (Springer Series in Surface Sciences 11, 1987)

[15] M. W. Finnis and V. Heine, *J. Phys. F* **4**, L37 (1974).

[16] M. S. Daw and M. I. Baskes, *Phys. Rev. Lett.* **50**, 1285 (1983). M. S. Daw and M. I. Baskes, *Phys. Rev. B* **29**, 6443 (1984).

[17] K. W. Jacobsen and J. K. Nørskov, *Phys. Rev. Lett.* **59**, 2764 (1987).

[18] A. P. Baddorf, I.-W. Lyo, E. W. Plummer, and H. L. Davis, *J. Vac. Sci. Techn. A* **5**, 782 (1987).

[19] J. Friedel in *The Physics of Metals* edited by J. M .Ziman (Cambridge Univ. Press, Cambridge, England, 1978) Vol I.

[20] D. G. Pettifor, *Phys. Rev. Lett.* **42**, 846 (1979).

[21] H. L. Skriver, *Phys. Rev. B* **31**, 1909 (1985).

[22] D. G. Pettifor, *J. Phys. C* **5**, 97 (1972).

[23] V. Heine, in *Solid State Physics*, edited by H. Ehrenreich, F. Seitz, and D. Turnbull (Academic, New York, 1980) Vol 35.

[24] J. A. Moriarty, *J. Phys. F* **5**, 873 (1975).

[25] O. K. Andersen and O. Jepsen, *Phys. Rev. Lett.* **53**, 2571 (1984).

[26] V. L. Moruzzi, J. F. Janak, and A. R. Williams, *Calculated Electronic Properites of Metals* (Pergamon, New York 1978).

[27] D. M. Newns, *Phys. Rev.* **178**, 1123 (1969).

[28] R. Feidenhansĺ, F. Grey, R. L. Johnson, S. G. J. Mochrie, J. Bohr, and M. Nielsen, preprint.

[29] J. K. Nørskov and K. W. Jacobsen, to be published.

[30] M. W. Finnis and J. E. Sinclair, *Phil. Mag. A.* **50**, 45 (1984).

[31] F. Ercolessi, E. Tosatti, and M. Parinello, *Phys. Rev. Lett.* **57**, 719 (1986).

[32] W. R. Tyson and W. A. Miller, *Surf. Sci.* **62**, 267 (1977).

[33] C. Kittel, *Introduction to Solid State Physics*, Wiley and Sons, NY (1971). G. Simmons and H. Wang, *Single Crystal Elastic Constants and Calculated Aggregate Properties: A Handbook*, MIT Press, Cambridge Massachusetts (1971).

The Embedded Atom Method: A Review*

M.S. Daw

Theoretical Division, Sandia National Laboratories,
Livermore, CA94551, USA

Abstract. The Embedded Atom Method (EAM) has become a popular replacement for interatomic pair potentials in total energy calculations of metallic systems. The EAM relates the embedding energy of an atom in an electron gas to the cohesive energy of a metallic solid. The resulting model is no more computationally demanding than simple interatomic potentials, but incorporates some important many-atom interactions in the expression for the total energy. In this paper, we discuss several points of interest:

(1) the EAM is heuristically derived from the density functional for the energy;
(2) the practice of obtaining semi-empirical functions is discussed;
(3) the EAM form is interpreted in terms of many-atom interactions; and,
(4) several applications are reviewed.

1. Introduction

The modern study of realistic defects in metals (such as grain boundaries, dislocations, high-index or defected surfaces) requires the use of a fairly simple model which represents a reasonable approximation to metallic properties. Until a few years ago, the common approach was to assume a pairwise interaction between atoms. It was realized long ago that these pair potential schemes leave out much of the essential physics by neglecting the effect of coordination on bond strength. But it was not until recently that a practical scheme for improving these calculations was devised.

Daw and Baskes[1, 2] have proposed a new framework for calculating the energetics of metals which they call the Embedded Atom Method (EAM). In this approach, the energy of the metal is viewed as the energy obtained by embedding an atom into the local electron density provided by the remaining atoms of the system. In addition, there is an electrostatic interaction. The ansatz they used is

$$E_{coh} = \sum_i G_i\Big(\sum_{j\neq i} \rho_j^a(R_{ij})\Big) + \frac{1}{2}\sum_{i,j(j\neq i)} U_{ij}(R_{ij}) \qquad (1)$$

* Work supported by U S. Department of Energy, Office of Basic Energy Sciences, Division of Materials Sciences.

Springer Proceedings in Physics, Vol. 48 — Many-Atom Interactions in Solids
Editors: R.M. Nieminen · M.J. Puska · M.J. Manninen

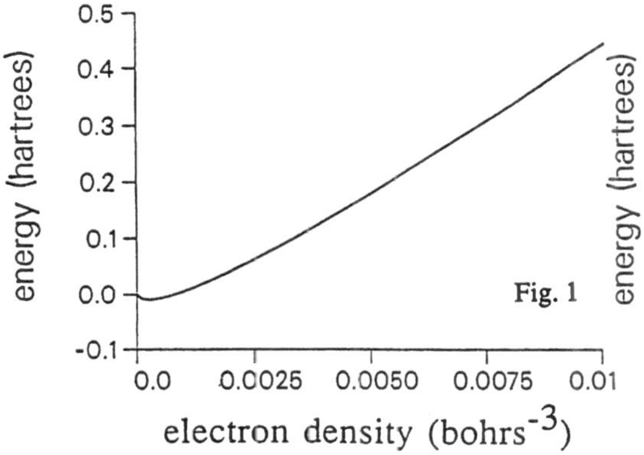

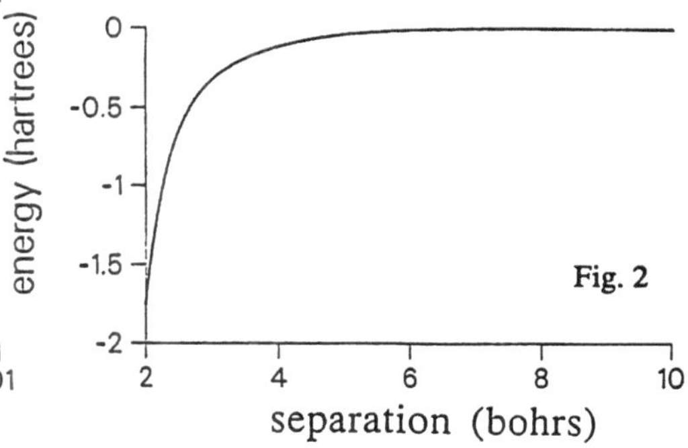

Fig. 1. The embedding energy as a function of the background electron density, from Ref. [40].

Fig. 2. The electrostatic interaction between atoms, from Ref. [40].

where G is the embedding energy, ρ^a is the spherically averaged atomic electron density, and U is an electrostatic, two-atom interaction. The embedding energy is defined as the interaction of the atom with the background electron gas, and is illustrated in Fig. 1. The background density for each atom in Eq. 1 is determined by evaluating at its nucleus the superposition of atomic density tails from the other atoms. The electrostatic pair interaction is illustrated in Fig. 2. Daw and Baskes determined the functions empirically, though in this paper we will illustrate how these functions can be determined from first principles (within the local density approximation).

The EAM combines the computational simplicity needed for large-scale systems with a physical picture which includes many-atom effects and avoids some of the problems of the pair potential scheme. This method has been applied successfully to such problems as phonons[3], liquid metals[4], defects[2, 5, 6], grain boundary structure[7, 8, 9], alloys[5, 6, 10, 11], interdiffusion in alloys[12, 13], fracture[14, 15, 16], surface structure[2, 5, 6, 17, 18, 19], surface adsorbate ordering[20, 21, 22], segregation to surfaces and grain boundaries[5, 6, 23, 24, 25, 26], surface order-disorder transitions[17], surface ordered alloys[18, 27], and surface phonons[28, 29].

This paper will summarize some of the related work of several people: primarily Drs. Mike Baskes and Stephen Foiles (of Sandia Livermore), and also Drs. Jim Adams (now at U. Illinois), Jeff Nelson (now at Sandia Albuquerque), and Erik Sowa (now at Lawrence Livermore).

2. Background

The EAM is antedated by the suggestion of Friedel[30] that the solution of hydrogen in copper was related to the solution of hydrogen in jellium. He was led to the investigation of the Friedel oscillations, i.e., impurity-induced oscillations in the jellium density.

More recently, detailed calculations of the impurity-in-jellium problem were carried out within the local density approximation (LDA) by Puska, Nieminen, and Manninen[31]. The solutions give the self-consistent impurity-in-jellium charge density and the embedding energies as functions of the background gas density.

The impurity-in-jellium work was the basis of the independent work of Nørskov and Lang (Effective Medium[32]) and of Stott and Zaremba (Quasiatom[33]). Nørskov and Lang showed in detail that the heat of solution of a light, interstitial impurity (e.g., H and He) in metals could be calculated by replacing the host with a suitable effective medium, which in this case was jellium. The heat of solution was then related to the embedding function. Nørskov and his co-workers have had great success in calculating from first principles the heats of solution[34] and heats of chemisorption[35] of hydrogen in metals. The optimal density of the jellium was determined by weighting the background metallic density by the Hartree potential of the metal ion. Stott and Zaremba arrived at a similar concept, based on viewing the impurity as a quasiatom in a nearly uniform electron gas.

The basis of the EAM is the realization that the cohesive energy of a metallic system can be expressed in terms of embedding energies. Daw and Baskes[1, 2] made this important generalization by proposing to view the cohesive energy of a metallic solid as comprised of the embedding energy plus electrostatic interactions. In this view, each atom in the metal is embedded into the electron gas created by the other atoms. Atoms near a defect such as a surface are embedded into an electron gas of different density profile than atoms in the bulk. The energy is apportioned so that the cohesive energy is manifestly symmetrical in the atomic index. They suggested the ansatz (Eq. 1) which will be discussed here. They then obtained the functions empirically by fitting to properties of the bulk metals. The generality of the functions was tested by applying them to surfaces and other defects. This generalization allowed calculations of complex metallic structures to be done within the approximate embedding energy framework. The EAM is thus a significant improvement in simplified total energy calculations for metallic systems.

Even more recently, Jacobsen, Nørskov, and Puska[36], Manninen[37], and Kress and DePristo[38, 39] re-examined the ansatz used in the EAM with arguments based on the effective medium approach. Jacobsen *et al.* demonstrated how the cohesive energy of a metallic system could be related to the embedding energies, with corrections accounting for the d-d hybridization in the transition metals. Their approach showed that with the neglect of the d-d hybridization (valid for simple metals and also presumably for early and late transition metals), the EAM expression is recovered. The density of the effective medium was taken to be an unweighted average of the background density over the Wigner-Seitz cell of the atom. Kress and DePristo suggested using as a weighting function the electron density of the atom itself, so that the background density is related to the overlap of charge densities. A correction to the embedding function based on a local energy functional was used in their Corrected Effective Medium Method to correct for inaccuracies in the definition of the optimal effective medium density.

It is also possible to derive the EAM form from approximations made within density functional theory[40]. This work will be discussed in this paper.

Several other methods, equivalent to the EAM, have been proposed since the original work of Daw and Baskes. These methods all take the form of Eq. 1, with differences due to specific parameterizations or functional forms. Finnis and Sinclair[41] proposed that the d-d hybridization in the second moment approximation could be described by Eq. 1. Their "N-body" potential has been applied to defects in BCC metals. The Finnis-Sinclair model is formally equivalent in the case of homonuclear metals to the EAM, though the physical motivation is quite different. For alloys, however, the Finnis-Sinclair model leads to a different form from the EAM. The "local volume forces" represents an alternative parameterization of the EAM by Chen, Voter, and Srolovitz[42], and has been applied to surface relaxations and grain boundaries in binary metallic alloys. The "glue model" of Ercolessi, Tosatti, and Parrinello[43, 44] also belongs to the EAM class of models, and has been applied mainly to surface reconstructions. The "equivalent crystal" model of Smith and Banerjea in its actual implementation is mathematically equivalent to the EAM, though again the motivation is different[45, 46].

3. Basic Theory

Our goal is to derive an approximate expression for the cohesive energy of a metallic system which is an explicit function of the positions of the atoms and which is simple to evaluate (i.e., Eq. 1). This has been discussed in detail in Ref. [40], and we will review the highlights here. We start with the electron density of the solid, $\rho(r)$, and the energy functional of the electron density, $E[\rho]$. We will assume that the kinetic, exchange, and correlation contributions to the energy, $G[\rho]$, can be described by

$$G[\rho] = \int \gamma(\rho(r), \nabla\rho(r), \nabla^2\rho(r), \ldots) \tag{2}$$

where γ is the energy density and is assumed to be a function of the local electron density and its derivatives. The integral is understood here and elsewhere to be over real space. The cohesive energy of a solid is then[47]

$$E_{coh} =$$

$$G[\rho] + \frac{1}{2}\sum_{i,j}{}' \frac{Z_i Z_j}{R_{ij}} - \sum_i \int \frac{Z_i \rho(r)}{|r-R_i|} + \frac{1}{2}\int\int \frac{\rho(r_1)\rho(r_2)}{|r_{12}|} - E_{atoms} \tag{3}$$

where the sums over i and j are over the nuclei of the solid, the primed sum indicates omission of the i=j term, Z_i and R_i are the charge and position of the i-th nucleus, and the integrals are over r (or r_1 and r_2). E_{atoms} is the collective energy of the isolated atoms.

In order to make Eq. 3 useful, we will express the electron density ρ explicitly in terms of the positions of the nuclei. We start by assuming that the electron density of the solid can be described as a linear superposition of the densities of the individual atoms ($\rho_s(r) \equiv \sum_i \rho_i^a(r-R_i)$). (The effect of charge redistribution is treated in detail in Ref. [40].) Substituting $\rho(r) = \rho_s(r)$ into Eq. 3 gives

$$E_{coh} = G[\sum_i \rho_i^a] - \sum_i G[\rho_i^a] + \frac{1}{2}\sum_{ij}{}' U_{ij}^a \tag{4}$$

with $U_{ij}^a \equiv \int\int \frac{n_i^a(r_1) n_j^a(r_2)}{|r_{12}|}$ and $n_i^a(r) \equiv \rho_i^a(r-R_i) - Z_i\,\delta(r-R_i)$. The first two terms in Eq. 4 involve the difference in kinetic, exchange, and correlation energies in going from the case of isolated atoms to the solid. The last term is the electrostatic energy of the overlapping charge distributions.

Consider the region around atom i. The electron density near atom i can be viewed as the sum of the density of atom i, ρ_i^a, and a background density due to the surrounding atoms. Let us define the background density for atom i to be $\rho_{b,i}(r) \equiv \sum_{j\neq i} \rho_j^a(r-R_j)$. In most of this region, the density ρ_i^a dominates in ρ_s; i.e., $\rho_{b,i}(r)$ is small and slowly varying compared to ρ_i^a. Thus it seems plausible to approximate $\rho_{b,i}(r)$ by a constant $\bar{\rho}_i$. Let us therefore define the embedding energy for an atom in an electron gas of some constant density $\bar{\rho}$ (neutralized by a positive background):

$$G_i(\bar{\rho}_i) \equiv G[\rho_i^a + \bar{\rho}_i] - G[\rho_i^a] - G[\bar{\rho}_i] \; . \qquad (5).$$

Using the embedding energy, we can rewrite the cohesive energy as

$$E_{coh} = \sum_i G_i(\bar{\rho}_i) + \frac{1}{2}\sum_{i,j} U_{ij}^a + E_{err} \tag{6}$$

where the error is

$$E_{err} = G[\sum_i \rho_i^a] - \sum_i G[\rho_i^a + \bar{\rho}_i] + \sum_i G[\bar{\rho}_i]$$

$$= \int_\infty [\ g(\sum_i \rho_i^a) - \sum_i g(\rho_i^a + \bar{\rho}_i) + \sum_i g(\bar{\rho}_i)\ ] \ . \qquad (7a).$$

Setting $E_{err} = 0$ provides a definition for the optimal background densities $\bar{\rho}_i$. Let us break up the integral into a sum of integrals over the regions around each atom, and set each individual integral to zero. For the sake of discussion, let us consider only the dominant part (in Ref. [40], the full solution is considered). This dominant part can be written

$$\int_{\Omega_i} g(\rho_i^a + \rho_{b,i}) \approx \int_{\Omega_i} g(\rho_i^a + \bar{\rho}_i) \qquad (7b)$$

and this expresses the relationship between $\bar{\rho}_i$ and $\rho_{b,i}$ more directly.

In the limit that the background density $\rho_{b,i}$ is very slowly varying compared to ρ_i^a, then $\rho_b(r) = \rho(0)$ and the solution to Eq. 7b is $\bar{\rho} = \rho(0)$. If we assume that the background density is largely constant with small variations, we can approximate the background density $\rho_b(r) \approx \rho(0) + \mathbf{r}\bullet\nabla\rho(0) + \frac{1}{2}\mathbf{rr}\bullet\bullet\nabla\nabla\rho(0)$, and expand in a Taylor's series involving powers of $\nabla\rho(0)$ and $\nabla\nabla\rho(0)$. We find that the lowest order corrections are

$$\bar{\rho} = \rho(0) + \alpha\nabla^2\rho(0) + \beta|\nabla\rho(0)|^2 \qquad (8)$$

with α and β being constants depending on the atom i and the function g. This relation is only approximate, due to the breakdown in the Taylor's series expansion near the edge of each atomic cell. However, Eq. 8 implies that the dominant contribution to the optimal background density is the density near the origin, with small corrections due to the gradient and curvature of the density.

In summary, the cohesive energy can be written $E_{coh} = \sum_i G_i(\bar{\rho}_i) + \frac{1}{2}\sum_{i,j} U_{ij}^a$, where U contains the two-atom electrostatic interactions and the embedding function is given by $G_i(\bar{\rho}_i) \equiv G[\rho_i^a + \bar{\rho}_i] - G[\rho_i^a] - G[\bar{\rho}_i]$. The constant background density is given by solving $E_{err} = 0$. This derivation is based on the assumption that the charge distribution in the solid is not very different from a superposition of atomic electron densities; thus this form is restricted to simple metals and early or late transition metals.

In Daw[40], the procedure outlined in this section was tested on a model system. A model energy functional based on the Thomas-Fermi-Dirac-von

Weizsäcker kinetic energy was developed and describes fcc nickel quite well. Within this model, the embedding function and electrostatic interactions were obtained, and the equation for the optimum electron density was solved in detail. A simple, linear relationship between $\bar{\rho}$ and $\rho(0)$ was found to hold, with some error attributed to gradient corrections as in Eq. 8. In the semi-empirical EAM, the assumption has been to set $\bar{\rho} = \rho(0)$.

This completes the heuristic derivation of the EAM. We note that the approximations in the derivation limit the range of validity to metallic systems with nearly empty or nearly filled d-bands. This would suggest that one should probably avoid metals near the center of the transition series and certainly avoid covalent materials. Also, any subtleties due to fermi surface or band structure effects are ignored in the EAM, so that any situations where these are important should be treated carefully.

4. Empirical Determination

The embedding function and pair interaction can be obtained from first principles, as is demonstrated in Ref. [40] with a model density functional for nickel (also see Ref. [36], for a different treatment). These first principles functions do reasonably well in describing the properties of nickel. However, in practice, we have taken a semi-empirical approach, where the fundamental theory has guided the fitting of the embedding function and pair interaction to basic bulk properties. In general, we have fitted to lattice constant, cohesive energy, elastic constants, and vacancy formation energy. We have also found it useful to require that the cohesive energy as a function of lattice constant follow the "universal binding curve" of Rose *et al.*[48] In one paper, we have also fitted to the dilute heats of alloying for binary alloys[5, 6]. In general, the fits are overdetermined, and the fact that we can achieve any kind of fit is in some sense a check of self-consistency. Functional forms have varied: in some cases, the functions are determined by general splines, in other cases, we have found it more useful to use analytic functions.

There is one fact to be aware of when doing the fitting. The following transformation does not change the total energy:

$$G_i(\rho) \rightarrow G_i(\rho) - c_i\rho$$

$$U_{ij}(R) \rightarrow U_{ij}(R) + c_i\rho_j^a(R) + c_j\rho_i^a(R).$$

Therefore, the embedding function cannot be determined to within a linear function. This is related to the fact that some of the electrostatic interaction can be grouped either as part of the pair term or as part of the embedding energy: the choice is arbitrary. This ambiguity also confuses the comparison of functions obtained by different workers. Some workers have required that U(R) be positive definite, which amounts to a particular choice of the constants in the transformation. Other workers have required that the embedding functions correspond to those obtained from atom-in-jellium calculations, which amounts to a different choice of the constants. However, the different choices of constants are arbitrary: very different looking sets of EAM functions may in fact give exactly the same energetics. Instead of comparing embedding functions and pair interactions directly, one should compare the effective pair and trio interactions (see the next section). These effective interactions are unaffected by the transformation, and therefore provide a way of comparing rather different looking sets of EAM functions.

In obtaining new functions, or in applying existing ones, there are several points to keep in mind:

(1) The functions should be smooth.
(2) The embedding function should have positive curvature. This ensures that the bond strength decreases with increasing coordination (see the section on many-atom interactions). This is a basic property of embedding functions, as can be demonstrated from fundamental principles[40].
(3) A direct consequence of the previous point is that metals with negative Cauchy discrepancy evidently have some basic properties which are neglected in the EAM. The approximations used in justifying the EAM break down in these cases. Therefore, it is wise to avoid such metals.

(3) Several workers[5, 6] have used the restriction of $U_{ij} = \sqrt{U_{ii}U_{jj}}$ (for positive definite U). This was used only to restrict the number of free parameters and is generally not a good approximation.
(4) Functions which have been fitted to the properties of two elemental metals with no regard for the alloying properties will be very lucky to describe the alloy properly. For example, Daw and Baskes[2] give functions for Ni and functions for Pd. These functions were fitted to the elemental properties but do not describe the Ni-Pd system well. On the other hand, the functions presented in Foiles, Baskes, and Daw[5] were fitted to the dilute heats of alloying and do well for Ni-Pd.
(5) Avoid covalent materials: the EAM does not describe the type of bonding exhibited by these materials[40].

5. Phenomenology

Although the EAM form is calculationally convenient, it is often conceptually useful to visualize effective N-atom interactions. One way to do this is to examine how the total energy changes when one makes small perturbations from a reference state.

We pick as a reference state a uniform solid. For this case, all the atoms have the same environment, so we define $\bar{\rho} = \sum_n \rho^a(R_n)$ and $\bar{U} = \sum_n U(R_n)$ where the sums are over the neighbors of a typical atom. Then the total energy of the uniform solid is $N(G(\bar{\rho}) + \frac{1}{2}\bar{U})$, where N is the number of atoms.

If we now allow a small distortion of the lattice, each atom will experience a slightly different electron density, and the embedding energy can be expanded in the small difference. The cohesive energy of the solid can then be written as a sum of N-body interactions.

For small distortions, the change in the energy within the EAM is equivalent to the change in the sum of effective two- and three-atom interactions[4]:

$$\psi_{ij}(R) \equiv$$

$$\frac{1}{2}\left(\left[U_{ij}(R) + 2G_i'(\bar{\rho}_i)\rho_j^a(R) + G_i''(\bar{\rho}_i)\rho_j^a(R)^2\right] + [(i \leftrightarrow j)] \right) \qquad (9a)$$

$$\chi_{ijk}(R_i,R_j,R_k) \equiv \left(G_i''(\bar{\rho}_i)\rho_j^a(R_{ij})\rho_k^a(R_{ik}) + (jki) + (kij) \right) \qquad (9b).$$

Higher-body interactions can also be obtained. The interactions in Eqs. 9 demonstrate that the effective interactions within the EAM are *environment-dependent*, in that the interaction between two atoms depends on the slopes of their embedding functions, which depend on the $\bar{\rho}$ for each atom. As one goes to the surface, the effective two-atom interaction becomes stronger and the bond-length shortens. Within the EAM, this is a result of the requirement that $G'' > 0$. In this way, the EAM describes the effect of coordination on bond strength: the less-coordinated surface atoms tend to have stronger bonds and shorter bond lengths. This in fact is the strength of the EAM, and is directly connected to the nonlinearity of the embedding function. Of course, this effect is neglected in the traditional pair potential formulation.

In Figures 3a and b, we have plotted the effective pair and trio interactions for three EAM sets of functions for Ni: the original semiempirical functions

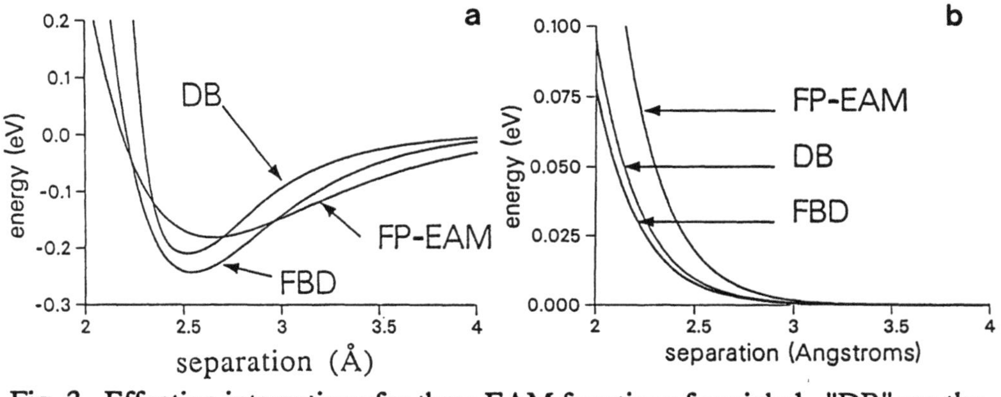

Fig. 3. Effective interactions for three EAM functions for nickel. "DB" are the semiempirical Daw-Baskes EAM functions from Ref. [1, 2]. "FBD" are the semiempirical Foiles-Baskes-Daw EAM functions from Ref. [5, 6]. "FP-EAM" are the first-principles EAM functions from Ref. [40]. (a) Effective two-body interaction in the fcc bulk. (b) Effective three-body interaction of atoms in an equilateral triangle, in the fcc bulk.

developed by Daw and Baskes[2], the semiempirical functions developed by Foiles, Daw, and Baskes[5], and the functions developed by Daw[40] from the first principles model. The results show that the effective energetics of the three functions are rather similar, and that the semiempirical functions are qualitatively quite similar to what one gets from more fundamental considerations.

6. Applications

We have implemented the EAM in four different types of calculations: (1) energy minimization, (2) molecular dynamics, (3) Monte Carlo, and (4) vibrational normal mode analysis. The current practical limitations of EAM calculations on a CRAY-XMP/24 allow molecular dynamics simulations of up to 35,000 atoms. Simulated times for molecular dynamics typically run up to tens of picoseconds. For typical problems, the calculational cost scales mostly like the number of atoms. Monte Carlo simulations based on the EAM are in practice capable of typically 10^5 iterations per atom. These simulations are most useful for "annealing" structures and obtaining transition temperatures or equilibrium concentration profiles. The dynamical matrix can be obtained analytically and the vibrational normal modes obtained by matrix diagonalization. Normal mode calculations are currently possible for unit cells of up to about 260 atoms.

We have applied the EAM to a large variety of problems involving intrinsic defects and surfaces of metals. From the EAM we obtain information on the structure, dynamics, phase transitions, and phonons. Reliable semiempirical functions have been obtained for Al, Ni, Pd, Pt, Cu, Ag, Au, and some alloys of these metals. Rather than go into detail in all of the subjects, we review briefly some of the applications in the literature, and refer the reader to the references for further information.

6a. Bulk properties

The lattice dynamical matrix was derived[3] for bulk phonons in fcc transition metals. The comparison with experiment showed that the EAM provides a reasonable description of elementary vibrational excitations in solids. The agreement near the zone center is to be expected, because the functions are fitted to the elastic constants. However, agreement over the full zone was obtained as well, and this was not guaranteed by the fitting.

The linear coefficients of thermal expansion of fcc metals were computed[49]. The results were in good agreement with experiment. These results demonstrated the importance of including the equation of state in the semi-empirical fitting, because without that additional constraint, anomalous thermal expansion coefficients may be obtained[50].

The Gibbs free energy of solid and liquid phases as well as the melting point were calculated using a combination of molecular dynamics and normal mode analysis[51]. The agreement with experiment was very encouraging.

Above the melting point, the static structure factor of various liquid transition metals were computed[4]. The results were in good agreement with experimental data. These calculations of basic bulk properties demonstrated that the EAM provides a good description of the interatomic interactions in metals.

The formation and diffusion of vacancies, di-vacancies, and self-interstitials have been analyzed in alloys of Cu, Ag, Au, Ni, Pd, and Pt, showing good agreement with experiment. The contribution of these processes to self-diffusion was analyzed by Adams, Foiles, and Wolfer[12].

6b. Surfaces

LEED analysis of several clean, unreconstructed metal surfaces is available. The relaxations have been compared to the theoretical predictions with good results[2, 5].

The energetics and structure of several fcc (110) surfaces were calculated[19, 52], demonstrating that for Pt and Au this surface was stabilized by a missing row reconstruction. The atomic relaxations were computed and compared in detail to experimental analysis, showing excellent agreement. Experimentally, it is known that the missing-row structure of Au disorders at 650K. EAM/Monte Carlo was used to study the structure and order-disorder transformation of Au and Pt(110) surfaces[17]. It was found that the missing-row structure of Au disordered at 570K, in very good agreement with experiment. Pt(110) was predicted to disorder at 750K, which prompted experiments currently in progress. Relaxations and vibrations were found to be essential for the accurate calculation of the critical temperature, implying that the bulk of the literature of lattice gas calculations must be carefully interpreted.

The interactions of metal adatoms were studied systematically by depositing Pt adatoms onto a Pt Field Ion Microscopy tip. The stability of various cluster geometries were studied as a function of the number of atoms in the cluster. Analogous calculations were done using the EAM, and detailed comparisons between theory and experiment showed excellent agreement[53].

The surface phonons of Cu and Ag low index surfaces have been computed. The theoretical results are in very good agreement with experiment for the (100) surface[29]. For the (111) surface, a strong debate has developed over the magnitude of the difference in the force constants between bulk and surfaces (see references in [29]). Analysis of helium scattering and EELS experiments appear to disagree substantially. The EAM results are consistent with the analysis of the results obtained with EELS[28].

6c. Grain boundaries

Internal interfaces have also been investigated with the EAM. Foiles has performed careful calculations of the structure of grain boundaries in gold. The atomic positions calculated for the $\Sigma 13$ (100) twist boundary are in excellent agreement with experiment[7]. The $\Sigma 5$ (210) tilt shows a very interesting reconstruction which doubles one of the periodic lengths in the plane of the boundary[8, 54]. He has also used the effective pair interaction (Eq. 9a) in similar calculations to compare to the full EAM results in order to examine the effects of the high-body interactions on grain boundary structure.

The elastic properties of a twist boundary in Cu has been analyzed by Adams, Foiles, and Wolfer [55] by studying the interlayer strain induced by small applied stresses. The results show marked changes in the "interlayer moduli", especially in a greatly reduced resistance to linear shear.

6d. Alloys: bulk and interfaces

Metallic alloys also represent a fruitful ground for testing the EAM and for performing calculations that cannot be reliably investigated with other methods. The EAM was used to calculate phase stability, phonons, point defect properties, and surface energies and relaxations for Ni_3Al[10, 11, 56]. Further calculations revealed that as stoichiometry was varied from the ideal, antisite defects formed in the bulk would segregate to grain boundaries. Furthermore, the segregation of Ni to the grain boundaries in the Ni-rich alloy caused a loss of compositional order in the grain boundaries. The problem of segregation to the surfaces of binary alloys was treated in a series of papers by Foiles[23, 24, 26, 57, 58, 59, 60]. The EAM/Monte Carlo simulations provided equilibrium concentration profiles. The results showed that the composition may vary nonmonotonically near the surface, in agreement with experimental studies. Calculations using the EAM showed that Au forms ordered surface layers on the low index faces of Cu[27]. These "ordered surface alloys" were demonstrated to exist in equilibrium with a bulk containing dilute amounts of Au. The results are in excellent agreement with experimental results for the structure of Au deposited on Cu.

Hydrogen in metals has also been a promising area of investigation using the EAM. The phase diagram of hydrogen on Pd(111) was calculated[22, 61], using EAM/Monte Carlo. The symmetry and critical temperatures of ordered phases were predicted in excellent agreement with experiment. A unique feature of the theoretical results was the prediction of occupation of subsurface sites by the hydrogen.

6e. Mechanical properties

Mechanical properties of metals has been the focus of a series of calculations using the EAM[1, 14, 15, 16, 62]. Molecular dynamics calculations of dislocation dynamics in fcc Ni were performed, showing the motion of an edge dislocation under applied stress. The results were compared quite favorably with continuum elasticity solutions of the same problem. Molecular dynamics studies were also performed of crack propagation in fcc Ni. Both brittle and ductile fracture were studied. Hydrogen atoms were added and demonstrated that hydrogen near the crack tip could reduce both the brittle fracture stress and the stress required to generate dislocations from the crack tip.

The structure and dynamics of an edge dislocation in Ni were compared to the predictions of elastic continuum theory in papers by Daw *et al.*[15, 16]. The computed structure of a superscrew dislocation in Ni_3Al compared favorably to available High Resolution Microscopy results[63].

The difficult problems of crack propagation and the question of ductility have been approached using the EAM. The results are greatly improved over earlier studies based on interatomic pair potentials[64].

6f. Summary

The wide range of problems to which the EAM has been applied demonstrate that the method describes ground state structural properties of fcc metals quite well. These benchmark calculations now provide us with the confidence to do more complicated systems.

7. CONCLUSIONS

The Embedded Atom expression for the cohesive energy of a solid can be derived from approximations to density functional theory. The energy can be divided into an embedding energy contribution with an electrostatic, two-atom correction. The procedure defines an optimal constant background density, ρ, for the embedding function, in terms of the actual background density $\rho_b(r)$ at each site. The EAM form is expected to be most applicable to simple metals and early or late transition metals The expressions developed here for the EAM can serve as a basis for the parameterization of the functions used in the semi-empirical technique. The EAM has been tested on a large variety of situations and has proven very versatile in describing interatomic interactions in a metallic environment. The EAM should replace the use of pair potentials in simulations of metallic systems.

Acknowledgements

Much of the work on the EAM and applications has been done in collaboration with Drs. S. M. Foiles and M. I. Baskes of this laboratory. Other collaborators have included Mr. C. L. Bisson, Drs. J. B. Adams, J. S. Nelson, E. C. Sowa, and W. G. Wolfer of this laboratory and Prof. R. D. Hatcher of Queens College. This work was supported by the U. S. Department of Energy, Office of Basic Energy Sciences, Division of Materials Science.

References

1. M. S. Daw and M. I. Baskes: Phys Rev Lett **50**, 1285 (1983)
2. M. S. Daw and M. I. Baskes: Phys Rev B **29**, 6443 (1984)
3. M. S. Daw and R. D. Hatcher: Solid State Commun **56**, 697 (1985)
4. S. M. Foiles: Phys Rev B **32**, 3409 (1985)
5. S. M. Foiles, M. I. Baskes and M. S. Daw: Phys Rev B **33**, 7983 (1986)
6. S. M. Foiles, M. I. Baskes and M. S. Daw: Phys Rev B **37**, 10387 (1988)
7. S. M. Foiles: Acta Met **in press**, (1989)
8. S. M. Foiles, M. I. Baskes and M. S. Daw: in Interfacial Structure, Properties and Design, ed. by M. H. Yoo, W. A. T. Clark and C. L. Brian (Materials Research Society, Pittsburgh, 1988)
9. S. M. Foiles: in Interfacial Structure, Properties and Design, ed. by M. H. Yoo, W. A. T. Clark and C. L. Briant (Materials Research Society, Pittsburgh, 1988)
10. S. M. Foiles and M. S. Daw: J of Metals **39**, 39 (1987)
11. S. M. Foiles: in High-Temperature Ordered Intermetallic Alloys II, ed. by N. S. Stoloff, C. C. Koch, C. T. Liu and O. Izumi (Materials Research Society, Pittsburgh, 1987)
12. J. B. Adams, S. M. Foiles and W. G. Wolfer: J Mater Research **4**, 102 (1989)
13. J. B. Adams, S. M. Foiles and W. G. Wolfer: in Atomistic Simulation of Materials, ed. by D. Srolovitz and V. Vitek (Plenum Publishing Corporation, Chicago, 1989)
14. M. I. Baskes, S. M. Foiles and M. S. Daw: J de Physique **C5**, 483 (1988)
15. M. S. Daw, M. I. Baskes, C. L. Bisson and W. G. Wolfer: in Modelling Environmental Effects on Crack Growth Processes (The Metallurgical Society, Metallurgical Society Fall Meeting in Toronto, Canada, 1985)
16. M. S. Daw and M. I. Baskes: in NATO Advanced Workshop on "Physics and Chemistry of Fracture" (Martinus Nijhoff, Dordrecht, Bad Reichenhall, West Germany, 1986)
17. M. S. Daw and S. M. Foiles: Phys Rev Lett **59**, 2756 (1987)
18. M. S. Daw and S. M. Foiles: in Second International Conference on the Structure of Surfaces (Springer-Verlag, Amsterdam, The Netherlands, 1987)
19. M. S. Daw and S. M. Foiles: J Vac Sci Tech. A **4**, 1412 (1986)
20. S. M. Foiles and M. S. Daw: J Vac Sci Tech **3**, 1565 (1985)
21. M. S. Daw and S. M. Foiles: in 1st International Conference on the Structure of Surfaces (Springer-Verlag, Berkeley, 1985)
22. T. E. Felter, S. M. Foiles, M. S. Daw and R. H. Stulen: Surf Sci **171**, L379 (1986)
23. S. M. Foiles: Phys Rev B **40**, 11502 (1989)

24. S. M. Foiles: in Surface Segregation and Related Phenomena, ed. by P. A. Dowben and A. Miller (CRC, in press)
25. S. M. Foiles, M. I. Baskes and M. S. Daw: in US-Japan Seminar on Electronic Structure and Lattice Defects in Alloys (Trans Tech Publications, Honolulu, 1987)
26. S. M. Foiles: J Vac Sci Technol A **5**, 889 (1987)
27. S. M. Foiles: Surf Sci **191**, 329 (1987)
28. J. S. Nelson, M. S. Daw and E. C. Sowa: Phys Rev B **40**, 1465 (1989)
29. J. S. Nelson, E. C. Sowa and M. S. Daw: Phys Rev Lett **61**, 1977 (1988)
30. J. Friedel: Phil Mag **43**, 153 (1952)
31. M. J. Puska, R. M. Nieminen and M. Manninen: Phys Rev B **24**, 3037 (1980)
32. J. K. Nørskov and N. D. Lang: Phys Rev B **21**, 2131 (1980)
33. M. J. Stott and E. Zaremba: Phys Rev B **22**, 1564 (1980)
34. J. K. Nørskov: Phys Rev B **26**, 2875 (1982)
35. P. Nordlander, S. Holloway and J. K. Nørskov: Surf. Sci. **136**, 59 (1984)
36. K. W. Jacobsen, J. K. Nørskov and M. J. Puska: Phys Rev B **35**, 7423 (1987)
37. M. Manninen: Phys Rev B **34**, 8486 (1986)
38. J. D. Kress and A. E. DePristo: J. Chem. Phys. **88**, 2596 (1988)
39. J. D. Kress and A. E. DePristo: J. Chem. Phys. **87**, 4700 (1987)
40. M. S. Daw: Phys Rev B **39**, 7441 (1989)
41. M. W. Finnis and J. E. Sinclair: Phil Mag A **50**, 45 (1984)
42. S. P. Chen, A. Voter and D. L. Srolovitz: Phys Rev Lett **57**, 1308 (1986)
43. F. Ercolessi, E. Tosatti and M. Parrinello: Surf Sci **177**, 314 (1986)
44. F. Ercolessi, E. Tosatti and M. Parrinello: Phys Rev Lett **57**, 719 (1986)
45. J. R. Smith and A. Banerjea: Phys Rev B **37**, 10411 (1988)
46. J. R. Smith and A. Banerjea: Phys Rev Lett **59**, 2451 (1987)
47. P. Hohenberg and W. Kohn: Phys Rev B **136**, 864 (1964)
48. J. H. Rose, J. R. Smith, F. Guinea and J. Ferrante: Phys Rev B **29**, 2963 (1984)
49. S. M. Foiles and M. S. Daw: Phys Rev B **38**, 12643 (1988)
50. M. Marchese, G. Jacucci and C. P. Flynn: Phil Mag Lett **57**, 25 (1988)
51. S. M. Foiles and J. B. Adams: Phys Rev B **40**, 5909 (1989)
52. S. M. Foiles: Surf Sci **191**, L779 (1987)
53. P. Schwoebel, S. M. Foiles and G. Kellogg: Phys Rev B **40**, 10639 (1989)
54. S. M. Foiles, M. I. Baskes and M. S. Daw: Mat Sci Forum **37**, 223 (1989)
55. J. B. Adams, S. M. Foiles and W. G. Wolfer: Phys Rev B **40**, 9479 (1989)
56. S. M. Foiles and M. S. Daw: J Mater Res **2**, 5 (1987)
57. S. M. Foiles: Phys Rev B **32**, 7685 (1985)
58. S. M. Foiles: J Vac Sci Technol A **4**, 761 (1986)
59. S. M. Foiles: in Computer-Based Microscopic Description of the Structure and Properties of Materials, ed. by J. Broughton, W. Krakow and S. T. Pantelides (Materials Research Society, Pittsburgh, 1986)

60. S. M. Foiles: in Physical and Chemical Properties of Thin Metal Overlayers and Alloy Surfaces, ed. by D. M. Zehner and D. W. Goodman (Materials Research Society, Pittsburgh, 1987)
61. M. S. Daw and S. M. Foiles: Phys Rev B **35**, 2128 (1987)
62. M. Baskes, M. S. Daw, S. M. Foiles and B. Dodson: MRS Bulletin **13**, 28 (1988)
63. M. H. Yoo and M. S. Daw: in "Beyond Pair Potentials", a Special Symposium to the World Materials Congress Chicago, 1988)
64. M. I. Baskes and M. S. Daw: , (in preparation)

From Exact to Approximate Theory: The Tight Binding Bond Model and Many-Body Potentials

D.G. Pettifor

Department of Mathematics, Imperial College of Science, Technology and Medicine, London SW7 2BZ, UK

Abstract. The approximate Tight Binding Bond (TBB) model is derived from first principles by working within the Harris-Foulkes approximation to density functional theory and using Anderson's chemical pseudopotentials. The TBB expression for the binding energy is then simplified still further by deriving a many-body potential for the bond order which gives the direct dependence of the bond strength on local atomic environment. The simplicity of the potential's analytic form will allow the Embedded Atom potentials to be extended to include the very important explicit three-body and four-body terms.

1. Introduction

This workshop is concerned with the development of reliable many-body potentials for atomistic simulation. It can be argued that the semi-empirical Tight Binding model [1, 2] is the simplest scheme for describing the energetics of semi-conductors and transition metals within a quantum mechanical framework. Assuming only one type of valence orbital per site, the total binding energy may be written [3, 4] in the form

$$U = U_{rep} + U_{bond} \tag{1}$$

where U_{rep} is a semi-empirical pairwise repulsive contribution and U_{bond} is the covalent bond energy which results from evaluating the local density of states $n_i(E)$ at site i within the two-centre, orthogonal Tight Binding approximation. That is,

$$U_{bond} = \Sigma_i \int^{E_F} (E - E_i)\, n_i(E)\, dE \tag{2}$$

where E_i is the effective atomic energy level at site i and E_F is the Fermi energy. In practice, eq. (1) gives the binding energies with respect to some *reference* free atom state which usually differs from the *true* atomic ground state due to, for example, the neglect of spin-polarization or the shift in atomic energy levels arising from the renormalization of the wave functions in the bonding situation.

The simple form of eq. (1) may be derived from first principles [5, 6, 7]. The pairwise nature of the repulsive term follows directly from the Harris-Foulkes approximation [5, 6] to density functional theory [8], whereas the Hückel-type two-centre orthogonal form of the matrix elements may be justified in principle within Anderson's [9] chemical pseudopotential theory. This derivation is outlined in section 2 and illustrated in detail for the particular case of hydrogen [10]. Section 3 stresses the importance of working with the bond energy rather than the band energy whenever the

Springer Proceedings in Physics, Vol. 48 Many-Atom Interactions in Solids
Editors: R.M. Nieminen · M.J. Puska · M.J. Manninen

repulsive term is not evaluated exactly but approximated semi-empirically. This is the reason why Pettifor and Podloucky [11] and Sutton et al [7] refer explicitly to the Tight Binding Bond (TBB) model.

Although the TBB model is a physically intuitive approximation to exact density functional theory, it still remains at least two orders of magnitude slower to implement than, for example, the Embedded Atom models of Daw and Baskes [12] and Finnis and Sinclair [13]. The form of these embedding potentials may be justified either within Effective Medium theory [14] or within the second-moment approximation to the local electronic density of states [13]. In section 4 the successes and failures of the second moment approximation are discussed with respect to the cohesion and structure of transition metals. In section 5 a new many-body potential for the Tight Binding bond order is presented which goes beyond the *explicit* 2 body terms of the Embedded Atom models, thereby providing for the first time a rational way to improve the accuracy and reliability of the latter potentials. In section 6 we conclude.

2. Derivation of the TBB Model

The total energy within density functional theory [8] can be written

$$U = U_{nn} + U_{dc} + U_{band} \tag{3}$$

where the first two terms are the nuclear-nuclear and double-counting contributions respectively. The last term is the bond energy which is obtained by summing over all the occupied states, namely

$$U_{band} = \sum_{\underline{k}} E_{\underline{k}}. \tag{4}$$

The $E_{\underline{k}}$ are the eigenvalues of the one-electron Schrödinger equation

$$[-\nabla^2 + V_H(\underline{r}) + V_{xc}(\underline{r})]\, \psi_{\underline{k}}(\underline{r}) = E_{\underline{k}}\, \psi_{\underline{k}}(\underline{r}) \tag{5}$$

where V_H and V_{xc} are the Hartree and exchange-correlation potentials corresponding to the ground state charge density

$$\rho_{gs}(\underline{r}) = \sum_{\underline{k}} |\psi_{\underline{k}}(\underline{r})|^2. \tag{6}$$

The TBB expression for the binding energy, eq. (1), may be derived from first principles for the simplest case of a system of hydrogen atoms [10] by using the Harris-Foulkes approximation [5, 6] to density functional theory and Anderson's chemical pseudopotentials [9]. Using the stationary property of the energy with respect to the true ground state charge density ρ_{gs}, Harris [5] and Foulkes [6] replaced ρ_{gs} by an approximate charge density, say

$$\hat{\rho}_{gs}(\underline{r}) = \sum_{\underline{R}} \rho_{atom}(\underline{r}-\underline{R}), \tag{7}$$

so that the density functional energy could be written exactly to first order in $(\rho_{gs} - \hat{\rho}_{gs})$ as

$$U = U_{nn} + \hat{U}_{dc} + \hat{U}_{band} \tag{8}$$

where the last two terms are the double-counting and band contributions evaluated using the approximate charge density $\hat{\rho}_{gs}$ as input. It follows from eq. (7) that the Hartree part of the double-counting term is now pairwise in nature because there is no bond charge present in $\hat{\rho}_{gs}$. The exchange-correlation part is well approximated by a pairwise form [7, 10], so that as first pointed out by Foulkes [5] $U_{nn} + \hat{U}_{dc}$ is describeable by a *pair-potential* as assumed by the TB models [1-4].

The Hückel-type *two-centre* orthogonal form of the TBB model may be obtained within Anderson's chemical pseudopotential theory [9]. Defining localized orbitals $|\phi_A\rangle$ by

$$\{(-\nabla^2 + V_A) + \sum_{B\neq A} [1 - |\phi_B\rangle\langle\phi_B|]\, V_B\}\, |\phi_A\rangle = E_A\, |\phi_A\rangle, \tag{9}$$

the secular equation takes the Hückel form

$$|(E_A - E)\, \delta_{AB} + \langle\phi_B|V_B|\phi_A\rangle(1 - \delta_{AB})| = 0 \tag{10}$$

where

$$E_A = \langle\phi_A|-\nabla^2 + V_A|\phi_A\rangle + \sum_{B\neq A} \langle\phi_B|V_B|\phi_A\rangle - \sum_{B\neq A} \langle\phi_A|\phi_B\rangle\langle\phi_B|V_B|\phi_A\rangle. \tag{11}$$

Thus, E_A comprises the free atom and crystal field terms as expected, plus the shift due to the non-orthogonality contribution.

Grouping the crystal-field term with the double-counting and nuclear-nuclear contribution in eq. (8) allows the total energy to be decomposed explicitly [7, 10] in terms of physically intuitive contributions as

$$U = U_{atom} + U_{es} + U_{xc} + U_{no} + U_{bond}, \tag{12}$$

where U_{es} is the electrostatic interaction between the overlapping atomic charge densities and U_{no} is the non-orthogonality contribution from eq. (11).

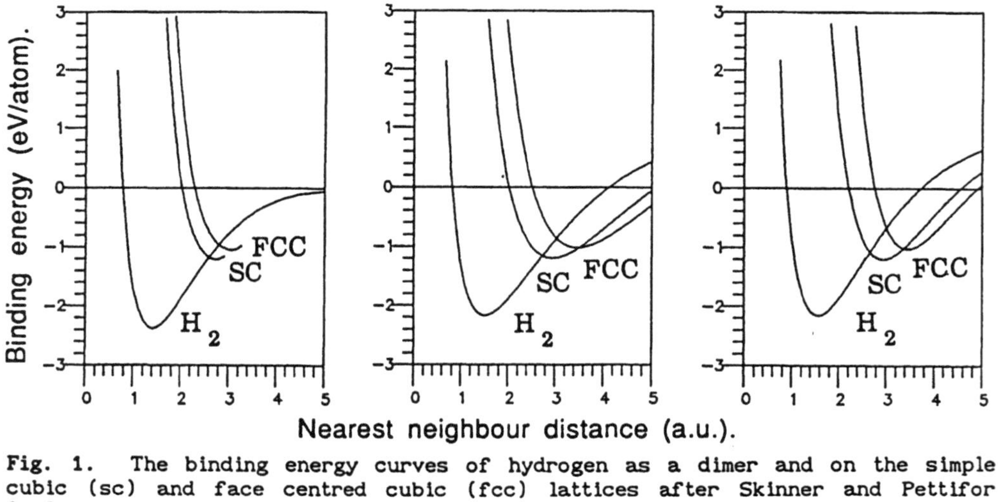

Fig. 1. The binding energy curves of hydrogen as a dimer and on the simple cubic (sc) and face centred cubic (fcc) lattices after Skinner and Pettifor [10].

Eq. (12) clearly shows that this binding energy is with respect to the reference atomic state defined by the localized orbitals $|\phi_A\rangle$, so that this must be taken into account when calculating the cohesive energy which is with respect to the true atomic ground state. The above derivation may be generalized [7] to the case where there is more than one type of valence orbital per atom, in which case the promotion energy also enters eq. (1).

Fig. 1 shows the binding energy curves for hydrogen as a dimer and on the infinite simple cubic (sc) and face centred cubic (fcc) lattices from Skinner and Pettifor [10]. The left-hand panel gives the ab initio results of Kolos and Wolniewicz [15] for H_2 and Min et al [16] for the bulk. The central panel shows the results of evaluating eq. (12) variationally assuming the localized orbitals have variational form $\exp(-\lambda r)$ and the approximate charge density $\hat{\rho}_{gs}$ is the sum of atomic densities $\exp(-2\lambda r)$. In order that all the integrals are kept analytic the Xα form [17] was taken [10] for the exchange and correlation with α chosen to give the correct binding energy for the simple cubic lattice at equilibrium, namely $\alpha = 0.871$. As is well-known the non-spin-polarized Local Density Functional calculations go off to the wrong limit as $R\to\infty$. This would be circumvented by allowing a self-consistent spin-polarized solution to develop for R>3.2a.u. (see ref. 18). We see that the variational TBB expression reproduces the correct ordering of the cohesive energy and equilibrium nearest neighbour separation between the dimer, simple cubic and face centred cubic lattices.

The right-hand panel shows the results of evaluating eq. (12) assuming a *transferable* pair potential and bond integral, which are chosen by freezing the localized orbital $\exp(-\lambda r)$ and atomic charge density $\exp(-2\lambda r)$ with their values obtained from the variational calculation for the simple cubic lattice at equilibrium, namely $\lambda_0 = 1.043$. We see that because of the variational nature of the central panel's results, the fixed λ calculations reproduce the cohesive energy and equilibrium nearest neighbour distance to first order in $(\lambda_0 - \lambda_{eq})$ but not the curvature about equilibrium. This appears to be the price which has to be paid for using short-range [9] or screened [19] TB parameters which are dependent on the local environment. It is not possible to reproduce accurately the bulk moduli of different structure types with a

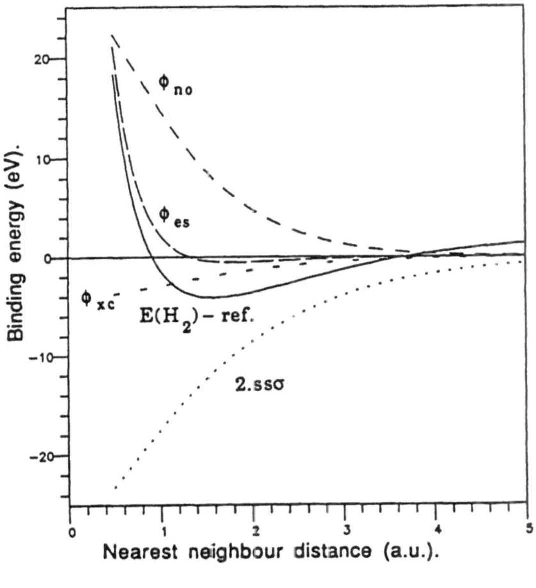

Fig. 2 The electrostatic (es), non-orthogonality (no), exchange-correlation (xc), and bond integral (ssσ) contributions for hydrogen with $\lambda = 1.043$ (after Skinner and Pettifor [10]).

universal set of *short-range* transferable two-centre hopping integrals. Nevertheless, fig. 1 shows that the transferable two-centre TBB model works remarkably well for hydrogen.

Fig. 2 shows the electrostatic (es), the exchange-correlation (xc) and the non-orthogonality (no) contributions to the pair potential in addition to the ssσ bond integral curve for the case where λ has been fixed at 1.043. Explicit analytic expressions for these individual terms are given by Skinner and Pettifor [10]. We see that at the equilibrium separation of the hydrogen dimer, the electrostatic contribution plays very little role. The repulsion is provided by the non-orthogonality term, the attraction by the bond and exchange-correlation energies. Thus, fig. 2 provides a detailed break down of the pair potential which enters the TBB model for hydrogen. In practice, at very large distances this so-called "repulsive" potential becomes attractive due to the long range exchange-correlation tail.

3. Why Bond Not Band?

It is customary within the TB community to refer to the second term in eq. (1) as the *band* energy. This associates it directly with the last term in eq. (8) so that the semi-empirical pairwise repulsive term represents the first two terms in eq. (8), namely $U_{nn} + \hat{U}_{dc}$. This grouping has no direct physical significance, the double-counting term entering to guarantee correct book-keeping. In the previous section, however, we associated the second term in eq. (1) with the *bond* energy by writing

$$U_{band} = \Sigma_i \int^{E_F} E\, n_i(E)\, dE = \Sigma_i N_i E_i + U_{bond}. \tag{13}$$

The site-diagonal contribution was then grouped with the double-counting and ion-ion contributions to give the physically intuitive pair potential term $U_{es} + U_{xc} + U_{no}$ in eq. (12).

Why do we prefer to use the bond energy rather than the band energy? Clearly if all the terms were calculated *exactly* it would make no difference to the total energy how the cake was divided up (although we would still argue that the bond description is preferable as all the pieces have physically transparent labels). On the other hand, if the pairwise contribution is *approximated* by an empirical repulsive term then it is crucial which description is adapted as we will now demonstrate.

The Harris-Foulkes approximation is only exact to first order. Problems such as evaluating the heats of formation or the bulk modulus are second order and demand an explicit statement about charge transfer and the corresponding re-alignment of the site-diagonal energies E_i. A model which ignores site shifts leads to serious inconsistencies not only in the bulk modulus of metals (see, for ex., fig. 1 of Sutton et al [7]) but also in alloy behaviour (see, for ex., section 7 of Pettifor [20]). The simplest way to remove these inconsistencies is to assume local charge neutrality by adjusting the individual atomic energy levels until each site is neutral [7, 11, 20]. This change in the local on-site potential leads directly to a first order change in the *band* energy $\Sigma_i N_i \Delta E_i$ from eq. (13). However, the force theorem of Pettifor [21, 22] and Andersen [23] requires that this first order change in the band energy due to the renormalization of the potential should be cancelled exactly by a similar change in the double-counting term, namely ½ $\Sigma_i 2N_i \Delta E_i$. If the double-counting term is not evaluated directly but is approximated empirically then this cancellation does not take place and the force theorem is no longer satisfied. The *bond* energy description, on the other hand, satisfies the force theorem as it does not contain the site-diagonal energies explicitly.

Thus the TBB model with local charge neutrality is the simplest semi-empirical Tight Binding scheme which is not only consistent with the first order force theorem but also provides a consistent description of second order properties such as the bulk modulus [7] or heats of formation [20].

4. The Second Moment Approximation

Finnis and Sinclair [13] have reduced the TBB expression for the binding energy into the form of an embedding potential by evaluating the bond energy within the second moment approximation [4, 24]. Assuming a constant local density of states of width W_i, the d bond energy associated with a transition element at site i may be written [3] from eq. (2) as

$$U^{i}_{bond} = -\frac{1}{20} W_i N_d (10 - N_d) \tag{14}$$

where N_d is the number of valence d electrons. But W_i may be related to the hopping integrals through the second moment of the local density of states μ^i_2 as follows. By definition, choosing E_i as the energy zero,

$$\mu^i_2 = \int^{\infty}_{-\infty} E^2 n_i(E)\, dE \tag{15}$$

$$= 10\,(W_i^2/12) \tag{16}$$

where the prefactor 10 accounts for spin and the 5-fold degeneracy of the d band. However, it follows from eq. (15) that

$$\mu^i_2 = 2 \sum_m \langle\phi_{im}|H^2|\phi_{im}\rangle \tag{17}$$

where the prefactor 2 accounts for the spin and the m sum runs over the 5 atomic d orbitals. Thus,

$$\mu^i_2 = 2 \sum_{j,mm'} \langle\phi_{im}|H|\phi_{jm'}\rangle\langle\phi_{jm'}|H|\phi_{im}\rangle$$

$$= 2 \sum_{j\neq i} (dd\sigma^2 + 2dd\pi^2 + 2dd\delta^2)_{R=R_{ij}}$$

$$= 10 \sum_{j\neq i} h^2(R_{ij}). \tag{18}$$

It follows from eqs. (16) and (18) that

$$\frac{1}{12} W_i^2 = \sum_{j\neq i} h^2(R_{ij}) \tag{19}$$

where h is the root mean square hopping integral given by

$$h^2 = (dd\sigma^2 + 2dd\pi^2 + 2dd\delta^2)/5.$$

Substituting eqs. (19) and (14) into eq. (1) the TBB expression for the binding energy may be written [13] within the second moment approximation as

$$U = \tfrac{1}{2} \sum_{i,j}{}' \Phi(R_{ij}) + \sum_i F(\rho_i) \tag{20}$$

where Φ is the repulsive pair potential and F is the embedding function

$$F(\rho) = -A\rho^{1/2} \tag{21}$$

with A a constant. Thus, eq. (21) corresponds to a particular choice for the embedding potential [12] in which the local "charge density" is given by

$$\rho_i = \sum_j{}' h^2(R_{ij}). \tag{22}$$

The TBB μ_2 embedding function shows the well known square root saturation which is discussed in more detail elsewhere in this volume [25].

The Embedded Atom potentials have been particularly successful in modelling the fcc noble metals and aluminium [26], and their alloys [26, 27]. They go beyond the pair potential description by including the Cauchy discrepancy automatically within their formalism since for a fcc nearest neighbour model [28], for example,

$$C_{12} - C_{44} = (\text{positive constant})\ F''. \tag{23}$$

Nevertheless, even for these elements there are problems with the accurate fitting of the vacancy formation energy, surface energy, and the elastic anisotropy [28]. Moreover the neighbouring fcc transition metals Rh and Ir cannot be fitted with realistic embedding potentials since experimentally $C_{12} - C_{44} < 0$ which implies from eq. (23) an embedding function with negative curvature. In addition, bcc transition metals and tetrahedrally co-ordinated semi-conductors provide a further challenge to the Embedded Atom method (see, for example, Baskes [29]).

The advantage of the TBB description is that it provides a logical procedure for improving upon the Embedded Atom potentials by going beyond the second moment approximation and including the higher moments μ_3, μ_4, μ_5, ... This allows a systematic study of the factors influencing, for example, structural stability (see [11] and references therein). We, therefore, begin with a study of the consequences of a very simple, but physically realistic second moment model for the transition metals, before improving upon it in the next section.

Consider the following TBB model [20] in which the repulsive pair potential varies exponentially as

$$\Phi(R) = a\, N_d^2 \exp(-2\, k\, R) \tag{24}$$

and the hopping or bond integral behaves as

$$h(R) = b\, N_d \exp(-k\, R) \tag{25}$$

where a and b are constants for a given transition metal series. The prefactors N_d^2 and N_d are suggested by the respective dependence of $\Phi(R)$ and h(R) on the atomic charge density. The explicit influence [30] of the single valence s electron per atom is neglected, its nett repulsive contribution being included implicitly in $\Phi(R)$. Assuming only nearest neighbour hopping and substituting eq. (24) and (25) into eq. (20), we find the following equilibrium expressions for the band width, nearest-neighbour distance, cohesive energy, and bulk modulus:

$$W = (3b^2/5a)\, N_d\, (10 - N_d) \tag{26}$$

$$R_{eq} = k^{-1} \ln \{10a\, \sqrt{z}/[\sqrt{3}\, b\, (10 - N_d)]\} \tag{27}$$

$$U_{coh} = (3b^2/200a)[N_d(10 - N_d)]^2 \tag{28}$$

and

$$B = (2\sqrt{2}\, k^2/9\, R_{eq})\, U_{coh} \tag{29}$$

where z is the co-ordination number.

Fig. 3 compares the theoretical predictions with the experimental values across the 4d series, assuming one valence s electron per atom [31]. The "experimental" values of the band width are taken from the first-principles calculations in table 2 of ref. [31]. We have found that the observed equilibrium atomic volumes are fitted by values of the inverse decay length k which vary almost linearly across the series [32]. We have taken

$$k = 0.435 + 0.075\, N_d\ . \tag{30}$$

The ratio $b^2/a = \frac{2}{3}$ eV is obtained from eq. (26) by choosing a band width W = 10 eV for Mo with $N_d = 5$. Fitting the observed Mo Wigner-Seitz radius of 2.92 a.u. gives from eq. (27) a/b = 18.0. Thus, a = 216 eV and b = 12 eV for the 4d series.

We see from fig. 3 that the *trends* across the 4d series for the cohesive energy, equilibrium lattice spacing, and bulk modulus are well represented by this simple model. However, the predicted cohesive energy is about 2 eV/atom too small due to the neglect of sd hybridization. The predictions become

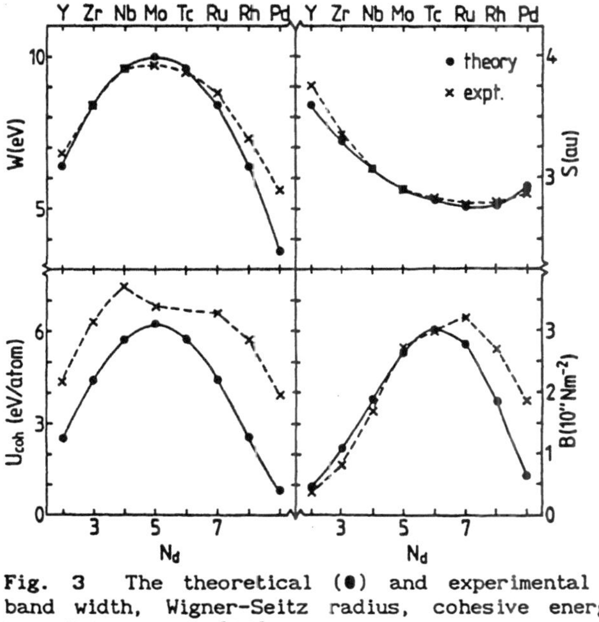

Fig. 3 The theoretical (●) and experimental (x) values of the equilibrium band width, Wigner-Seitz radius, cohesive energy, and bulk modulus of the 4d transition metals [20]

poorer towards the noble metal end of the series as the strength of the d bond weakens and the influence of the sp electrons become more noticeable. The equilibrium nearest neighbour distance depends on the co-ordination number z through eq. (27). This may be written in terms of the bond number $n = N_d/z$ which gives the number of electrons per bond contributed by a given atom. It takes the identical form to that proposed by Pauling [33], namely

$$R_{eq}(n) = R_{eq}(1) - (1.15/k)\log_{10}n \tag{31}$$

where $R_{eq}(1)$ is the bond length associated with a single pair of electrons. For Mo $k = 0.81$ a.u.$^{-1}$ so that the prefactor $(1.15/k)$ takes the value 0.75Å, which compares with Pauling's empirical value of 0.6Å for the metallic bond.

The most important feature of this second moment model is that the cohesive energy is structure independent since the co-ordination number z does not enter eq. (28). The origin of this result may be traced back to the form of the binding energy within this nearest neighbour model, namely

$$U(R) = \alpha\, z \exp(-2kR) - \beta \sqrt{z} \exp(-kR) \tag{32}$$

where R is the nearest neighbour distance and α, β are constants. Then, the equilibrium condition $dU/dR = 0$ gives immediately

$$\exp(-k R_{eq}) = (\beta/2\alpha)/\sqrt{z}, \tag{33}$$

so that substituting into eq. (32) the equilibrium binding energy is given by

$$U(R_{eq}) = \frac{1}{4}\beta^2/\alpha - \frac{1}{2}\beta^2/\alpha = -\frac{1}{4}\beta^2/\alpha. \tag{34}$$

That is, the binding energy curves for different nearest neighbour structure types (such as dimeric ($z = 1$), diamond ($z = 4$), simple cubic ($z = 6$), hexagonal ($z = 8$), and face centred cubic ($z = 12$)) all display the same minimum energy but at different separations.

Thus the second moment approximation is unable to *predict* structural stability. This is total consistent with the structural energy difference theorem [34] where the higher moments μ_3, μ_4, and μ_5 have been shown to control the structural stability of sp-bonded elements [35], transition elements [36], and pd bonded AB compounds [11]. Of course, $\Phi(R)$ and $h(R)$ may be *fitted* to give the observed ground state structure, but it must be appreciated that the square root embedding function, eq. (21), does not contain the microscopic factors controlling structural stability.

An unexpected consequence of this second moment model is that it predicts a zero vacancy formation energy [37]. It follows from eq. (1) that the vacancy formation energy may be written

$$\Delta U^{vac} = \Delta U^{vac}_{rep} + \Delta U^{vac}_{bond} \tag{35}$$

where the first contribution is pairwise so that [25]

$$\Delta U^{vac}_{rep} \simeq -U_{rep} \tag{36}$$

and the second contribution is square root saturated so that [25]

$$\Delta U^{vac}_{bond} \simeq -\frac{1}{2} U_{bond}. \tag{37}$$

Therefore, ignoring atomic relaxation, we have from eqs. (34)-(37) that

$$\Delta U^{vac} \simeq -U_{rep}(R_{eq}) - \frac{1}{2} U_{bond}(R_{eq})$$

$$= -\frac{1}{4} \beta^2/\alpha + \frac{1}{4} \beta^2/\alpha = 0. \qquad (38)$$

That is, whereas pair potentials predict a value of the vacancy formation energy equal to the cohesive energy, the second moment approximation predicts a value of zero. In reality, the experimental values lie approximately midway between! This zero result tends to counter one's intuition that the vacancy formation energy is in some way related to the bulk modulus or compressibility. In fact, Johnson [28] has recently shown that the vacancy formation energy is most closely related to the average shear modulus G, which in the above simple model is also zero. (Note that the second moment model predicting G = 0 is consistent with the fact that, unlike the bulk modulus, the sign of the shear modulus for a given lattice oscillates as a function of band filling; see, for example fig. 3 of Finnis et al.[38].)

This last example of the vacancy illustrates the danger of modelling the atomistic properties of transition metals with embedding potentials which do not contain the necessary higher moments for structural prediction. The embedding potentials predict that the first nearest neighbour shell of atoms relaxes inwards about a vacancy in bcc metals such as Nb or Mo, whereas more accurate TBB calculations by Ohta et al.[39] show a slight outward relaxation. This latter result is consistent with the known behaviour of the first and second nearest neighbour force constants reflected in the phonon spectra [38, 40].

5. A Many Body Potential for the Bond Order

We have seen that the second moment approximation is not sufficient for structural prediction. In this section we extend the embedding potential by including higher moment contributions. In this volume Moriarty [41] demonstrates very beautifully how the third and fourth moment terms lead to explicit three-body and four-body contributions in the perturbative expansion

$$U = U_0 + \frac{1}{2} \sum_{ij}' \Phi_2 (ij) + \frac{1}{6} \sum_{ijk}' \Phi_3 (ijk) + \frac{1}{24} \sum_{ijkl}' \Phi_4 (ijkl) + .. \qquad (39)$$

Here we adopt a slightly different approach by focussing on a many-body expansion for the bond order which is perhaps more in the spirit of embedding potentials.

The bond energy may be broken down in terms of contributions from individual pairs of bonds [7, 42, 43] by writing eq. (2) for s-valent systems as

$$U_{bond} = \frac{1}{2} \sum_{i,j}' h(R_{ij}) \Theta_{ij} \qquad (40)$$

where Θ_{ij} is the bond order of the ij-th bond. It is defined to be the difference between the number of electrons in the bonding $\frac{1}{\sqrt{2}}|i+j\rangle$ and anti-bonding $\frac{1}{\sqrt{2}}|i-j\rangle$ states. Although eq. (40) looks like a pair potential, the bond order Θ_{ij} is dependent on the surrounding atomic environment. For the isolated dimer with one valence s electron per atom the bond order of the singlet state is clearly 2. We will see that the bond order is reduced if the isolated dimer is embedded in a surrounding atomic matrix.

The analytic dependence of the bond order on its surroundings may be derived [44] by using the recursion method [45] to write the bond order as an integral over the difference of two continued fractions:

$$\Theta_{ij} = -\frac{2}{\pi} \Im m \int^{E_F} [G_{00}^{+}(E) - G_{00}^{-}(E)]\, dE \tag{41}$$

where the factor 2 outside the integral sign accounts for spin degeneracy. $\Im m$ is the imaginary part of the bonding/anti-bonding Green functions which are given by

$$G_{00}^{\pm}(E) = \frac{1}{2} \langle i \pm j|(E - H)^{-1}|i \pm j\rangle = \frac{1}{\left|(E-a_0^{\pm})\right.} - \frac{(b_1^{\pm})^2}{\left|(E-a_1^{\pm})\right.} - \dots \tag{42}$$

The first three recursion coefficients take the values

$$a_0^{\pm} = \pm\, h(R_{ij}) \tag{43}$$

$$(b_1^{\pm})^2 = \frac{1}{2} (\hat{\mu}_2^{\,i} + \hat{\mu}_2^{\,j}) \pm \sum_{k\neq i,j} h(R_{ik})h(R_{kj}) \tag{44}$$

$$(b_1^{\pm})^2\, a_1^{\pm} = \frac{1}{2} (\hat{\mu}_3^{\,i} + \hat{\mu}_3^{\,j}) \pm \sum_{k,l\neq i,j} h(R_{ik})h(R_{kl})h(R_{lj}) \tag{45}$$

where $\hat{\mu}_2^{\,i(j)}$ and $\hat{\mu}_3^{\,i(j)}$ correspond to the second and third moments about the site i(j) in which the direct path linking the bond ij is excluded from the path counting summation.

The first level of the continued fraction (42), namely $1/(E-a_0^{\pm})$, corresponds to the bonding/anti-bonding Green function for an isolated dimer which has eigenvalues of $\pm\, h(R_{ij})$ with respect to the free atomic energy level. The second level of the continued fraction begins to reflect the influence of the first shell of neighbours about the bond on the bonding/anti-bonding Green function. We see that the three- and four-membered *ring* contributions are solely responsible for the differences between the bonding and anti-bonding recursion coefficients given by eqs. (44) and (45).

A many-body form for the bond order may be derived by using perturbation theory to expand the bonding/anti-bonding Green function $G^{\pm}$ about the average Green function $G^0 = \frac{1}{2}(G^+ + G^-)$. Assuming that the latter is characterized by the Hamiltonian corresponding to the semi-infinite linear chain [45] defined by the recursion coefficients a_n^0, b_n^0 of G_{00}^0, we have, following Turchi and Ducastelle [36], from the first order Dyson equation

$$G_{00}^{\pm} = G_{00}^{0} + \sum_{n=0}^{\infty} (\delta a_n \pm \delta\alpha_n)\, G_{0n}^{0}\, G_{n0}^{0} + 2\sum_{n=1}^{\infty} (\delta b_n \pm \delta\beta_n)\, G_{0(n-1)}^{0}\, G_{n0}^{0} \tag{46}$$

where $a_n^{\pm} = a_n^0 + \delta a_n \pm \delta\alpha_n$, $b_n = b_n^0 + \delta b_n \pm \delta\beta_n$. (Note that the recursion coefficients a_n^0, b_n^0 of the average Green function are not in general equal to the average recursion coefficients $(a_n^+ + a_n^-)/2$, $(b_n^+ + b_n^-)/2$). Substituting into eq (41), the bond order becomes

$$\Theta = -4[\sum_{n=0}^{\infty} \chi_{0n,n0}(E_F)\, \delta\alpha_n + 2\sum_{n=1}^{\infty} \chi_{0(n-1),n0}(E_F)\, \delta\beta_n] \tag{47}$$

where the response functions $\chi_{0m,n0}(E_F)$ are defined by

$$\chi_{0m,n0}(E_F) = \frac{1}{\pi} \Im m \int^{E_F} G^{o}_{0m}(E)\, G^{0}_{n0}(E)\, dE. \tag{48}$$

The response functions depend on the average Green function G^0. However, since the perturbations in the recursion coefficients δa_n, δb_n do not enter into the bond order to first order, we can approximate G^0 by the Green function corresponding to the semi-infinite linear chain with zero diagonal elements $a^0_n = 0$ and constant off-diagonal elements $b^0_n = b^0_1$. Writing $b^0_1 = b$ for simplicity, we have

$$b = \left\{\left[\sum_{k\neq i} h^2(R_{ik}) + \sum_{k\neq j} h^2(R_{jk})\right]/2\right\}^{1/2} \tag{49}$$

where b is negative corresponding to $h = ss\sigma < 0$. Then substituting [36] $bG^0_{0n} = \exp[i(n+1)\phi]$ where $\cos\phi = E/2b$ into eq. (48), we have $\chi_{0m,n0} = \hat{\chi}_{0m,n0}/|b|$ where the reduced susceptibility

$$\hat{\chi}_{0m,n0}(N) = \frac{1}{\pi}\left[\frac{\sin(m+n+1)\phi_F}{(m+n+1)} - \frac{\sin(m+n+3)\phi_F}{(m+n+3)}\right] \tag{50}$$

with $\phi_F = \cos^{-1}(E_F/2b)$. ϕ_F is fixed by the number of electrons per atom N through [7]

$$N = (2\phi_F/\pi)[1 - (\sin 2\phi_F)/2\phi_F]. \tag{51}$$

Fig. 4 shows the behaviour of the first three reduced response functions $\hat{\chi}$ as a function of the number of electrons per atom N.

Finally, the many-body character of the bond order may be displayed explicitly by substituting the values of $\delta\alpha_0$, $\delta\alpha_1$, and $\delta\beta_1$ from eqs. (43), (44) and (45) into eq. (47). To first order, using $\mu_3/\mu_2^{3/2} \ll 1$, we find

$$\Theta_{ij} = 4\left\{\hat{\chi}_{00,00}(N)\,[h(R_{ij})]/b + \hat{\chi}_{00,10}(N)\left[\sum_{k\neq i,j} h(R_{ik})\,h(R_{kj})\right]/b^2\right.$$

$$\left. + \hat{\chi}_{01,10}(N)\left[\sum_{k,l\neq i,j} h(R_{ik})\,h(R_{kl})\,h(R_{lj})\right]/b^3 + \dots\right\} \tag{52}$$

This important result gives the analytic dependence of the bond order on the surrounding environment. The first term is similar to an embedding potential in that this contribution may be written

$$\Theta^{(2)}_{ij} = -4\,\hat{\chi}_{00,00}(N)\,h(R_{ij})\,F^{(2)}[(\rho_i + \rho_j)/2] \tag{53}$$

where the embedding function

$$F^{(2)}(\rho) = -\rho^{-1/2} \tag{54}$$

with the local "charge density" given as before by eq. (22). For nearest neighbour interactions only, $\Theta^{(2)}_{ij} \propto 1/\sqrt{z}$, a well-known result [43] within the

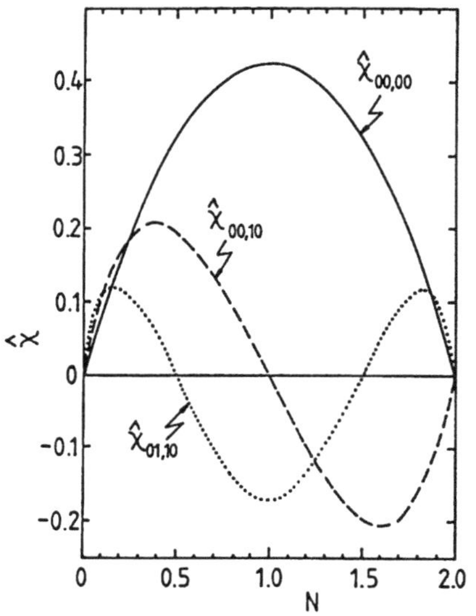

Fig. 4 The reduced susceptibilities as a function of the number of electrons per atom.

second moment approximation, following directly from the covalent energy per bond in eq. (32). It accounts naturally for the back-bond strengthening factor which Chelikowsky et al.[46] found necessary to include for an accurate representation of both bulk and cluster silicon geometries.

The renormalization of the effective pair-wise bond strength with change in environment also explains why the usual pair potential contribution to the vacancy formation energy, namely U_{coh} of eq. (36), is reduced within the second moment approximation to $\frac{1}{2}$ U_{coh} of eq. (37) if the repulsive contribution is neglected. It follows from eqs. (40) and (53) that the bond contribution to the vacancy formation energy may be written

$$\Delta U_{bond}^{vac} = - U_{bond} + 4\, \hat{\chi}_{00,00}\, h\, z(z-1)\{[((z-1)+z)/2]^{-\frac{1}{2}} - [z]^{-\frac{1}{2}}\} \tag{55}$$

where

$$U_{bond} = 2\, \hat{\chi}_{00,00}\, h \sqrt{z}. \tag{56}$$

The first term in eq. (55) reflects the usual pairwise contribution to the vacancy formation energy due to breaking bonds. The second term reflects the change in the strength of the bonds due to the presence of the vacancy. It may be simplified to give

$$\Delta U_{bond}^{vac} \simeq - U_{bond} + \hat{\chi}_{00,00}\, h \sqrt{z} \tag{57}$$

$$= - U_{bond} + \frac{1}{2} U_{bond} = - \frac{1}{2} U_{bond} \tag{58}$$

which agrees with eq. (37).

Only explicit 2-body sums are involved in $\Theta_{ij}^{(2)}$. The importance of the 3-body and 4-body contributions to the bond order may be demonstrated by considering the topology of four atom s-valent clusters. The total bond energies of the one-dimensional linear (ℓ), two-dimensional square (s), and three-dimensional tetrahedral (t) clusters are shown in the left hand panel of

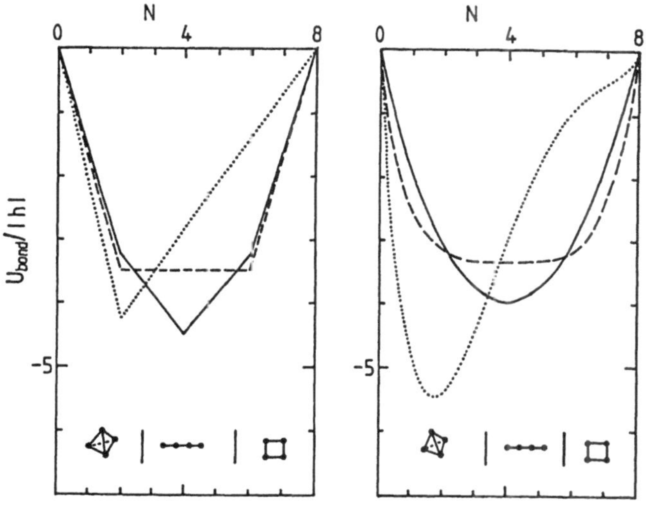

Fig. 5 The total bond energy as a function of the total number of electrons for a linear (full curves), square (dashed), and tetrahedral (dotted) cluster of four atoms. The left hand panel gives the Molecular Orbital result, whereas the right hand panel shows the predictions of the many-body potential for the bond order.

fig. 5 as a function of the total number of valence electrons. The *total* bond energy may be obtained directly from eq. (2) by occupying the nearest neighbour two-centre Hückel eigenvalues namely $[\pm(\sqrt{5} \pm 1)/2]h_\ell$ for the linear, $[0, 0, \pm 2]h_s$ for the square, and $[-1, -1, -1, 3]h_t$ for the tetrahedral clusters respectively. The relative strengths of the hopping integrals $h_\ell : h_s : h_t$ were set by assuming that the equilibrium bond lengths of the individual clusters are such as to give identical second moments in the spirit of the structural energy difference theorem [34]. That is, $\mu_2 = 6h_\ell^2 = 8h_s^2 = 12h_t^2$ or $h_t = h/\sqrt{2}$, $h_s = \sqrt{3}\,h/2$, $h_\ell = h$. We see that the curves predict a change in structural stability from tetrahedral to linear to square as the total number of electrons in the cluster increases. Identical curves would, of course, be obtained by evaluating the *individual* bond energies directly from eq. (40) using the appropriate recursion coefficients given in table 1. Note that the four-atom linear chain has two non-equivalent bonds, which we have labelled the inner and outer.

The right-hand panel of fig. 5 shows the predictions of the many-body potential approximation to the bond order. The linear cluster has no three-membered or four-membered rings, so that its behaviour is controlled by the response function $\hat{\chi}_{00,00}$ in fig. 4. However, each bond in the square cluster is associated with one four-membered ring so that there is an additional contribution from $\hat{\chi}_{01,10}$. This leads to the oscillatory behaviour of the stability from square to linear to square observed in fig. 5. On the other hand, each bond in the tetrahedral cluster is associated with two three-membered rings and two four membered rings. The three-membered rings introduce the $\hat{\chi}_{00,10}$ response function which skews the binding curve towards small values of the electron number. It is the presence of these three-membered rings which favours the close-packed structures over the more open structures for monovalent, divalent and trivalent sp-bonded elements [35]

Table 1. Bond order recursion coefficients for the linear, square, and tetrahedral clusters.

Recursion Coefficient	Linear Inner	Linear Outer	Square	Tetrahedral
$a_0^{\pm}$	$\pm h_\ell$	$\pm h_\ell$	$\pm h_s$	$\pm h_t$
$a_1^{\pm}$	0	0	$\pm h_s$	$(\frac{1}{2} \pm \frac{1}{2})h_t$
$a_2^{\pm}$	-	$\mp\frac{1}{3}h_\ell$	-	-
$a_3^{\pm}$	-	$\mp\frac{2}{3}h_\ell$	-	-
$b_1^{\pm 2}$	h_ℓ^2	$\frac{1}{2}\, h_\ell^2$	h_s^2	$(2 \pm 2)h_t^2$
$b_2^{\pm 2}$	0	$\frac{3}{2}h_\ell^2$	0	0
$b_3^{\pm 2}$	-	$\frac{2}{9}h_\ell^2$	-	-
$b_4^{\pm 2}$	-	0	-	-

and which probably causes the metal-metal bond to be strengthened in the presence of hydrogen impurities for less than half-filled d bands, but weakened otherwise [47]. Higher order contributions to eq. (52) may be important in controlling, for example, the Jahn-Teller distortion in clusters or the relative stability of the sp- and sd-bonded cubic and hexagonal close packed lattices. Nevertheless the convergence of the perturbation series may be guaranteed by introducing temperature to damp down the long range Green functions [48].

This many-body potential for the bond order may be generalized to the case of sp-(or sd-)valent systems. Assuming that the crystal field terms produce identical shifts for the s and p atomic energies on a given site [7], the total energy of eq. (12) will contain the additional promotion energy contribution, namely

$$U_{prom} = (\varepsilon_p^0 - \varepsilon_s^0)\, \Delta N, \tag{59}$$

where ε_s^0, ε_p^0 are the atomic energy levels corresponding to the *reference* free atom state with N_s^0, N_p^0 s, p valence electrons respectively. ΔN gives the change in p occupancy on going from the reference free atom state to a given bonding situation, that is $\Delta N = N_p - N_p^0$.

Now define a new bond energy $(U_{bond})_{\Delta N_{s,p}=0}$ such that $\varepsilon_{sp} = \varepsilon_p - \varepsilon_s$ has been adjusted on each site to guarantee N_s and N_p retain the reference values N_s^0 and N_p^0 respectively. Then the difference in this bond energy between two situations 1 and 2 may be written [11]

$$(\Delta U_{bond})_{\Delta N_{s,p}=0} = U^{(2)}_{bond}(\varepsilon^{(2)}_{sp}) - U^{(1)}_{bond}(\varepsilon^{(1)}_{sp})$$

$$= U^{(2)}_{band}(\varepsilon^{(2)}_{s}, \varepsilon^{(2)}_{p}) - U^{(1)}_{band}(\varepsilon^{(1)}_{s}, \varepsilon^{(1)}_{p}) - \Delta\varepsilon_s N_s^{\,0} - \Delta\varepsilon_p N_p^{\,0}. \qquad (60)$$

But, to first order,

$$U^{(2)}_{band}(\varepsilon^{(2)}_{s}, \varepsilon^{(2)}_{p}) = U^{(2)}_{band}(\varepsilon^{(1)}_{s}, \varepsilon^{(1)}_{p}) + \Delta\varepsilon_s N_s^{\,0} + \Delta\varepsilon_p N_p^{\,0}. \qquad (61)$$

Therefore, substituting eq. (61) into eq. (60), we have

$$(\Delta U_{bond})_{\Delta N_{s,p}=0} = (\Delta U_{band})_{\Delta\varepsilon_{s,p}=0} \qquad (62)$$

$$= (\Delta U_{prom} + \Delta U_{bond})_{\Delta\varepsilon_{s,p}=0}. \qquad (63)$$

Hence, the TBB expression of eq. (1) is perfectly general provided the bond energy is evaluated under the constraint that the numbers of s and p electrons on each site are kept fixed and equal to their reference values $N_s^{\,0}$ and $N_p^{\,0}$ respectively through the adjustment of ε_{sp}.

In general, the bond energy can be written [7]

$$U_{bond} = \tfrac{1}{2} \sum_{\substack{im,jn\\ i\neq j}} h_{im,jn}\, \Theta_{jn,im} \qquad (64)$$

where m(n) label the orbitals on site i(j). For the particular case of $sp\sigma = (|ss\sigma| pp\sigma)^{1/2}$ the ij-th bond energy reduces to the form

$$U^{ij}_{bond} = (ss\sigma_{ij} - pp\sigma_{ij})\, \Theta_{j\sigma,i\sigma} + 2pp\pi_{ij}\, \Theta_{j\pi,i\pi} \qquad (65)$$

where the normalized orbitals $|i\sigma\rangle$ and $|j\sigma\rangle$ are defined by

$$|i\sigma\rangle = (\sqrt{s}\,|is\rangle + \sqrt{p}\,|iz\rangle)/\sqrt{s+p} \qquad (66)$$

and

$$|j\sigma\rangle = (\sqrt{s}\,|js\rangle - \sqrt{p}\,|jz\rangle)/\sqrt{s+p} \qquad (67)$$

with $s = |ss\sigma|$, $p = pp\sigma$ choosing the z-axis along R_{ij} and $\Theta_{j\pi,i\pi} = (\Theta_{jx,ix} + \Theta_{jy,iy})/2$. For silicon in the diamond structure there are no three-, four-, or five- membered rings so that $\Theta^{(2)}_{im,jn}$ dominates. It follows from eqs. (49) and (52) that

$$\Theta^{(2)}_{im,jn} \propto \{ [\langle im|H^2|im\rangle + \langle jn|H^2|jn\rangle]/2\ \langle im|H|jn\rangle^2 \}^{-1/2}, \qquad (68)$$

where from table 1 of Slater and Koster [49] for $\varepsilon_{sp} = 0$ we have

$$\langle is|H^2|is\rangle = \sum_{k\neq i} (ss\sigma^2_{ik} + sp\sigma^2_{ik}), \qquad (69)$$

$$\langle iz|H^2|iz\rangle = \sum_{k\neq i} [(sp\sigma^2_{ik} + pp\sigma^2_{ik}) \cos^2\theta_{jik} + pp\pi^2_{ik} \sin^2\theta_{jik}], \tag{70}$$

$$\langle is|H^2|iz\rangle = \sum_{k\neq i} - sp\sigma_{ik}(ss\sigma_{ik} - pp\sigma_{ik}) \cos\theta_{jik}, \tag{71}$$

and

$$\langle ix|H^2|ix\rangle + \langle iy|H^2|iy\rangle = \sum_{k\neq i} \begin{bmatrix} (sp\sigma^2_{ik} + pp\sigma^2_{ik}) \sin^2\theta_{jik} \\ + pp\pi^2_{ik}\ (1 + \cos^2\theta_{jik}) \end{bmatrix} \tag{72}$$

with θ_{jik} the angle between the bonds ij and ik. Substituting eqs. (69)-(71) into (68) we find

$$\Theta^{(2)}_{i\sigma,j\sigma} \propto \{1 + \sum_{k\neq i,j} [G^\sigma_{ij}(R_{ik}, \theta_{jik}) + G^\sigma_{ji}(R_{jk}, \theta_{ijk})]/2\}^{-½} \tag{73}$$

where

$$G^\sigma_{ij}(R_{ik}, \theta_{jik}) = A^\sigma_{ij}(R_{ik}) + B^\sigma_{ij}(R_{ik}) \cos\theta_{jik} + C^\sigma_{ij}(R_{ik}) \cos 2\theta_{jik} \tag{74}$$

with

$$\left.\begin{matrix} A^\sigma_{ij} \\ B^\sigma_{ij} \\ C^\sigma_{ij} \end{matrix}\right\} = \left.\begin{matrix} a^\sigma_{ij} \\ b^\sigma_{ij} \\ c^\sigma_{ij} \end{matrix}\right\} \times (|ss\sigma_{ij}| + pp\sigma_{ij})^{-3} \tag{75}$$

such that

$$a^\sigma_{ij}(R_{ik}) = |ss\sigma_{ij}|(ss\sigma^2_{ik} + sp\sigma^2_{ik}) + ½\ pp\sigma_{ij}(sp\sigma^2_{ik} + pp\sigma^2_{ik} + pp\pi^2_{ik}), \tag{76}$$

$$b^\sigma_{ij}(R_{ik}) = -2\ sp\sigma_{ij}\ sp\sigma_{ik}(ss\sigma_{ik} - pp\sigma_{ik}), \tag{77}$$

and

$$c^\sigma_{ij}(R_{ik}) = ½\ pp\sigma_{ij}(sp\sigma^2_{ik} + pp\sigma^2_{ik} - pp\pi^2_{ik}). \tag{78}$$

We have, therefore, derived in eqs. (73) and (74) a bond order which has a form and angular dependence which may be compared with those proposed by Chelikowsky et al [46] and Tersoff [50]. We should note that for the case where the s and p hopping integrals show the same distance dependence, say h(R), G^σ_{ij} factorizes as

$$G^\sigma_{ij}(R_{ik}, \theta_{jik}) = h^{-2}(R_{ij})h^2(R_{ik})(a + b \cos\theta_{jik} + c \cos 2\theta_{jik}). \tag{79}$$

A particular short-range analytic form for h(R) has been given recently by Goodwin et al [51] which reproduces silicon clusters and bulk.

For sd-bonded transition-metals the ij-th bond energy may be written for the case $sd\sigma = -(ss\sigma . dd\sigma)^{½}$ as

$$U^{ij}_{bond} = (ss\sigma_{ij} + dd\sigma_{ij})\,\Theta_{j\sigma,i\sigma} + 2dd\Pi_{ij}\,\Theta_{j\pi,i\pi} + 2dd\delta_{ij}\,\Theta_{j\delta,i\delta} \tag{80}$$

where the normalized orbitals $|i\sigma\rangle$ and $|j\sigma\rangle$ are defined by

$$|i\sigma\rangle = (\sqrt{s}\,|is\rangle + \sqrt{d}\,|i(3z^2 - r^2)\rangle)/\sqrt{s+d} \tag{81}$$

and

$$|j\sigma\rangle = (\sqrt{s}\,|js\rangle + \sqrt{d}\,|j(3z^2 - r^2)\rangle)/\sqrt{s+d} \tag{82}$$

with $s = |ss\sigma|$, $d = |dd\sigma|$ and

$$\Theta_{j\pi,i\pi} = (\Theta_{j(xz),i(xz)} + \Theta_{j(yz),i(yz)})/2 \tag{83}$$

$$\Theta_{j\delta,i\delta} = (\Theta_{j(xy),i(xy)} + \Theta_{j(x^2-y^2),i(x^2-y^2)})/2. \tag{84}$$

For $\varepsilon_{sd} = 0$ we have

$$\langle is|H^2|is\rangle = \sum_{k\neq i} (ss\sigma^2_{ik} + sd\sigma^2_{ik}), \tag{85}$$

$$\langle i(3z^2 - r^2)|H^2|i(3z^2 - r^2)\rangle = \sum_{k\neq i} [\tfrac{1}{4}(3n^2 - 1)^2(sd\sigma^2_{ik} + dd\sigma^2_{ik}) + 3n^2(1 - n^2)\,dd\Pi^2_{ik} + \tfrac{3}{4}(1 - n^2)^2\,dd\delta^2_{ik}], \tag{86}$$

$$\langle is|H^2|i(3z^2 - r^2)\rangle = \sum_{k\neq i} \tfrac{1}{2}(3n^2 - 1)\,sd\sigma_{ik}\,(ss\sigma_{ik} + dd\sigma_{ik}), \tag{87}$$

$$\langle i(xz)|H^2|i(xz)\rangle + \langle i(yz)|H^2|i(yz)\rangle = \sum_{k\neq i} [3n^2(1 - n^2)(sd\sigma^2_{ik} + dd\sigma^2_{ik}) + (1 - 3n^2 + 4n^4)\,dd\Pi^2_{ik} + (1 - n^4)\,dd\delta^2_{ik}] \tag{88}$$

and

$$\langle i(x^2 - y^2)|H^2|i(x^2 - y^2)\rangle + \langle i(xy)|H^2|i(xy)\rangle = \sum_{k\neq i} [\tfrac{3}{4}(1-n^2)^2(sd\sigma^2_{ik} + dd\sigma^2_{ik}) + (1 - n^4)\,dd\Pi^2_{ik} + \tfrac{1}{4}(1 + 6n^2 + n^4)\,dd\delta^2_{ik}] \tag{89}$$

where $n = \cos\theta_{jik}$. From this it follows that the explicit 2-body contribution to the σ bond order may be written as eq. (73) with

$$G^{\sigma}_{ij}(R_{ik}, \theta_{jik}) = D^{\sigma}_{ij}(R_{ik}) + E^{\sigma}_{ij}(R_{ik}) \cos 2\theta_{jik} + F^{\sigma}_{ij}(R_{ik}) \cos 4\theta_{jik} \tag{90}$$

with

$$\left.\begin{matrix} D^{\sigma}_{ij} \\ E^{\sigma}_{ij} \\ F^{\sigma}_{ij} \end{matrix}\right\} = \left.\begin{matrix} d^{\sigma}_{ij} \\ e^{\sigma}_{ij} \\ f^{\sigma}_{ij} \end{matrix}\right\} \times (|ss\sigma_{ij}| + |dd\sigma_{ij}|)^{-3} \tag{91}$$

such that

$$d^{\sigma}_{ij}(R_{ik}) = |ss\sigma_{ij}|(ss\sigma^2_{ik} + sd\sigma^2_{ik}) + \frac{1}{2}|sd\sigma_{ij}|sd\sigma_{ik}(ss\sigma_{ik} + dd\sigma_{ik})$$

$$+ \frac{1}{32}|dd\sigma_{ij}|(11sd\sigma^2_{ik} + 11dd\sigma^2_{ik} + 12dd\pi^2_{ik} + 9dd\delta^2_{ik}), \qquad (92)$$

$$e^{\sigma}_{ij}(R_{ik}) = \frac{3}{2}|sd\sigma_{ij}|sd\sigma_{ik}(ss\sigma_{ik} + dd\sigma_{ik})$$

$$+ \frac{3}{8}|dd\sigma_{ij}|(sd\sigma^2_{ik} + dd\sigma^2_{ik} - dd\delta^2_{ik}), \qquad (93)$$

and

$$f^{\sigma}_{ij}(R_{ik}) = \frac{3}{32}|dd\sigma_{ij}|(3\, sd\sigma^2_{ik} + 3\, dd\sigma^2_{ik} - 4\, dd\pi^2_{ik} + dd\delta^2_{ik}). \qquad (94)$$

It remains to test the predictions of these many-body angularly dependent bond order potentials against the results of previous Tight Binding calculations [35, 36, 51].

6. Conclusion

We have seen that it is possible to derive the simple Tight Binding Bond (TBB) model from first principles by working within the Harris-Foulkes approximation [5, 6] to density functional theory and using Anderson's chemical pseudopotentials [9]. The TB bond contribution was reduced to that of an embedding potential through the second moment approximation to the local electronic density of states, as suggested by Finnis and Sinclair [13]. The higher moment contributions were, however, argued to be essential for *predicting* structural stability. A many-body potential for the bond order was derived which included the higher moment contributions explicitly. Its simple analytic form, eq. (52) , will allow the Embedded Atom potentials to be extended to include *explicit* three-body terms, for example. These could help remove problems in the accurate fitting of the surface energy, the vacancy formation energy, and the elastic anisotropy of fcc metals discussed in this volume [14, 28]. These potentials may be generalized to the case of sp-valent elements such as silicon [51] or sd-valent elements such as molybdenum [41], although actual calculations have not yet been performed.

Acknowledgements

I should like to thank the organizers of this workshop for such a stimulating meeting. In particular, I should like to thank Anders Carlsson for pointing out that the vacancy formation energy is approximately zero within the simple second moment model presented in section 4. I am grateful to Dr. Masato Aoki for checking the many-body bond order expressions and pointing out mistakes in the original eqs. (89) and (93). I should like to thank José Cressoni, Mike Finnis, Leif Goodwin, Andrew Skinner and Adrian Sutton for many illuminating discussions. I should also like to acknowledge the support of the US Department of Energy under their Energy Conversion and Utilization Technologies (ECUT) Materials Program, subcontract No. 19X-55992V, through Martin Marietta Systems, Inc.

References

1. W.A. Harrison, "Electronic Sructure and the Properties of Solids", Freeman, San Francisco, (1980); and references therein.

2. V. Heine, R. Haydock, D. Bullett and M.J. Kelly, Sol. St. Phys. 35 (1980); and references therein.

3. J. Friedel in "The Physics of Metals", Cambridge University Press, Cambridge, p. 494 (1969).

4. F. Ducastelle, J. de Physique 31, 1055 (1970).

5. W.M.C. Foulkes, College Dissertation (1985); Ph.D. thesis, University of Cambridge (1987).

6. J. Harris, Phys. Rev. B31, 1770 (1985).

7. A.P. Sutton, M.W. Finnis, D.G. Pettifor and Y. Ohta, J. Phys. C21, 35 (1988).

8. P. Hohenberg and W. Kohn, Phys. Rev. 136, 864 (1964); W. Kohn and L.J. Sham, Phys. Rev. 140, 1133 (1965).

9. P.W. Anderson, Phys. Rev. Lett. 21, 13 (1968).

10. A.J. Skinner and D.G. Pettifor, J. Phys. (Con. Mat.), (submitted).

11. D.G. Pettifor and R. Podloucky, J. Phys. C19, 315 (1986).

12. M.S. Daw and M.I. Baskes, Phys. Rev. B29, 6443 (1984).

13. M.W. Finnis and J.E. Sinclair, Phil. Mag. A50, 45 (1984).

14. K.W. Jacobsen, J.K. Nørskov and M.J. Puska, Phys. Rev. B35, 7423 (1987); and this volume.

15. W. Kolos and L. Wolniewicz, J. Chem. Phys. 41, 12 (1964).

16. B.I. Min, H.J.F. Jansen and A.J. Freeman, Phys. Rev. B30, 5076 (1984).

17. J.C. Slater, "Quantum Theory of Molecules and Solids", Vol. 4, McGraw-Hill (1974).

18. O. Gunnarson and B.I. Lundquist, Phys. Rev. B13, 10 (1976).

19. O.K. Andersen and O. Jepsen, Phys. Rev. Lett. 53, 2571 (1984).

20. D.G. Pettifor, Sol. St. Phys. 40, 43 (1987).

21. D.G. Pettifor, Commun. Phys. 1, 141 (1976).

22. D.G. Pettifor, J. Chem. Phys. 69, 2930 (1978).

23. O.K. Andersen, §5.3 of A.R. Mackintosh and O.K. Andersen in "Electrons at the Fermi Surface", ed. M. Springford, Cambridge University Press, Cambridge (1980).

24. M. Cyrot and F. Cyrot-Lackmann, J. Phys. F6, 2257 (1976).

25. V. Heine, this volume.

26. M.S. Daw, this volume.

27. S.P. Chen, A.F. Voter, R.C. Albers, A.M. Boring and P.J. Hay, Scripta Met. 23, 217 (1989).

28. R.A. Johnson, Phys. Rev. B37, 3924 (1988); and this volume.

29. M.I. Baskes, Phys. Rev. Lett. 59, 2666 (1987).

30. D.G. Pettifor, J. Phys. F8, 219 (1978).

31. D.G. Pettifor, J. Phys. F7, 613 (1977).

32. D.G. Pettifor in "Physical Metallurgy", eds. R.W. Cahn and P. Haasen, North-Holland Publ., Amsterdam, Ch. 3 (1983).

33. L. Pauling, "The Nature of the Chemical Bond", 3rd ed. Cornell Univ. Press, Ithaca (1960).

34. D.G. Pettifor, J. Phys. C19, 285 (1986).

35. J.C. Cressoni, Ph.D. thesis, Imperial College, London (1989).

36. P. Turchi and F. Ducastelle, in "The Recursion Method and its Applications", eds. D.G. Pettifor and D.L. Weaire, Sol. St. Sciences 58, 104, (1985).

37. A. Carlsson, private communication (1989).

38. M.W. Finnis, K.L. Kear, and D.G. Pettifor, Phys. Rev. Lett. 52, 291 (1984).

39. Y. Ohta, M.W. Finnis, D.G. Pettifor and A.P. Sutton, J. Phys. F17, L273 (1987).

40. S.H. Chen and B.N. Brockhouse, Sol. St. Commun. 2, 73 (1964).

41. J.A. Moriarty, Phys. Rev. B38, 3199 (1988); and this volume.

42. C.A. Coulson, Proc. Roy. Soc. A169, 413 (1939).

43. G.C. Abell, Phys. Rev. B31, 6184 (1985).

44. D.G. Pettifor (to be published).

45. R. Haydock, V. Heine and M.J. Kelly, J. Phys. C5, 2845 (1972).

46. J.R. Chelikowsky, J.C. Phillips, M. Kamal and M. Strauss, Phys. Rev. Lett. 56, 632 (1989).

47. H. Adachi and S. Imoto, J. Phys. Soc. Jpn. 46, 1194 (1979).

48. D.G. Pettifor, A.J. Skinner and R.A. Davies, in Proc. of Int. Conf. on "Atomistic Simulation of Materials: Beyond Pair Potentials", eds. V. Vitek and D. Srolovitz, Plenum, New York (1989).

49. J.C. Slater and G. F. Koster, Phys. Rev. 94, 1498 (1954).

50. J. Tersoff, Phys. Rev. B37, 6991 (1988).

51. L. Goodwin, A.J. Skinner and D.G. Pettifor, Europhys. Lett. 9, 701 (1989); L. Goodwin and A.J. Skinner, this volume.

Implications of the Embedded-Atom Method Format

R.A. Johnson

Materials Science Department, Thornton Hall, University of Virginia,
Charlottesville, VA 22901, USA

Abstract. The implications of format of the governing equations of the embedded-atom method, or several alternate theories which obtain the same mathematical formulas, has been studied. Transformation properties which lead to the concept of an effective two-body potential are discussed relative to both monatomic and alloy models with a new form for alloy potentials proposed. Correspondence with earlier empirical models is made through the effective two-body potential and it is used to clarify the role of many-body forces and the curvature of the embedding function relative to the Cauchy condition. Also some general relations between defect energies and the elastic properties of metals are given. Simple analytic models for fcc metals using nearest-neighbor interactions, bcc metals with first- and second-neighbor interactions, and hcp metals with interactions through third neighbors are developed and used to further study the relationships between physical input parameters and other properties of metals.

1. Introduction

Computer calculations based on atomistic models have been used to study metals ever since the availability of digital computers for scientific research in the late 1950's. An atomistic model is simply an equation for the energy of a material as a function of the position of all its constituent atoms. In practice, often only the difference in energy between two atomic configurations is of interest, which leads to considerable simplifications in the calculations.

The selection of an interatomic potential, i.e., the way in which atoms interact, is of course the fundamental input into the development of an atomistic model. Numerous approaches have been used leading to a long bibliography [1-5] and a full complement of controversy. All choices involve approximations, so there is always a trade-off and no ideal way of treating this problem. If the model is too simple it will not have enough flexibility to accurately portray real metals, while more fundamental approaches tend to be too unwieldy to carry out the desired calculations with sufficient efficiency.

The detailed requirements for a model shift with time as computational power increases, but the intent remains the same: a model must be general enough to carry out a broad range of computer experiments, accurate enough to yield a reasonable expectation that the results will be physically meaningful, and simple enough to do so within your computer budget.

Empirical and semi-empirical atomistic models based on two-body potentials, and global volume dependence in some cases, were well developed by the mid 1960's and were reasonably successful in treating a number of bulk defect properties in transition metals. In all such models, the analytic form for the potentials is given, and the actual parameters are obtained by fitting to experimental data such as the lattice constant, the elastic moduli, and the cohesive energy or the vacancy formation energy. There was a dilemma here, however: models fit to the cohesive energy always gave too large a vacancy formation energy by a factor of two to three, and models fit to the vacancy formation energy either gave too small a cohesive energy by the same factor or could not yield a cohesive energy.

There was little improvement until the 1980's, when two new procedures for designing atomistic models for metals were proposed: the Finnis and Sinclair (FS) model [6,7] and the Embedded-Atom Method (EAM) by Daw and Baskes [8,9]. The essential band character of metallic cohesion is incorporated in the FS model through the second-moment approximation to tight-binding theory in which the cohesive energy per atom varies as $\sqrt{z}$, with z the atomic coordination. The application of the FS model has been primarily to BCC metals [6,10-12]. The EAM is based on density-functional theory as derived by Nørskov and Lang (effective-medium approach) [13] and Stott and Zaremba (quasiatom approach) [14]. Here the energy required to place a small impurity atom in a lattice is taken solely as a function of the electron density at that particular site. Each atomic species therefore has a unique energy function which is in turn a function of just the electron density. For example, Puska *et al.* [15] have calculated this function for a number of light atoms using a constant electron density (jellium) formulation. Daw and Baskes [6,7] extended this idea to a general lattice model which permits variation in atomic positions by also including a two-body central interatomic potential. Thus it has been possible to include lattice relaxation and cover a broad range of metals, impurities, and alloys, including bulk, surface, and liquid properties [16-23]. The basis of the EAM is discussed in detail in a recent paper by Daw [24].

Although the theoretical development is very different, the format of the governing equations in both the FS and EAM models is the same. Jacobsen *et al.* [25] have recently derived an expression for the total energy of a set of atoms based on effective medium theory (EMT). With approximations to render their results amenable to defect calculations, the same format is obtained. Indeed, equations which have the same effect as this format were unsuccessfully sought in the mid 1960's to provide local volume dependence for the empirical models. Thus, the emphasis of the present paper is towards understanding the implications of this "standard" format relative to the properties of metals and, for convenience, it will be referred to as the EAM format. In addition to studying the general formulation, a set of functions and fitting schemes has been designed to provide as simple EAM models as possible. Although not as precise as more detailed models, these simple models maintain the basic character of the EAM and can be investigated analytically.

2. Formulation

The EAM [8,9,24] involves the calculation of the total energy of a set of atoms as a sum of an energy from inserting each atom in the electron density arising from all the other atoms and an energy from two-body interactions. The fundamental concept is that the embedding energy for a particular species of atom is a unique function of electron density, independent of the source of that electron density. The electron density at any site is then taken as a linear superposition of spherically symmetric electron densities from contributing atoms. In this respect there is a difference between the EAM and the FS model [6,7], in which the equivalent terms (overlap integrals) are dependent on the type of atom at each end of the interaction. This difference manifests itself in alloy calculations, so the models are only equivalent mathematically for monatomic materials.

The basic equations of the EAM and FS have the form

$$E = \sum (F_i + \Phi_i) , \tag{1}$$

$$\rho_i = \sum' f_j(r_{ij}) , \tag{2}$$

$$\phi_i = \frac{1}{2} \sum' \phi_{ij}(r_{ij}) , \tag{3}$$

where E is the total internal energy, ρ_i is the electron density at atom i due to all other atoms, $f_j(r)$ is the electron density of atom j as a function of distance from its center, r_{ij} is the separation distance between atoms i and j, $F_i(\rho)$ is the energy to embed atom i in electron density ρ, and $\phi_{ij}(r)$ is a two-body central potential between atoms i and j. Furthermore, $f(r)$ is only determined to within a scaling factor from fitting to monatomic properties and the relative scaling factor between elements must be specified for alloy models.

In summary, for an EAM model an embedding function $F(\rho)$ and an atomic electron density function $f(r)$ must be specified for each atomic species and a two-body potential $\phi(r)$ specified for each possible combination of atomic species. For an FS model, the $f(r)$ term must also be specified for each possible combination of atomic species.

2.1. Model Transformations

A monatomic EAM model is invariant to a transformation in which a term linear in the electron density is added to or subtracted from the embedding function and an appropriate adjustment is made to the two-body potential [6,9,18,26-29]. To illustrate this point, define transformed embedding function $G(\rho)$ and two-body potential $\psi(r)$ by

$$G(\rho) = F(\rho) + k\rho \, , \tag{4}$$

$$\psi(r) = \phi(r) - 2kf(r) \, . \tag{5}$$

The total energy using these functions is the same as that using the initial functions:

$$E = \sum [G_i + \Psi_i] = \sum [F_i + \Phi_i] \, . \tag{6}$$

Thus, the reported forms for monatomic EAM model functions involve an arbitrary choice and, with transformation, models which appear to be very different are in fact found to be similar. The effect of this transformation on alloy models is more complex and is discussed below. In some models the embedding function is determined by purely numerical methods and in others an analytic form is used. The general analytic form

$$F(\rho) = F_0\left[\frac{n}{n-m}\left(\frac{\rho}{\rho_e}\right)^m - \frac{m}{n-m}\left(\frac{\rho}{\rho_e}\right)^n\right] + F_1\left(\frac{\rho}{\rho_e}\right) \tag{7}$$

is proposed for the embedding function. Due to the transformation invariance pointed out above, EAM models are independent of the choice of F_1. Eq. (7) reduces to the analytic form in the FS model with $m = 1/2$, $n = 1$, and $F_1 = F_0$, to that given by Baskes [30] for silicon with $m = 1$, $n = 1$, and $F_1 = -F_0$, while the universal form recently suggested by Banerjea and Smith [31] is obtained with $m = n$ and $F_1 = 0$. For these last two correspondencies, the limit

$$\lim_{m=n}\left[\frac{n}{n-m}\left(\frac{\rho}{\rho_e}\right)^m - \frac{m}{n-m}\left(\frac{\rho}{\rho_e}\right)^n\right] = \left[1 - \ln\left(\frac{\rho}{\rho_e}\right)^n\right]\left(\frac{\rho}{\rho_e}\right)^n \tag{8}$$

is required, where subscript e implies evaluation at equilibrium.

2.2. Effective Two-Body Potential

The concept of an effective two-body has been discussed in the literature [6,8,18,26,28,32]. The two-body potential becomes the effective two-body potential when the transformation of the embedding function and two-body potential is made such that the slope of the embedding function is zero when evaluated at the equilibrium electron density for the perfect crystal. For the suggested form for the embedding function, Eq.(7), this means $F_1 = 0$. I will refer to a model with the slope of the embedding function at equilibrium equal to zero and the two-body potential the effective two-body potential as being in normalized form. It is very useful to transform models to the normalized form for comparative purposes.

For defect calculations in the bulk, all atoms remain quite tightly packed, i.e., the electron density at atom sites does not vary much from the equilibrium value. Since the slope of the embedding function is zero in a normalized model and the curvature of the embedding function is not large (discussed below), there is little contribution from the embedding function to defect energies. Thus, the empirical two-body potentials [4] which were surprisingly successful in many bulk calculations in essence are the effective two-body potentials of the EAM. Since the energy per atom in equilibrium is $-E_c$ and, from Eqs. (1) and (3) is also $F_e + \Phi_e$, $\Phi_e \approx -E_{uv}$ (the unrelaxed vacancy formation energy) and $F_e = -(E_c - E_{uv})$ for a normalized model. Since the vacancy relaxation energy is only a small fraction of the vacancy formation energy, and the vacancy formation energy one-half to one-third the cohesive energy, the EAM provides the solution to the dilemma mentioned in the introduction: the empirical two-body potentials should fit to the vacancy formation energy and should not be expected to give a realistic cohesive energy.

For a normalized model, certain physical properties depend on the effective two-body potential and others on the embedding function. For the potential,

$$2E_{uv} \approx -\sum N_i \phi_{i,e} , \tag{9}$$

$$0 = \sum N_i r_{i,e} \phi'_{i,e} , \tag{10}$$

$$30\Omega G = \sum N_i r^2_{i,e} \phi''_{i,e} , \tag{11}$$

with the sums over shells of symmetric atoms, N_i the number of atoms in the ith shell, Ω the atomic volume, G the Voigt average shear modulus, and primes indicating differentiation. For a cubic crystal, $C = C_{44}$, $C' = (C_{11}-C_{12})/2$, and $G = (3C+2C')/5$. C and C' are the limiting values of the shear modulus, i.e., one is the maximum and the other the minimum. The first of these relations comes from the discussion in the preceding paragraph, the second from equilibrium, and the third shows that shear (a volume conserving deformation) is independent of the embedding function. For the embedding function,

$$E_c - E_{uv} \approx -F_e , \tag{12}$$

$$0 = F'_e , \tag{13}$$

$$3\Omega(3B - 5G) = F''_e \left(\sum N_i r_{i,e} f'_{i,e}\right)^2 , \quad (14)$$

where the prime on F means differentiation with respect to ρ. The first of these relations comes from the discussion in the preceding paragraph, the second from equilibrium, and the third is related to the Cauchy condition. In a normalized model the equilibrium condition is decoupled between the two-body potential and the embedding function. The Cauchy condition gives relationships between elastic constants which must hold for any lattice model employing only two-body forces. Real materials do not satisfy the Cauchy condition, and so many-body forces are required to describe them. For a cubic material, the Cauchy condition is $3B = 5G$, i.e., it implies $F_e'' = 0$ for an EAM model so that changes in energy depend solely on the two-body potential. To the author's knowledge, $3B > 5G$ for all cubic metals except Cr. Thus the curvature of the embedding function is positive for all cubic metals except Cr and the form for the embedding function given by Eq. (7) is applicable.

2.3. Alloys

For an EAM alloy model, an embedding function $F(\rho)$ and an atomic electron density function $f(r)$ must be specified for each atomic species, and a two-body potential $\phi(r)$ specified for each possible combination of atomic species. As mentioned above, $f(r)$ must also be specified for each possible combination of atomic species for an FS model. Since the electron density at any location is taken as a linear superposition of atomic electron densities, and since the embedding energy is assumed to be independent of the source of the electron density, these two functions can be taken directly from monatomic models. The electron density functions are only determined to within a scaling factor from fitting to monatomic properties, however, so the relative scaling factor between elements must be specified for an alloy model.

For a binary alloy with a and b type atoms, ϕ^{aa} and ϕ^{bb} are given by the respective monatomic models, and ϕ^{ab} and ϕ^{ba} are assumed to be equal. The alloy potential could be determined separately from the monatomic models if sufficient experimental fitting data were available, but the assumption is commonly made that it is a function of the monatomic potentials. For example, Foiles *et al.* [20] tested both geometric and arithmetic averages, and reported results using geometric averages.

As discussed in section 2.1 above, monatomic EAM models are invariant to a specific form of transformation. A problem with using a functional relationship for the alloy potentials in terms of the monatomic potentials is that arbitrary choices in monatomic models change the alloy potentials. Indeed, the geometric average is imaginary if one of the monatomic two-body potentials is negative while the other is positive. This problem did not occur in the Sandia alloy calculations [20] because their two-body potentials were always taken as positive. Certainly a geometric average cannot be used for EAM models in normalized form because the effective two-body potentials are negative in some range.

For the transformation like that in Eqs. (4) and (5) to the monatomic embedding functions and potentials in an alloy, i.e., for

$$G^{a}(\rho) = F^{a}(\rho) + k^{a}\rho \,, \tag{15}$$

$$G^{b}(\rho) = F^{b}(\rho) + k^{b}\rho \,, \tag{16}$$

$$\psi^{aa}(r) = \phi^{aa}(r) - 2k^{a}f^{a}(r) \,, \tag{17}$$

$$\psi^{bb}(r) = \phi^{bb}(r) - 2k^{b}f^{b}(r) \,, \tag{18}$$

the condition on the cross terms for invariance of the model to the transformation is [29]

$$\psi^{ab}(r) = \phi^{ab}(r) - k^{a}f^{a}(r) - k^{b}f^{b}(r) \,. \tag{19}$$

If the alloy potentials are taken as linear functions of the monatomic potentials, then this condition requires the form

$$\phi^{ab}(r) = \frac{1}{2}\left[\frac{f^{b}(r)}{f^{a}(r)}\,\phi^{aa}(r) + \frac{f^{a}(r)}{f^{b}(r)}\,\phi^{bb}(r)\right] . \tag{20}$$

No nonlinear form which maintains the transformation invariance has been found. To avoid singularities, the electron density function for any metal cannot be zero at a distance at which the two-body potential for that metal is non-zero, i.e., the cutoff distance for the electron density function must be equal to or greater than that for the potential.

3. Models

The procedures for obtaining the parameters to describe specific materials are discussed in this section. Most parameters are obtained by fitting to experimental data, but theoretical input has also been used. There is not very much information available, and traditionally the equilibrium lattice constant and the elastic moduli have been used. In earlier models, either the cohesive energy or the vacancy formation energy were used, but both can be fitted with the EAM. In general the fitting schemes are quite complex, but simple analytic models have been developed [26-29,32,33] which, although not quite as accurate, display the parameter dependence for both fitted and calculated properties.

3.1 FCC

The original work by Daw and Baskes [9] applied to fcc metals. The atomic electron densities were taken as the spherically averaged free-atom densities calculated from Hartree-Fock theory, with an additional adjustable parameter to account for an admixture of different electronic configurations. These densities are rapidly converging monatonically decreasing functions and can be approximated by exponential or power law equations. For a nearest-neighbor model, either can be used, but with more distant neighbors, power-law dependence is more convenient. For later fitting, it is desirable to be able to invert the functional relationship for ρ and r so that the lattice constant can be written as a function of ρ for a homogeneous change in lattice constant. If $f(r)$ is assumed to have the form $f(r) = f_e(a_e/r)^\beta$, then $a = a_e(\rho/\rho_e)^{-1/\beta}$ with $\rho_e = f_e\,\Sigma(1/x_i)^\beta$ and $r_i = x_i a$. The summation is over all neighbors to a given site and depends only on lattice structure. This form was used for a bcc model [31]. For a monatomic calculation the parameter f_e cancels, so only the parameter β remains. For a nearest neighbor fcc model [26,27], $f(r) = f_e\exp[-\beta(r/r_e-1)]$ gives $a = a_e[1-(1/\beta)\ln(\rho/\rho_e)]$. Although the inversion was not as convenient, this form was also used for longer ranged fcc and hcp models [33,34].

Daw and Baskes [6] used a complex fitting scheme for the two-body potentials for fcc calculations. This has been simplified to Born-Mayer exponential repulsion, $\phi(r) = \phi_e\exp[-\gamma(r/r_e-1)]$, in the nearest-neighbor fcc model [26,27]. Here it was desirable to have the same functional form for $f(r)$ and $\phi(r)$.

In early EAM calculations, the embedding function was also determined by a complex fitting procedure. Foiles [19] introduced a straight-forward scheme which was also used in the simple analytic fcc model [26,27]. Rose *et al.* [34] have shown that, for a broad range of materials, the energy as a function of lattice constant (or any length parameter in the lattice) is well approximated by a universal equation of state in the form of a Rydberg function:

$$E(a) = -E_c\left[1 + \alpha\left[\frac{a}{a_e} - 1\right]\right] \exp\left[-\alpha\left[\frac{a}{a_e} - 1\right]\right] \qquad (21)$$

with $\alpha^2 = 9\Omega B_e/E_c$. This form assures agreement with equilibrium, the cohesive energy, and the bulk modulus. Using the inversion formulas from the previous paragraph and Eqs. (1) and (3), the embedding function is determined as a function of ρ:

$$F(\rho) = -E_c\left[1 - \left[\frac{\alpha}{\beta}\right]\ln\left[\frac{\rho}{\rho_e}\right]\right]\left[\frac{\rho}{\rho_e}\right]^{\alpha/\beta} - \Phi_e\left[\frac{\rho}{\rho_e}\right]^{\gamma/\beta}, \qquad (22)$$

where $\rho_e = 12f_e$ and $\Phi_e = 6\phi_e$. This form is very close to that given in Eqs. (7) and (8).

In addition to the atomic volume, cohesive energy, and bulk modulus required for the universal equation of state, this nearest-neighbor model requires parameters β, γ, and ϕ_e. For Cu, the spherically-averaged Hartree-Fock atomic calculations are reasonably well approximated with $\beta = 6$ (indeed this value is not far off for all the fcc metals tested), and then γ and ϕ_e are given by the Voigt average shear modulus and (approximately) the vacancy formation energy.

Instead of the traditional C_{11}, C_{12}, and C_{44} elastic constants, it is better to use the bulk modulus B (how the metal responds to volume change), the Voigt average shear modulus G (how, on average, the metal responds to volume conserving shear), and the anisotropy ratio A (how the shear varies with orientation) for understanding the elastic properties of cubic materials. The anisotropy ratio of a cubic material is the ratio of the extremal values of the shear modulus: $A = C/C'$. Both B and G are used as inputs with this simple analytic nearest neighbor model, but $A = 2$ and is not adjustable. Extending the model to more neighbors permits variation in A and there is a tendency for A to increase somewhat. Experimentally A is greater than 2 for all fcc metals except Pt (1.6) and Al (1.2). It is very difficult to force any EAM model to fit the almost isotropic Al value.

If this simple analytic model is normalized, the effective two-body potential is

$$\phi_{\text{eff}} = \phi_e\left\{ \exp\left[-\gamma\left(\frac{r}{r_e} - 1\right)\right] - \left(\frac{\gamma}{\beta}\right)\exp\left[-\beta\left(\frac{r}{r_e} - 1\right)\right]\right\} . \tag{23}$$

This form is similar to a Morse potential, except that ϕ_{eff} is one contribution to the energy, whereas the Morse potential is intended to provide the total energy. Furthermore, the dependence of ϕ_e on β and γ is such that ϕ_{eff} is unchanged by interchanging β and γ, but ϕ_e changes sign, i.e., the two-body potential becomes attractive rather than repulsive.

As mentioned earlier, the EMT approach of Jacobsen *et al.* [25] reduces to the EAM format when simplifying approximations are made. Although the format is the same, there are differences in interpretation between EAM and EMT. In EMT the two-body potential arises from atomic-core--electron-gas electrostatic attraction, whereas it is taken as repulsive in EAM. Also, except in the perfect-crystal case, the electron density is not a simple superposition of contributions from neighboring atoms in EMT, but such a superposition raised to a power close to but not equal to one. This occurs due to a charge conservation self-consistency requirement using an atomic sphere approximation in which the spheres overlap but are not space filling. As shown above, the same effective potential can be obtained with either a purely repulsive or attractive potential. Since the parameters are obtained by fitting to experimental results, whether the two-body potential is attractive or repulsive has little effect on the resultant model.

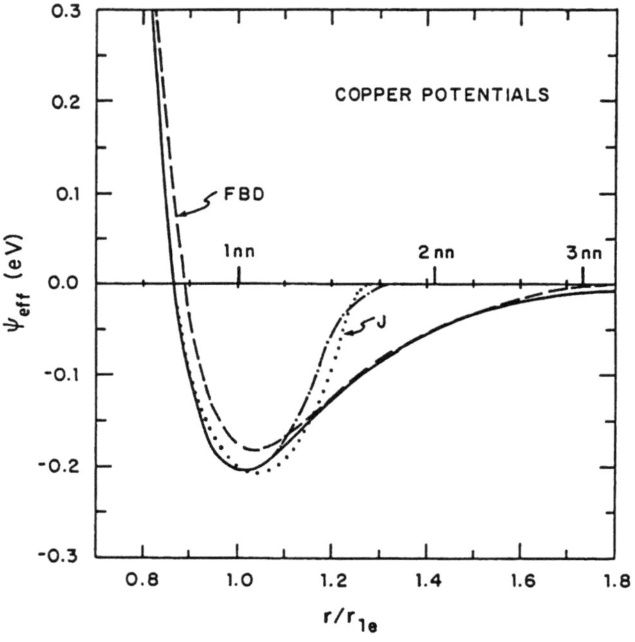

FIG. 1. Representative effective two-body potentials for copper. The solid line is from Ref [26] without any cutoff and the dash-dot curve shows the effect of a smooth cutoff to 85 percent of the distance between first-and second-neighbor distances. The dashed curve is from a numerical EAM model for copper by Foiles *et al.* [20] and the dotted curve is a short-ranged empirical two-body potential for copper [35]. The minimum of the empirical potential is not at r/r_{1e} = 1 because global volume dependence was included in the metal.

For any calculation in which there is significant relaxation, a cutoff distance must be given. With the analytic nearest-neighbor model, best results are obtained if it is taken at about 85 percent of the distance between first and second neighbors, with the potential smoothly cut off rather than truncated. An effective two-body potential for copper is compared to an early empirical two-body potential [35] in Fig. 1. A recent Cu potential from Foiles *et al.* [20], as well as the present Cu potential without the cutoff are also shown.

Calculations with this model yield the following general conclusions. Since there is little relaxation near a vacancy, the vacancy formation energy depends primarily on the effective two-body potential, which in turn depends on G, not on B or E_c. This leads to the surprising result that the shear modulus, rather than the bulk modulus or the cohesive energy, is the dominant parameter in determining the vacancy formation energy. Although B plays a larger role in surface energy calculations, to lowest order the surface energy is proportional to the vacancy formation energy. The proportionality factor is such that, in general, EAM models that give a good accounting of the vacancy formation energy will tend to give too small surface energies compared to the best available experimental data.

Overall, the parameter dependence in defect calculations may be summarized as follows [27]:

(a) the cohesive energy has a negligible effect on defect energies,

b) the bulk modulus has a minor effect on defect energies,

(c) the variation of defect energies with the shear modulus is approximately linear,

(d) there is little variation in defect energies with β and γ if their average is held constant,

(e) the unnormalized two-body potential can be attractive or repulsive with little effect on calculated defect energies, and

(f) vacancy energies are decreased and interstitial energies increased by increasing the average of β and γ.

3.2. BCC

A short-range bcc model requires both eight first and six second neighbor interactions. In the nearest neighbor fcc model best results are obtained if the cutoff distance is taken at about 85 percent of the distance between first and second neighbors. If this fcc model is applied to the bcc structure holding the atomic volume constant, the cutoff is about midway between the second- and third-neighbor distance. The resulting bcc shear elastic moduli have a very high anisotropy ratio: $A = 42$. The bcc value of C is just 25 percent greater than the fcc value while the value of C' is very small. Thus the bcc model is almost unstable to C' shear. In general, fcc transition and noble metals have anisotropy ratios in the range of 2 to 3, while it varies from 0.5 for bcc Nb to 2.4 for bcc iron. The anisotropy ratio for bcc alkali metals is very high: 7.2 for Na and 8.5 for Li. With appropriate parameter variation, the fcc analytic model can be adjusted to match the alkali metals, but cannot give an adequate fit to the bcc transition metals.

The difficulty in fitting the bcc metals can be understood easily by considering the normalized form [32]. The shear constants then just depend on the effective two-body potential. The nearest-neighbor atoms only contribute to C, not to C', while the second-neighbor atoms only contribute to C', not to C. Because of the exponentially decreasing functions, the curvature of the two-body potential is much less at the second-neighbor distance than at the first-neighbor distance. This leads to a greater contribution to C than to C' from the curvature. Furthermore, the negative slope of the two-body potential at the first-neighbor distance increases C, while its positive slope at the second-neighbor distance decreases C'. The net effect yields a large C

and a very small C'. Extending the model to more distant neighbors while maintaining the exponential forms for $f(r)$ and $\phi(r)$ does not help because the additional contributions to the elastic moduli are negative.

It turns out that the bcc shear properties can be matched with a cubic equation in the range of first- and second-neighbor distances together with a cutoff between second and third neighbors [32]. A power-law dependence was used for the electron density function, and Eq. (8) was used for the embedding function with $m = n$.

The fcc model did yield stability of the fcc lattice relative to the bcc lattice. The fitting to bcc elastic constants with the present model in turn gave bcc stability relative to the fcc lattice. This again can be understood by considering the normalized form. If the effective two-body potential has a sharp well, its contribution to the cohesive energy per atom in the fcc lattice (in a nearest-neighbor model the nearest neighbor distance is at the minimum of the effective two-body potential) will be greater than that for the bcc lattice and yield a stable fcc lattice. A sharp well favors fcc stability. The requirement of attaining a realistically large C' shear modulus causes the two-body potential to be broad in the range of first-and second-neighbor distances in the bcc lattice and provides bcc stability. The FS model has been applied primarily to bcc metals and, if transformed to the normalized form, shows a broad well shape of the two-body potential discussed here, as illustrated in Fig. 2.

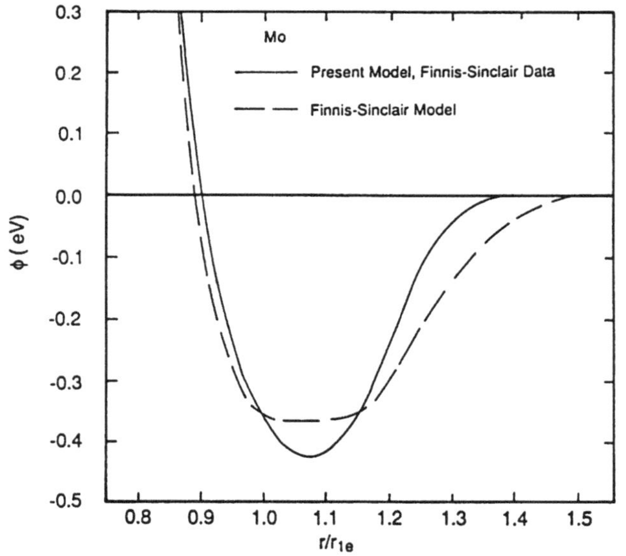

FIG. 2. Comparison of the effective two-body potential using a power-law dependence for the electron density function and a cubic equation for the two-body potential [32] (solid line) with that determined from the FS model [6] (dashed line).

In the discussion of stability, close-packed (cp) should actually be used rather than fcc because a short-range model cannot discriminate between fcc and hcp with an ideal *c/a* ratio. The basic parameter dependence relative to bcc and cp stability are [32]:

(a) increasing the anisotropy ratio favors cp stability over bcc, although the variation is relatively minor for $0.5 < A < 10.0$,

(b) increasing the *B/G* ratio provides for greater relative bcc stability, although the effect is not strong,

(c) increasing both *B* and *G* relative to E_c yields a significant increase in cp stability,

(d) increasing the unrelaxed vacancy formation energy relative to the cohesive energy strongly favors bcc stability,

(e) combining effects (c) and (d), which is equivalent to decreasing E_c while holding the other parameters constant, gives cancellation between the two, with the net effect being some favoring of the bcc structure, and

(f) relative stability is almost independent of the electron density parameter β.

3.3. HCP

There has been little application of the EAM to hcp metals [33,36]. In a recent study [36], a simple model was devised with the cutoff distance between third- and fourth-neighbor distances in the hcp lattice to distinguish stability between the hcp and the fcc structure. Defect and stability calculations were carried out varying the *c/a* ratio while holding the model input parameters constant (the atomic volume, the cohesive energy, and the Voigt average shear and bulk moduli). The physical parameters of zirconium were used as a prototype, but the results are independent of this choice. The electron density function was taken as an exponential, the two-body potential as an exponential plus a constant, and the embedding function a simple version of Eq. (7). The smooth cutoff applied to both $f(r)$ and $\phi(r)$ had little effect on the exponential terms as they were very small, but was necessary for the constant term in $\phi(r)$.

With the other physical parameters used as input held constant, a stable (or metastable) hcp model was determined for each choice of *c/a*. The stability of the model relative to fcc and bcc was then calculated and relaxation calculations for various point defects were carried out. A rather restricted range of hcp stability was found: for $c/a \geq 1.63$ (ideal), the fcc lattice was more stable, while for $c/a \leq 1.58$, the bcc lattice was more stable. Of 25 hcp metals, the *c/a* ratio of 22 of them is within this range, with Be just below it and Zn and Cd far above it. The three metals outside the stability range have a negative Cauchy pressure, i.e., they require a negative curvature to the embedding function at the equilibrium electron density.

As the c/a ratio decreases from ideal, stacking-fault energies increase, which favors prism slip over basal slip. Also, divacancy and trivacancy binding becomes more anisotropic favoring planar defects and the planar B_O self-interstitial configuration becomes more stable than the three-dimensional C configuration. The activation energy for self diffusion is independent of the variation of the c/a ratio and is isotropic.

3.4. FCC Alloys

Foiles *et al.* [20] have developed fcc alloy models based on the EAM using numerical fitting to tabulate the necessary EAM functions in a manner similar to their monatomic models [9,17]. For the six metals Ni, Pd, Pt, Cu, Ag, and Au they tried both geometric and arithmetic averaging for the alloy potentials and used all the standard physical inputs as well as the heats of solution (22 of the 30 possible values are available). The EAM parameters were varied to yield the best overall fit to all the data and numerous calculations were carried out with reasonable results and the method has been extended, e.g., to a detailed investigation of Ni Al alloys [22].

Again to obtain insight into the parameter dependence associated with alloying and to test the alloy potential form in Eq. (20), simple analytic nearest-neighbor models similar to the one discussed in section 3.1 were developed [29]. As before, exponentials were used for both $f(r)$ and $\phi(r)$, and the universal equation of state given by Rose *et al.* [34] was employed to determine the embedding function. The required model parameters are f_e, ϕ_e, α, β, and γ, and the input physical quantities to determine them are Ω, E_c, E_{uv}, B, and G. The relations $f_e = E_c/\Omega$ and $\phi_e = E_c/6$ were found to provide realistic results.

With an approximation for relaxation, agreement with the available data for heats of solution involving combinations of Ni, Pt, Cu, Ag, and Au was surprisingly good. These results are shown in Fig. 3, with the

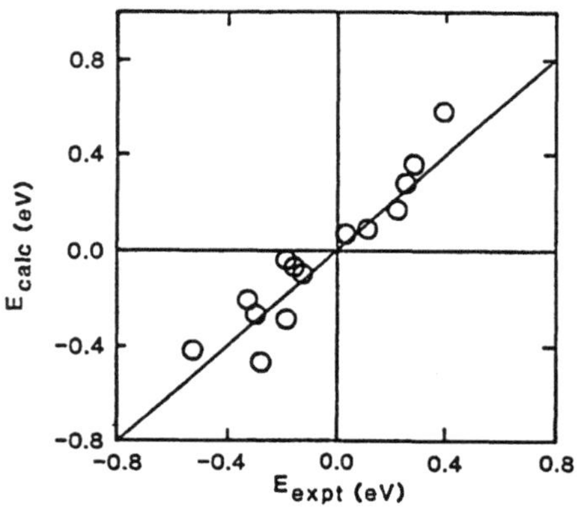

FIG. 3. Comparison of the calculated and experimental heats of solution for all data available for binary alloys of Cu, Ag, Au, Ni, and Pt [29]. The calculated values involving Pd averaged almost 0.4 eV too high.

correct sign obtained in all cases. Calculated heats of solution involving Pd did not have this excellent accord with experiment, but did show some uniformity in that they were too high in all cases with an average error of almost 0.4 eV.

Since there are no adjustable parameters in this model, the effect of variations in the five input quantities could be easily studied. The details associated with the relationship between the physical parameters of the solute and solvent and the heat of solution are complex, but some general qualitative features of the pattern were found.

The lattice constant and the cohesive energy are the most significant parameters. If they are equal for the solute and solvent, the heat of solution $E_s = 0$, independent of the other parameters. If the size of the solute is greater than while the cohesive energy is less than the corresponding quantities in the solvent, or vice versa, the heats of solution are positive, independent of the other parameters. However, if the size of the solute and its cohesive energy are both greater or both less than the corresponding solvent quantities, the heats of solution are negative or, in several cases, positive but small. The effect of the vacancy formation energy of the solute is not very regular, but it contributes to a more negative heat of solution when the heat of solution is initially negative if it varies in the opposite sense as the cohesive energy. Increasing the bulk modulus of either the solute or the solvent increases the heat of solution, while increasing the shear modulus of either generally decreases the heat of solution.

The disagreement in heats of solution relative to Pd cannot be understood through this parameter dependence study. All the physical values used as input for Pd are near the average for the six metals, i.e., there is nothing evidently unique about Pd. Any adjustment which improves agreement with one of the other metals makes the agreement worse relative to some other metal.

4. Summary

The invariance of EAM models to a transformation of the embedding function and the two-body potential is discussed. If the embedding function is transformed so that it has zero slope at the equilibrium electron density, the two-body potential becomes the effective two-body potential. This "normalized" form provides a correspondence with earlier empirical models, and is used to clarify the role of the two-body potential and the embedding function relative to the vacancy formation energy and the cohesive energy. Also, transformation to the normalized form shows that models which appear to be very different actually can be functionally very similar. The effective two-body potential is independent of the bulk modulus and the cohesive energy, while its magnitude provides the primary contribution to the vacancy formation energy, its slope is associated with equilibrium, and its curvature is proportional to the Voigt average shear modulus in a cubic metal. The magnitude of the embedding function contributes most of the cohesive energy, its slope at the equilibrium electron density is zero by definition, and its curvature is zero if the Cauchy condition is satisfied, and is proportional to the Cauchy pressure. The role of volume dependence in defect energies is minor if the electron density does not change too much, but becomes significant, for example, at surfaces where the coordination is severely altered.

There is not much experimental information upon which to construct alloy potentials. If alloy models are assumed to have the same transformation invariance as monatomic models and if alloy potentials are assumed to be functions of the potentials for the corresponding monatomic metals, then a unique form for alloy potentials is found. Although similar to an arithmetic average, the weighting factors depend on the electron density functions, are functions of position, and do not sum to one.

A series of simple analytic models have been constructed for fcc, bcc, and hcp metals. The electron densities and two body potentials were taken as having an exponential dependence in the close-packed fcc and hcp metals, while a power-law dependence was used for the electron density and a cubic equation for the potential in the bcc metals. In the nearest-neighbor fcc calculations, defect energies (vacancy and self interstitial) were found to have negligible dependence on the cohesive energy, vary only slightly with the bulk modulus, but to be linearly proportional to the average shear modulus. Defect energies were insensitive to variations in the exponent parameters in the electron density and two-body potential if their average was held constant, and consequently to whether the two-body potential was attractive or repulsive. Vacancy energies decreased while interstitial energies increased with an increase in the average of the exponent parameters. Extension of the short-range fcc model to include alloy potentials using the unique form mentioned above gave reasonable values for heats of solution of binary alloys of Cu. Ag, Au, Ni, and Pt, but the heats of solution of binary alloys of these metals with Pd were consistently too high.

A first- and second-neighbor model for bcc metals showed that the effective two-body potential had to be broader than that for fcc metals. In general it was found that increasing the elastic moduli, increasing the elastic anisotropy ratio, and decreasing the vacancy formation energy favor of a close-packed lattice (either fcc or hcp) relative to the bcc lattice.

The effect of the *c/a* ratio on lattice stability and point defect properties in hcp metals was studied using third-neighbor interactions. The range of *c/a* ratios for hcp stability was from 1.58 to 1.63, with fcc more stable at larger ratios and bcc more stable at smaller ratios. The activation energy for self diffusion was approximately isotropic and was independent of variation in the *c/a* ratio. The smaller the *c/a* ratio, the more "planar" defects tended to be with, e.g., the di and trivacancy energies becoming more anisotropic and the planar B_0 self-interstitial more stable than the three-dimensional *C* configuration.

Acknowledgements

It is a pleasure to acknowledge collaboration with D. J. Oh at the University of Virginia (now at North Carolina State University) and stimulating discussions with M. I. Baskes, M. S. Daw, and S. M. Foiles at Sandia National Laboratory, Livermore. Support from the U. S. Department of Energy (Office of Basic Energy Sciences, Division of Materials Science) under Grant Number DE-FG05-86ER45246 is gratefully acknowledged.

References

[1] *Calculation of the Properties of Vacancies and Interstitials*, Natl. Bur. Stand. (U.S.) Publ. No. 287 (U.S. GPO, Washington, D.C., 1966).
[2] *Interatomic Potentials and Simulation of Lattice Defects*, edited by P. C. Gehlen, J. R. Beeler, Jr., and R. I. Jaffee (Plenum, New York, 1972).
[3] I.M. Torrens, *Interatomic Potentials* (Academic, New York, 1972).
[4] R. A. Johnson, J. Phys. F **3**, 295 (1973).
[5] *Interatomic Potentials and Crystalline Defects*, edited by J. K. Lee (Metallurgical Society, AIME, Warrendale, PA, 1981).
[6] M. W. Finnis and J. E. Sinclair, Philos. Mag. A **50**, 45 (1984).
[7] M. W. Finnis and J. E. Sinclair, Philos. Mag. A **53**, 161 (1986).
[8] M. S. Daw and M. I. Baskes, Phys. Rev. Lett. **50**, 1285 (1983).
[9] M. S. Daw and M. I. Baskes, Phys. Rev. B **29**, 6443 (1984).
[10] G. J. Ackland and M. W. Finnis, Philos. Mag. A **54**, 301 (1986).
[11] G. J. Ackland and R. Thetford, Philos. Mag. A **56**, 15 (1987).
[12] G. J. Ackland, G. Tichy, V. Vitek, and M. W. Finnis, Philos. Mag. A **56**, 735 (1987).
[13] J. K. Nørskov and N. D. Lang, Phys. Rev. B **21**, 2131 (1980).
[14] M. J. Stott and E. Zaremba, Phys. Rev. B **22**, 1564 (1980).
[15] M. J. Puska, R. M. Nieminen, and M. Manninen, Phys. Rev. B **24**, 3037 (1981).
[16] M. I. Baskes, J. Nucl Mater. **128&129**, 676 (1984).
[17] S. M. Foiles, Phys. Rev. B **3232**, 3409 (1985).
[18] M. S. Daw and R. D. Hatcher, Solid State Commun. **56**, 697 (1985).
[19] S. M. Foiles, Phys. Rev. B **32**, 7685 (1985).
[20] S. M. Foiles, M. I. Baskes, and M. S. Daw, Phys. Rev. B **33**, 7983 (1986).
[21] M. S. Daw, Surf. Sci. **166**, L161 (1986).
[22] S. M. Foiles and M. S. Daw, J. Mater. Res. **2**, 5 (1987).
[23] S. M. Foiles and M. S. Daw, Phys. Rev. B **38**, 12643 (1988).
[24] M. S. Daw, Phys. Rev. B **39**, 7441 (1989).

[25] K. W. Jacobsen, J. K. Nørskov, and M. J. Puska, Phys. Rev. B **35**, 7423 (1987).
[26] R. A. Johnson, Phys. Rev. B **37**, 3924 (1988).
[27] R. A. Johnson, Phys. Rev. B **37**, 6121 (1988).
[28] D. J. Oh and R. A. Johnson, in *Atomistic Simulation of Materials: Beyond Pair Potentials*, edited by D. J. Srolovitz (Plenum, New York, 1989).
[29] R. A. Johnson, Phys. Rev. B **39**, 12554 (1989).
[30] M. I. Baskes, Phys. Rev. Lett. **59**, 2666 (1987).
[31] A. Banerjea and J. R. Smith, Phys. Rev. B **37**, 6632 (1988).
[32] R. A. Johnson, J. Mater. Res. **4** (1989).
[33] D. J. Oh and R. A. Johnson, J. Mater. Res. **3**, 471 (1988).
[34] J. H. Rose, J. R. Smith, F. Guinea, and J Ferrante, Phys. Rev. B **29**, 2963 (1984).
[35] R. A. Johnson, Radiat. Eff. **2**, 1 (1969).
[36] D. J. Oh and R. A. Johnson, J. Nucl. Mater.

Molecular Dynamics Simulations of Materials Phenomena

U. Landman, W.D. Luedtke, and R.N. Barnett

School of Physics, Georgia Institute of Technology,
Atlanta, GA 30332, USA

1. INTRODUCTION

Basic understanding of the structure and dynamics of materials and their properties often requires knowledge on a microscopic level of the underlying energetics and interaction mechanisms, whose consequences we observe and measure. Answers to material science problems are in principle possible, embodied in solutions to the Schrödinger equation subject to the appropriate boundary conditions. However, the full implementation of such a program is impossible for most (one may dare say all) material science and condensed matter systems and we must resort to various approximations and simplifications. In many situations the level of complexity of the model, which is necessary in order to describe faithfully the physical phenomena, is such that analytical approaches fail to provide a solution. In these situations, which include the majority of material systems and phenomena, the use of computer-based methods [1-10] is essential.

Computer simulations (in the classical [1-10] as well as in the quantum [4,11-19] regimes) where the evolution of a physical system is simulated, with refined temporal and spatial resolution, via a direct numerical solution to the model equations are in a sense computer experiments which open new avenues in investigations of the microscopic origins of materials phenomena. These methods alleviate certain of the major difficulties which hamper other theoretical approaches, particularly for complex systems such as those characterized by a large number of degrees of freedom, lack of symmetry, non-linearities and complicated interactions. In addition to comparisons with experimental data, computer simulations can be used as a source of physical information which is not accessible to laboratory experiments, and in some instances the computer experiment itself serves as a testing ground for theoretical concepts.

In this short review we focus on classical computer simulations which employ realistic non pair-wise interaction potentials (for recent reviews and references of quantum simulations see [4,11-19]). Our aim is to demonstrate the level of reality which can be achieved via computer simulations of materials systems and the wealth of information which they provide about the fundamental mechanisms and processes which govern the properties and response of materials systems.

2. CLASSICAL MOLECULAR DYNAMICS SIMULATIONS AND INTERACTION POTENTIALS

The classical molecular-dynamics (MD) method [1-10] consists of a numerical generation of the phase-space trajectories for a system of N particles, (and, for extended systems, employing periodic boundary

Springer Proceedings in Physics, Vol. 48 Many-Atom Interactions in Solids
Editors: R.M. Nieminen · M.J. Puska · M.J. Manninen

conditions), interacting via a potential function $V(\vec{r}_1,\ldots,\vec{r}_N)$, where $\vec{r}_i$ is the coordinate of particle i. Formulations have been developed which allow simulations of various equilibrium ensembles, such as constant volume microcanonical (E,Ω,N), constant pressure $(\mathcal{H}PN)$, or constant temperature canonical (T,Ω,N) ensembles.

The starting point of a MD simulation is a well-defined microscopic description of the physical system, in terms of a Hamiltonian or a Lagrangian from which the equations of motions are derived and the phase-space trajectories which are generated can be used in studies of equilibrium and non-equilibrium phenomena.

The correspondence between the simulation model and the physical system depends to a large extent on the nature of the potential energy employed in the simulations. In general the potential energy of a system of N atoms is a function of the atomic coordinates and may also contain a contribution from a term which does not depend explicitly on the atomic coordinates but on the density (such as in metallic systems). Thus in general the potential energy may be expanded in a series of n-body interatomic potentials

$$V = \phi_o + \Sigma\phi_1(i) + \Sigma\phi_2(ij) + \Sigma\phi_3(ijk) + \ldots + \phi_N(ijkl,\ldots), \qquad (1)$$

where ϕ_n is the n-body interaction potential which is a function of the positions of n atoms ijk...... . The sums in Eq. (1) are over all combinations (excluding redundant contributions) of n atoms in the system.

The potentials, ϕ_n, and the number of terms which are retained in a practical application depend on the nature of the system under investigation. Thus for example rare gases and ionic systems can be adequately described in terms of pair potentials [20], which depend only upon the distance between interacting pairs, (such potentials include the 6-12 Lennard Jones potentials for rare gases and potentials which include electrostatic interactions and Born-Meyer repulsive terms for ionic systems [21]).

For systems characterized by covalent bonding (such as semiconductors) one must go beyond the pair interactions due to the directionality of the localization of charge in the bonding regions. Potentials which include 3-body terms have been used for sometime [22] for the description of equilibrium structures of crystalline semiconductors and recently generalizations for arbitrary atomic locations have been developed [23-25]. In our work on silicon we have employed the potential developed by Stillinger and Weber (SW) [23] which has the form

$$\phi_2(r_{ij}) = A[Br_{ij}^{-P} - 1]g_\beta(r_{ij}) \quad , \qquad (2a)$$

$$\phi_3(r_i,r_j,r_k) = U_{jik} + U_{ijk} + U_{ikj} \quad , \qquad (2b)$$

$$U_{jik} \equiv \lambda\, g_\gamma(r_{ij})g_\gamma(r_{ik})[\cos\theta_{jik} + 1/3]^2 \quad , \qquad (2c)$$

$$g_\gamma(r) = \exp[\gamma/(r-a)] \quad . \qquad (2d)$$

The parameters of the potential have been determined [23] via empirical fits to properties of silicon in the crystalline and liquid phases. We have employed these potential in studies of liquid phase and vapor phase epitaxial growth of silicon and in investigations of the structural dynamics and energetics of amorphous silicon.

For metallic systems it is well known [26] that the cohesive energy contains density (volume) dependent contributions. Consequently, the cohesive energy of a simple metal can be written as

$$V = E_\Omega + \frac{1}{2} \sum_{i,j}' \phi_2(r_{ij};r_s) \; ; \tag{3a}$$

$$E_\Omega = N\, E_{el}(r_s) + N\phi_1(r_s) \tag{3b}$$

where Ω is the volume of the system, $r_s = (3/4\pi\rho)^{1/3}$, E_Ω is the volume energy expressed in terms of the energy of the uniform electron gas, $E_{el}(r_s)$, and the single particle contribution $\phi_1(r_s)$, and $\phi_2(r_{ij},r_s)$ is a density dependent effective pair-potential [26]. ϕ_1 and ϕ_2 are derived from pseudo-potential theory [26] and their specific form depends upon the choice of ionic pseudopotentials (model, local or non-local). Clearly, due to the explicit and implicit dependence of V on the density of the system, simulations of metal system phenomena which may involve changes in volume or structural transformations require a method in which the volume of the system varies dynamically [27,28]. This formalism was used by us in investigations of liquid metals (Mg) and metallic glasses ($Ca_{67}\,Mg_{33}$).

Alternatively, based on the philosophy of density-functional theory [29], a description of metallic systems which is ammenable to molecular dynamics simulations is provided by the effective medium theory (EMT) [30] or the related embedded atom method (EAM) [31]. The basic feature of these methods is that the effect of the surroundings on each atom in the system can be described in terms of the average electron density which other atoms in the system provide around the atom in question. The electronic structure problem is then converted to that of the embeddement of an atom in a homogeneous electron gas, which can be described in terms of a universal density dependent energy function. Thus the density dependent term gives rise to many-body interactions.

In the EAM the cohesive energy E_{coh} of the metal is written as

$$E_{coh} = \sum_i \{F_i[\sum_{j\neq i} \rho_j^a(R_{ij}) + \frac{1}{2} \sum_{j\neq i} \phi_{ij}(R_{ij})]\} \tag{4}$$

where ρ^a is the spherically averaged atomic electron density and R_{ij} is the distance between atoms i and j. In EAM the functions F and ϕ are determined by choosing functional forms which meet certain general requirements, and fitting parameters in these functions to a number of bulk equilibrium properties of the solid, such as lattice constant, heat of sublimation, elastic constants, vacancy-formation energy, etc.

Invoking the adiabatic (Born-Oppenheimer) approximation we have used the EMT method in MD simulations of premelting phenomena in aluminum [32] and the EAM in investigations of the crystal-to-melt and melt-to-vapor interfaces of nickel [33], disordering and premelting of Ni(110) [34], the formation and properties of $Ag_{55}Cu_{45}$ metallic glass [34] and adhesion and microindentation in Ni/Au systems [35].

3. SAMPLE RESULTS

Amorphous Silicon

The difficulty in studying matter in the amorphous state originates from the lack of periodicity which hinders the use of conventional condensed-matter structural determination and modeling methods leading to the development of various models [36-40] in which disordered structures are hand built or computer generated according to well-defined algorithms.

Recently, several MD simulations of the preparation of amorphous silicon [41-44] (and germanium [45]) via cooling of a melt have been performed. While the SW potential which was parameterized [23] on the basis of solid and liquid phase data for Si yields an adequate description of these aggregation states (bulk [23] and surface phenomena [46], point defects [47], and studies of equilibrium [48,49] and non-equilibrium [50] solid-melt interfaces of silicon), it appears [41-45] that upon quenching this potential (as well as other proposed potentials [43]) produces a range of degenerate disordered structures, and that the difference of the configurational entropies of the low-temperature liquid and amorphous phases is in error resulting in a too small value of the heat of fusion for the liquid to amorphous transition.

In an attempt to alleviate these difficulties indirect methods for preparing amorphous silicon, using MD simulations, have been developed [41-45]. In one of the preparation procedures which we have explored [44] the system is first prepared at high temperature and then cooled using an increased three-body potential term (see Eq. 2c), thus enhancing the tendency toward tetrahedral coordination. Once the path which the system traverses in its configurational space has been redirected in this manner, restoring the value of the coefficient (λ) of the three-body potential term to its original SW value (λ=21) results in an amorphous material which exhibits structural, dynamical, and thermodynamical properties in agreement with experimental data. Furthermore, upon fast heating, the so-prepared amorphous silicon melts via a first-order phase transition at a temperature which is 230 K below the crystalline melting temperature in agreement with experimental measurements. When the behavior of fully relaxed configurations of the amorphous material at several temperatures is analyzed we find that the nature of the transition and the latent heat associated with it maintains, but melting starts at a lower temperature.

While the amrophous sample prepared in the above indirect manner can be utilized for studies of the properties of the amorphous phase under various conditions, and in investigations of transformations of the amorphous phase, a principal question remains: do the above documented failures to achieve the amorphous state via direct quenching of the melt [41-45] reflect an inherent defect of the interaction potentials which were employed, or is the difficulty of a kinetic origin, i.e., related to the quench rates used. To resolve this question (which is often raised, in various variants, in the context of MD simulations, and in particular those involving nonequilibrium phenomena) we have performed [51] lengthy simulations in which a SW silicon bulk melt (588 particles) was cooled at a rate smaller than has been attempted in all previous MD studies. These simulations employed a periodically replicated calculational cell possessing full dynamical freedom to change shape and volume [52] under zero external pressure. Cooling of the system was achieved by reducing the temperature at selected stages of the simulation over a 2000 Δt (Δt = 1.15 x 10^{-3} psec) interval by 0.005ϵ (ϵ = 50 kcal/mole) followed by prolonged relaxations at constant energy.

The evolution versus temperature of the energy and volume of the system during the slow cooling process (the overall time span is 3.447 x 10^{-9} sec) and during heating of the system prepared via the indirect route [44] are shown in Fig. 1 (to convert to T in K multiply the reduced temperature by 2.5173 x 10^4). To demonstrate the slow rate by which the system explores the accessible phase-space we show in Fig. 2 the temporal relaxation of system properties for the time span 4.75 x

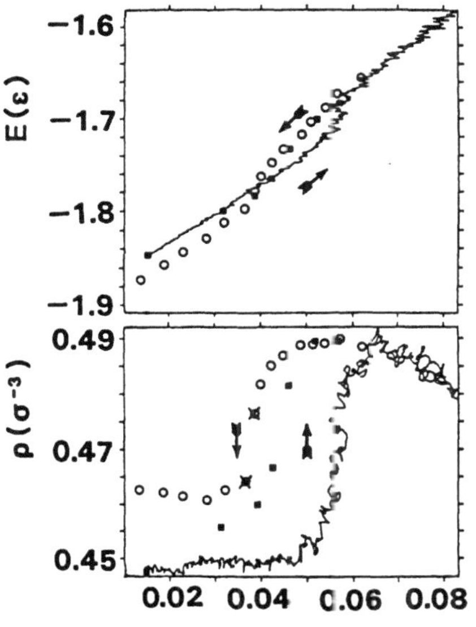

Figure 1: Total energy per particle E and density ρ of the slowly cooled melt (open circles) and those obtained by heating of the indirectly prepared (Ref. 44) amorphous Si. Solid lines correspond to rapid heating and the solid squares to slow heating allowing for more complete relaxation at the corresponding temperatures, both exhibiting the characteristics of a first-order transition, with melting occurring at a lower temperature for the slow-heating rate. ϵ = 50 kcal/mole, density in units of σ^{-3} (σ = 2.0951Å).

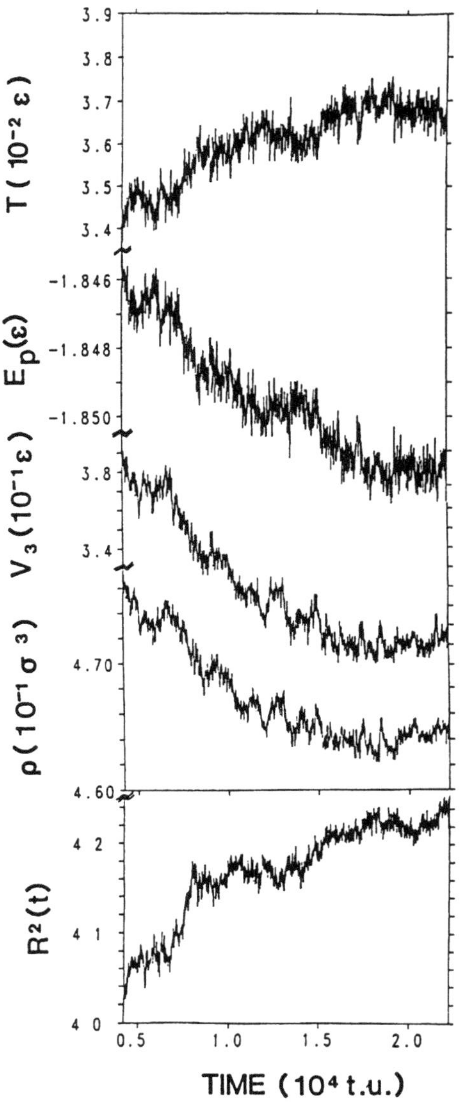

Figure 2: Kinetic temperature (T), potential energy (E_p), three-body potential contribution (V_3), density (ρ), and measure of particles' mobility, $R^2(t) = N^{-1} \Sigma_{i=1}^{N} [\vec{r}_i(t)-\vec{r}_i(0)]^2$, during the simulation of slowly cooled silicon, in the time interval 4.75×10^3 t.u. $\leq t \leq 2.25\times10^4$ t.u., (i.e., 1.2×10^6 integration time steps), where the time unit t.u. = 7.66×10^{-14} sec. This time interval corresponds to the transition between the two states of the system marked by x's in Fig. 1. ϵ = 50 kcal/mole and σ = 2.0951Å.

$10^3 \leq t \leq 2.25 \times 10^4$ t.u., where t.u. = 7.66×10^{-14} sec, corresponding to the transition between the two points marked by x's on the density plot.

The quality of the amorphous material obtained in these simulations [51] can be assessed via inspection of the structure factors and density of states (obtained via Fourier transformation of the velocity auto-correlation function) and their good correspondence with experimental data, given in Figs. 3 and 4, respectively. Other structural characteristics obtained in these simulations compare favorably with our previous results [44] and are in accord with experimental data. The average coordination at 300 K and 800 K are 4.21 and 4.31, respectively, compared to 4.14 and 4.26 for the a' (indirect preparation [44]) sample; the average bonding angles and their standard deviations for the two temperatures are 107.8 ± 16.0^o and 107.1 ± 18.3^o, as compared to 108.4 ± 14.2^o and 107.3 ± 18.0^o for the a' sample [44]; the average bonding angles corresponding to 4-fold coordinated sites at the two temperatures are 109.0 ± 10.3^o and 108.9 ± 11.1^o as compared to 109.0 ± 11.3^o and 108.8 ± 12.0^o for the a' sample, and those for 5-fold coordinated sites are 105.2 ± 23.8 and 104.8 ± 24.7^o for the a' system; the density of the system is ~ 3% higher than that of the a' sample. In addition the amorphous silicon material obtained in these simulations [51] exhibits a lack of medium-range order, measured via the absence of correlations between dihedral angles associated with adjacent bonds.

These results demonstrate that the origin of previous failures to obtain amorphous silicon via direct cooling of the melt, using the SW potentials, is of kinetic origin. This conclusion may be extended as a general cautionary remark concerning investigations in the area of non-equilibrium phenomena involving phase transformations and "ill condensed phases" (i.e., amorphous and glassy states).

Solid-Melt Interfaces of Silicon and Liquid-Phase Epitaxial (LPE) Growth

Understanding the equilibrium properties of interfaces and of the non-equilibrium mechanisms, kinetics, and dynamics of crystal-growth processes and identification of the microscopic material properties and of the macroscopic control parameters related to the method of growth, which govern the growth processes and quality of the grown crystals, are of fundamental importance from both basic and applied perspectives.

Ample evidence exists that surfaces of crystalline silicon become faceted upon melting and the solid-melt interface of growing Si establishes itself on the (111) crystal plane [55]. Motivated by these observations comprehensive MD studies of the solid-melt interfaces of Si [48,49] and simulations of liquid-phase epitaxial growth of Si [50] have been performed, employing the SW potentials. It was discovered that the nature and morphology of the interphase-interface exhibits crystalline face anisotropy [48,49]. Thus while the Si(111) equilibrium solid-melt interface was found to be atomically smooth, the Si(100) interface is structured exhibiting facets established along (111) directions [49], in qualitative agreement with experiment [49]. Furthermore, it was found that faceting initiates upon melting and further refines upon achieving equilibrium at a temperature of 1665K (compared to the experimental value of 1683K).

Liquid-phase epitaxial growth [50] was initiated by driving the crystal-melt systems out of equilibrium via allowing heat conduction to the underlying substrate. It was found that for both interfaces the system initially supercools by about 150K (in agreement with some experimental estimates), followed by ordering and crystallization processes. On the (001) surface the growth processes yielded a perfect crystal at a growth velocity of ~ 18m/sec while growth on the (111) surface, driven by the same rate of heat removal, resulted in an imperfect crystal containing an assortment of defects (stacking faults and disordered region), at a growth velocity of ~14m/sec. A slower growth rate of the Si(111) system results in a grown crystal possessing a markedly higher degree of crystalline perfection (stacking faults and a very narrow region of disorder limited to the crystal-vacuum interface). The crystallization rates and their orientational dependences and the crystal face dependence of the maximum growth velocities for the formation of defect-free crystals found in the simulations are in agreement with those deduced from experimental measurements [50]. Furthermore the growth mode found in the simulations emphasizes the dynamical nature of the two interacting phases (solid and melt), the kinetics of ordering processes in the melt in the vicinity of the moving interface [50,56] and the interface morphology. While the (111) microfaceting remains during growth on Si(100), which proceeds in an almost continuous, monotonous manner, growth on Si(111) proceeds in a layerwise manner, with periods of crystallization interrupted by stages of no apparent movement of the interface. Close inspection reveals that during these stages dynamical self-annealing of defects occurs. As the defective region achieves a high level of structural perfection crystallization resumes [50].

Amorphous Film Growth Via Molecular Beam Epitaxy (MBE)

Thin film fabrication via direct condensation of vapor atoms onto a substrate at temperatures below the melting point has been used for over 100 years [57]. More recently low-energy beam deposition (generally below KeV) methods have emerged as important techniques for growth of electronic materials of specified properties [58]. From a basic science perspective studies of crystal growth processes provide opportunities for integrated approaches based on statistical mechanical formulations of nonequilibrium phenomena, phase transitions and nucleation kinetics, surface science techniques, solid state methodologies and computer simulations of materials phenomena.

While much experimental work has been done on vapor and molecular beam epitaxy [59-65] the available theoretical approaches to these problems are phenomenological in nature due to their complex many-body character [66]. On the other hand computer simulations, where the evolution of a physical system is simulated, with refined temperal and spatial resolution, alleviate certain of the difficulties which hamper other theoretical approaches. We note however, that such atomistic simulations of growth processes are difficult due to system size, physical time span and intensive computational requirements, which typically limit the spatial extent of the simulated systems to several nanometers (i.e., simulations of the dynamics of several thousand atoms, with appropriate periodic boundary conditions) and the physical time span to up to about a nanosecond. These considerations guide the design and assessment of the simulations and lead to judicious choices of the phenomena to be investigated.

The focus of previous [67-69] MD simulations of vapor-phase epitaxial growth of silicon has been on determining the conditions which are optimal for growth of ordered crystals. In contrast to the growth of materials characterized by spherically symmetric interatomic interaction potentials [70,71] (such as metals), these simulations have shown that in the case of silicon, epitaxial growth into a properly stacked, ordered, crystalline film occurs at high substrate temperatures, T_s, while at low T_s an amorphous film is grown. The focus of our investigations [72], on the other hand, is on the dynamics and mechanisms of growth of amorphous silicon films and on identification of certain deposition parameters (such as incidence geometry) which control the growth mode and the structural characteristics of the grown film. In addition we provide information about the structure of the interfacial region between the crystalline substrate and the epitaxial grown a-Si film.

To simulate the growth process, following equilibration of the crystalline substrate at room temperature (T = 300K), particles are released at a rate of one particle every N_r integration time steps, Δt (where $\Delta t = 1.15 \times 10^{-3}$ psec is the integration time-step), from a randomly chosen position on a planar source located above the substrate outside the range of interatomic interactions. The initial kinetic energy of the deposited particles is set to $3k_BT_{cm}/2$, where T_{cm} is the melting temperature of crystalline Si (T_{cm} = 1665K for the SW potential [48,50] compared to the experimental value of 1683K) and is equivalent to 0.145 eV, and their incidence direction is chosen according to the objective of the simulation.

In these simulations the initial calculational cell consists of 2 static crystalline Si layers and 3 dynamic crystalline layers which are positioned on top of the static substrate, exposing the (111) surface. The number of atoms in each of the crystalline layers is N_ℓ, and while two-dimensional periodic boundary conditions are imposed in the directions parallel to the surface plane the system is free in the Z ([111]) direction.

In order to explore the dependencies of the growth mechanisms and of the structural and dynamical properties of the grown films on the deposition geometry, we have performed simulations of a-Si epitaxial film growth on a Si(111) employing three different incidence geometries: (i) incidence along the normal to the substrate; $N_\ell = 49$, $N_r = 500$, (ii) deposition using an atomic beam directed at 60° from the normal to the substrate; $N_\ell = 96$, $N_r = 250$, and (iii) a random incidence geometry, where the impinging particles are released at random directions with respect to the substrate normal. In these simulations special precaution was taken in order to avoid excess heating of the surface region of the growing film due to interaction with the incident particles. In this way we modeled the main mechanisms and structure characteristics of a-Si film growth where limited surface diffusion plays an important role. We believe that the microstructural features which we observe are faithfully simulated since the physical size of the simulated system is much larger than the range of the interatomic interactions. Dynamical simulations of larger and more complex growth morphologies would require a much larger computational effort.

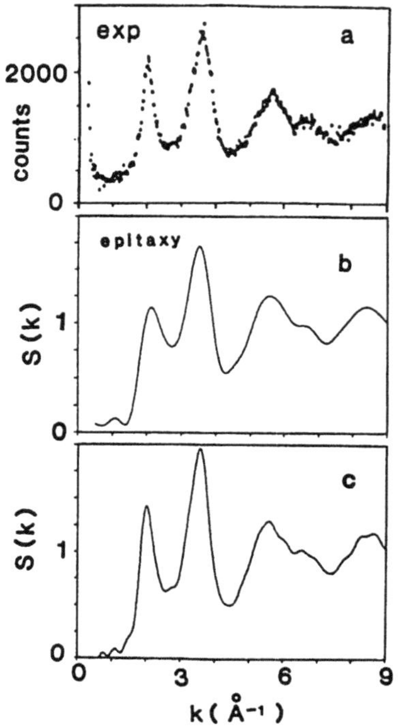

Figure 3: The static structure factor S(k) from experiment [53] (in a) and for the a-Si films obtained via deposition from the vapor (in b). The structure factors for the films grown under normal and random incidence deposition conditions are essentially identical. In addition, S(k) for a bulk a-Si system prepared by slow cooling from the melt [51] is shown for comparison in (c). The wave-vector, k, is in units of $Å^{-1}$.

tribution (ρ = 0.472 σ^{-3}, σ = 2.0951 Å) with 64% of the atoms possessing 4-fold coordination and 34% in 5-fold coordinated sites. The uniformity of the amorphous film grown for this geometry (i.e., the absence of well defined voids or columnar microstructure) is related [65] to the reduced tendency for the formation of such microstructures for small angles of incidence with respect to the normal to the substrate and to the short-range nature of the interaction potentials [23]. The static structure factor (S(k)) for the grown a-Si film (which was cooled to 300K after the termination of the growth process) agrees well with that obtained from diffraction experiments [53] (see Fig. 3), as well as with S(k) calculated for a bulk a-Si sample prepared previously [51] by slow cooling of the melt (see Fig. 3).

The density of vibrational modes, D(ω), for the film (see Fig. 4, dashed line), is in general correspondence with that obtained from experiment [54], although the level of agreement is inferior compared to D(ω) calculated for bulk a-Si, prepared via slow cooling of the melt [51].

(b) Deposition using a particle beam directed at 60° with respect to the surface normal results in a columnar microstructure, with inter-

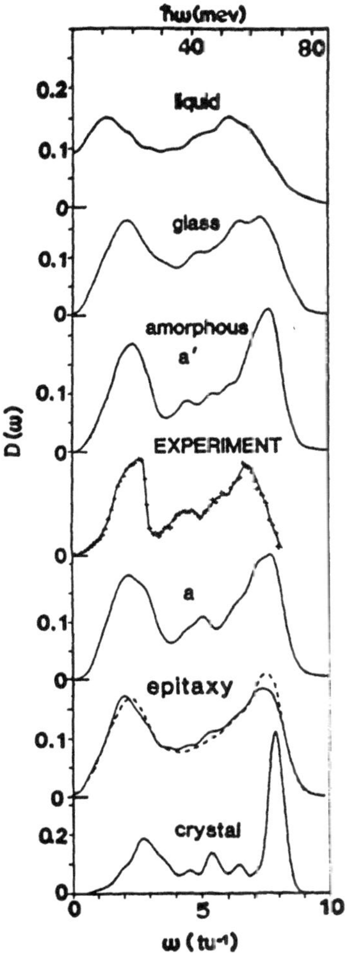

Figure 4: Density of vibrational states, $D(\omega)$ for the a-Si film grown epitaxially by deposition using a normal incident beam (dashed line) and via deposition at random incidence directions (solid line). In addition we show $D(\omega)$ determined experimentally [54], as well as the densities of states for bulk a-Si obtained by an indirect preparation procedure [44] (a') and via slow cooling from the melt [51] (a). The densities of states for simulated room temperature crystal and glass and for a liquid simulated at the melting point [44], are shown for comparison. The frequency ω is given in units of the inverse of the time unit, t.u. = 7.66×10^{-14} sec.

columnar voids. Using a column-orientation distribution function we determine that the distribution of orientations of the intercolumnar voids peaks at $\sim41^{\circ}$ in agreement with the value obtained by the "tangent rule". The pronounced tendency for development of columnar structure for this angle of incidence and the apparent agreement with the tangent

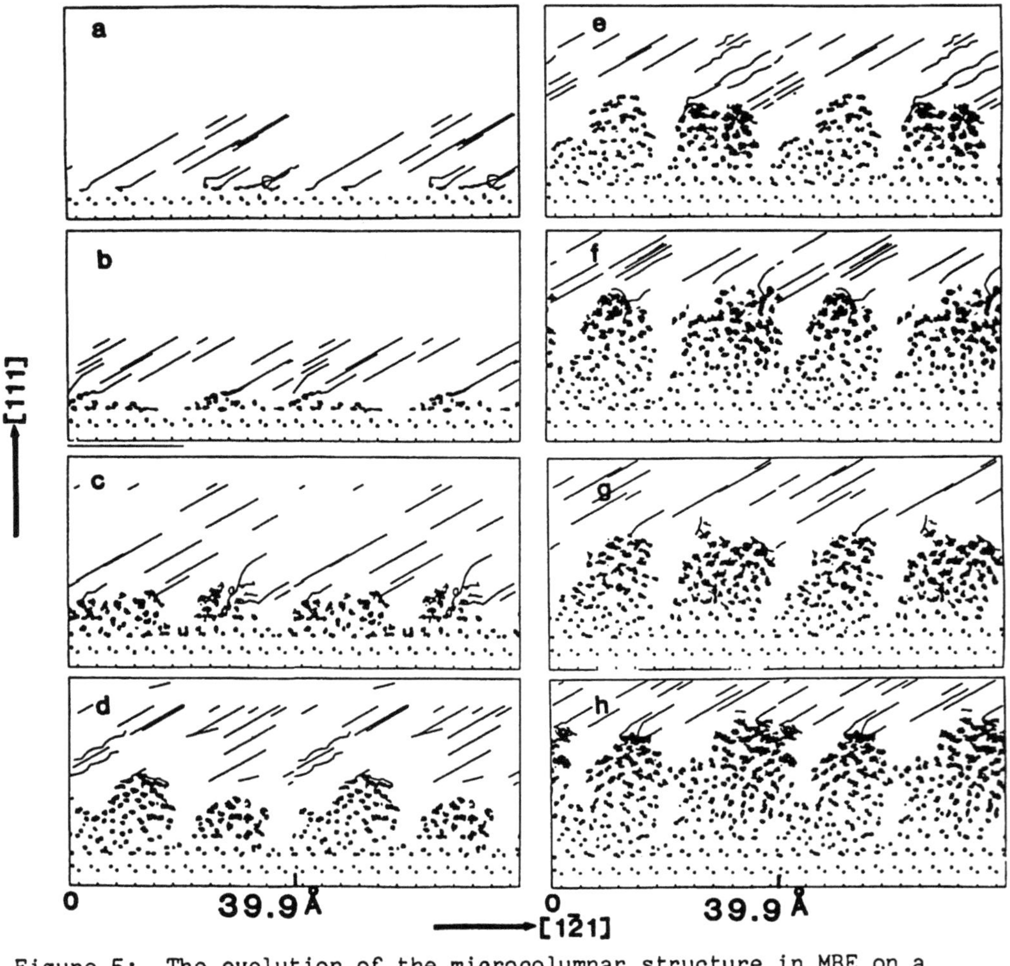

Figure 5: The evolution of the microcolumnar structure in MBE on a Si(111) substrate, using a beam directed at 60° from the normal to the surface. In each of the frames trajectories for a time span of 6.9 psec are shown, and the interval between successive frames is 57 psec. The view is along the [1$\bar{0}$1] direction. (a) Nucleation of two-dimensional crystalline islands. (b,c) Growth of three-dimensional clusters. (d-h) Growth of columnar microstructure, and the development of intercolumnar voids due to shadowing and deflection of incident particles' trajectories. The length of the calculational cell along the [1$\bar{2}$1] direction is 39.9Å, and it was duplicated along this direction for visual impression.

rule are related [65] to the small mobility of atoms deposited on the film and the short-range nature of the interaction potentials [23].

As seen from Fig. 5(a-c) the initial stage of growth of the film involves a stochastic nucleation of two-dimensional islands on the substrate which provide the nuclei for the growth of three dimensional clusters in a process which involves deflections of the trajectories of impinging particles. Further growth of these clusters shadows the intercluster regions and a film characterized by preferentially oriented density inhomogeneities, corresponding to a columnar microstructure, develops (see Figs. 5d-5h). We note the marked crystalline structure at the bases of the columns. Detailed inspection of the trajectories of deposited particles reveals that the factors underlying the growth of this structure are of geometrical (angle of incidence of the beam) and dynamical interaction origins (i.e., deflection of impinging particles' trajectories due to interactions with film atoms).

Columnar structures, i.e., preferential orientation of density inhomogeneities, in amorphous films occur on various length scales. Our discussion is limited by the nature of our simulations to microstructures, i.e., structural characteristics of microscopic (tens of angstroms) dimensions. In this context we remark that in view of the mechanism of the development of the observed microstructure, which as discussed above involves in the initial stages small localized islands and clusters from which the microstructure evolves via shadowing of the intercluster valleys, we conclude that for the conditions of our simulations the geometrical characteristics of the simulated microstructure are not influenced in a significant manner by size effects or the two-dimensional periodic boundary conditions imposed in the directions parallel to the surface plane.

(c) Random incidence deposition results in growth of an a-Si film containing micro-voids (whose diameter in our simulations is ~ 10Å). The density of the grown film (ρ = 0.457 σ^{-3}, σ = 2.0951Å) is lower (due to the voids) than that of the film grown via normal incidence (see (a)) and is comparable to that of the crystal (ρ = 0.455 σ^{-3}) at 300K. Other structural and dynamical characteristics of the film are similar to those of the film grown via normal incidence deposition.

(d) The structure of the interfacial film (regardless of the deposition geometry) in the vicinity of the crystalline substrate, exhibits over a region of ~ 8Å, partial crystalline and amorphous character, originating from crystalline intrusions embedded in an amorphous environment.

(e) Common to all the vapor phase epitaxial growth systems which we investigated in this paper are the fast rate (~ 2 psec) of dissipation of the kinetic energy of the low-energy incident particles upon attachment to the surface, and the lack of significant diffusion on the surface. These results are in contrast to those obtained for a smooth crystalline silicon surface [69], and are due to the enhancement of the collisional energy transfer rates of particles impinging on disordered or amorphous substrates and the strong local bonding, characteristic to covalent materials.

Metallic Systems

Simulations of metallic systems require special consideration of the nature of interactions and cohesion in these systems. These considerations led to the development of a new method of simulations [27,28] which allows for volume and shape variations and incorporates explicitly the dependence on the density of the "volume energy" and the effective pair potentials (see Eq. 3). Application of this method to studies of a liquid metal (Mg) yielded good agreement with experiments for several properties, including internal energy, density, structure factors, the adiabatic bulk modulus and trends in the diffusion, electrical resistivity and thermopower at several temperatures and pressures [27]. In addition, simulations of the formation and properties of a metallic glass ($Ca_{67}Mg_{33}$) [28], i.e., the solid phase formed by ultra rapid cooling of the liquid alloy, yielded results in good agreement with neutron scattering data for the static and dynamic structure factors, thus providing structural information and revealing the nature of elementary excitations in disordered media. Furthermore, we found that the accessible configurational energy of the glass possesses several nearly degenerate potential minima and that transitions between these accessible local minima involve a large displacement of single atoms accompanied by local structural relaxation of the environment [1,28].

More recently we have employed the EAM in investigations of the solid-to-liquid and liquid-to-vapor interfaces of Ni(111) and Ni(110) and in studies of the formation and properties of the metallic glass $Ag_{55}Cu_{45}$ via pressurization and/or temperature quenches [34]. The pair-distribution functions of these glasses agree with those extracted from x-ray experiments [73] and show little dependence on the formation procedure. Furthermore, in agreement with experiment [73] the glass pair-distribution function of the glass at 300K is rather similar to that of the molten alloy. However, closer inspection reveals that upon approaching the glass transition region and in the glassy state the material is characterized by a development of icosahedral local structures (~ 10% of the atoms are found to be at the center of icosahedral clusters) which are absent in the liquid alloy [34]. Finally, the glass transition temperature was found to increase linearly with applied pressure.

Although the melting of solids is one of the most common observations of a phase transition, the mechanism of melting is still an outstanding problem in condensed matter physics [74,75]. While the suggestion that melting nucleates at the surface of the solid and then propagates inward and the idea that the surface of a solid may become liquid at a temperature below the (nominal) bulk melting point, T_m, dates back over a century ago [76], a comprehensive understanding of the phenomena, at the microscopic level, is still lacking.

Surface melting which can be viewed as the wetting of the solid-gas interface by the liquid (or quasi-liquid, see below) upon approaching the triple point, provides the opportunity for investigating surface phase transformations and solid-liquid interfaces using modern surface science techniques. While numerous attempts to detect surface-initiated melting phenomena have been made [77-79], it is only recently that direct observations of surface melting have been made on a microscopic level and employing atomically clean, well-characterized surfaces [74,75].

The genesis of theoretical investigations of surface melting phenomena may be traced to the classic paper by Burton, Cabrera and Frank [80] who pointed out that above a characteristic temperature, T_R, the structure of the surface region of a crystalline solid may depart from the structure corresponding to the Terrace-Ledge-Kink (TLK) model due to a phase transition which they have termed as <u>surface melting</u>, and which has been called later the <u>roughening transition</u> (perhaps this is the cause of some confusion in the literature between the two terms

[74,75,79], which occasionally were taken to be synonymous). Surface melting, as distinguished from roughening, involves lack of registry of the atoms in the premelted surface region with respect to the underlying crystal, resulting in liquid-like properties of that surface region. However, the atoms in the premelted finite-thickness film are influenced by the presence of the underlying crystal. Therefore the surface-melt-layer at the initial stages of the process, should be regarded as a quasi-liquid exhibiting structural, dynamical and transport properties which are intermediate between those of the solid and the bulk liquid. It is the formation of the thin quasi-liquid layer (QLL), whose thickness grows as the temperature approaches T_m, which one terms appropriately as surface premelting [74].

While the Lindemann melting criterion [81], applied to the surface region has been used to rationalize the existence of surface premelting (due to the softening of vibrational modes near the surface, i.e., larger root-mean-square thermal displacements of surface atoms), the applicability of this criterion to surface melting shares the same deficiencies encountered in applying it to bulk solids.

Most theoretical approaches to surface melting are of thermodynamical [74,75,82-85] and/or phenomenological nature, including lattice dynamics and stability analysis [86-88], lattice-gas models (employing mean-field [89] and the cluster variation method [90], and Landau theory [91-93]. Microscopic descriptions of surface disordering and melting phenomena emerged from computer molecular dynamics (MD) simulations which until very recently [32,34] employed [94-98] simple pair-wise interatomic interactions in the form of Lennard-Jones potentials, which are appropriate for the description of rare gas solids and liquids, but are known to be inadequate for metals.

In our MD studies [34] we have used the many-body potentials obtained via the EAM, for Ni (for a previous investigation of premelting of Al(110) using the EMT, see reference [32]). From the several parameterization procedures which have been discussed [31], we have chosen the one described by Foiles [99] in his study of liquid transition metals. These potentials were used by us recently [33] in studies of the structure and energetics of the equilibrium crystal-to-melt interfaces of Ni(001) and Ni(111), and the melt-to-vapor interface of Nickel. Furthermore, from molecular dynamics simulations of Nickel, employing the above parameterized [99] EAM potentials we have determined that the melting temperature of the bulk metal is $1733^{o}K \pm 22^{o}K$, and the latent heat of melting is 0.19 eV/atom in good agreement with the experimental values [100] (T_m = 1725 K and ΔH_m = 0.182 eV/atom.

In our molecular dynamics simulations the semi-infinite system is modeled via a thick slab of interacting dynamic particles, exposing the (110) surface, which in addition interact with several crystalline layers of a static substrate in the desired crystallographic orientation [101]. Thus in these simulations periodic boundary conditions are imposed only in directions parallel to the surface plane. Due to the short range of the repulsive pair interactions and the finite-range of the atomic charge densities it is sufficient to represent the static substrate by 5 solid layers. The lattice constant of the static substrate, and thus the dimensions of the calculational cell in the directions parallel to the surface plane, is appropriate for the temperature of the study. From constant pressure simulations (P_{ext} = 0) of the bulk solid [52], i.e., 3-dimensional periodic boundary conditions, we have determined the coefficient of thermal expansion α = 1.2 x 10^{-5} K^{-1} which compares with the experimental [100] value (at T = 300 K) of 1.3 x 10^{-5} K^{-1}. The lattice constant at temperature T is given by $a(T) = a(T=0) \exp(\alpha T)$ where a(T=0) = 3.52 Å and T is in degrees Kelvin.

The structure of the system at various temperatures, and in particular the formation of the quasi-liquid region can be analyzed using the layer pair-distribution functions, $p_\ell(r_{||})$, shown in Figs. 6(a-f) for 1000K $\leq$ T $\leq$ 1733K. The equilibrium $p_\ell(r_{||})$ functions are calculated as

$$p_\ell(r_{||}) = \langle \frac{1}{n_\ell} \sum_{\substack{i,j\epsilon\ell \\ (i\neq j)}} \frac{1}{2\pi r_{||}} \delta(r_{ij,||} - r_{||})]\rangle \tag{5}$$

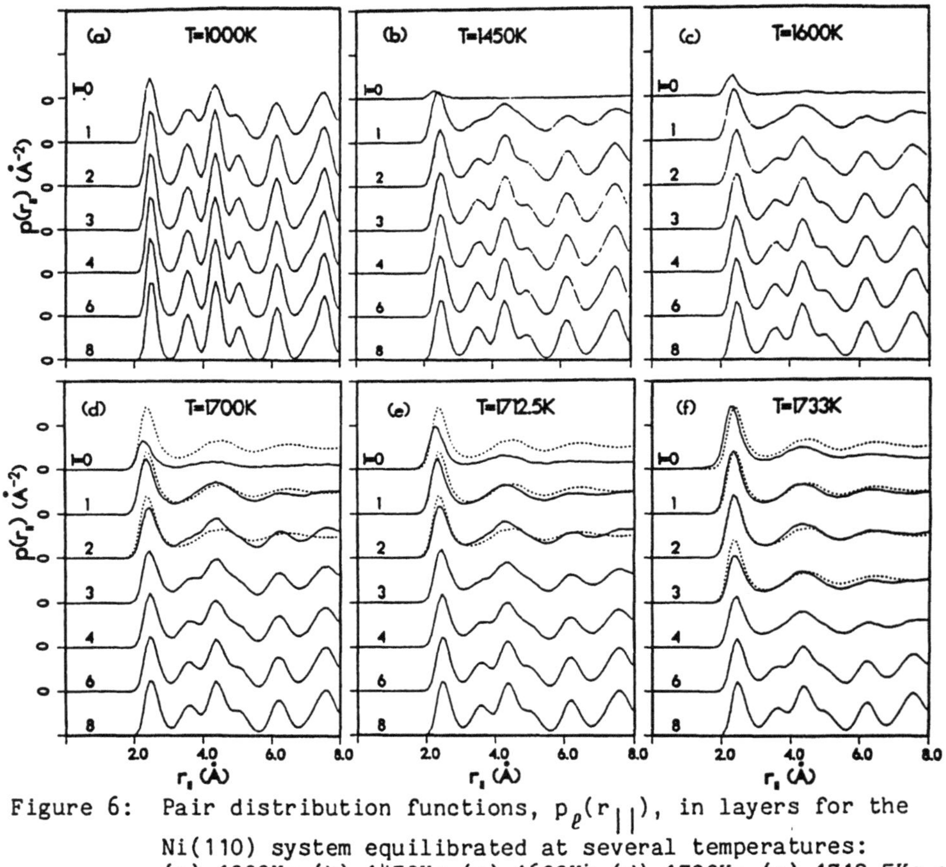

Figure 6: Pair distribution functions, $p_\ell(r_{||})$, in layers for the Ni(110) system equilibrated at several temperatures: (a) 1000K; (b) 1450K; (c) 1600K; (d) 1700K; (e) 1712.5K; and (f) 1733K. Layer ℓ = 0 corresponds to the adlayer which forms at T $\gtrsim$ 1450K. Note the gradual transition from solid to liquid character upon increasing temperature. The dashed line in d through f corresponds to the bulk liquid at the melting point [33]. Distance in units of Å.

where $r_{ij,||}$ is the component of $\vec{r}_i - \vec{r}_j$ parallel to the surface plane, n_ℓ is the instantaneous number of particles in layer ℓ, the sums extend over the particles in layer ℓ and the angular brackets denote averaging over time. For reference, the pair-distribution function for the bulk liquid at the melting point [33] (T = 1733) is shown (dashed line) for T $\geq$ 1700K. As seen from Figs. 6(a-f), the intralayer structure in the topmost layers of the system ($\ell \lesssim 3$) changes gradually from crystalline to liquid-like character as the temperature increases. In this context we recall [33] that the crystal to melt interfaces of metals, as well as other materials, are characterized by a "liquid-layered" transition region (or quasi-liquid region), extending typically over several layers, wherein structural and transport properties of the material are influenced by the crystalline field of the underlying substrate, exhibiting a gradual transition from solid to liquid properties. We note that for the adlayer (ℓ=0) at T < 1733K, the probability of finding particles with separations beyond the first-neighbor shell is small, indicating a tendency for clustering, which persists even for T $\simeq$ 1733K (compare the ratio between peak heights in $p_\ell(r_{||})$ for ℓ = 0 and in the bulk).

The structural data, presented as the natural logarithm of the squared magnitude of the layer structure factors corresponding to layers $0 \leq \ell \leq 8$ plotted versus temperature is shown in Fig. 7. The behavior of the structure factors versus temperature exhibits two regimes: (i) a low-temperature regime, T $\lesssim$ 1450K, characterized for all layers by a slow monotonic decrease of $\langle |S_\ell(\vec{g}_\alpha)|^2 \rangle$ upon increasing the temperature, and (ii) a high-temperature regime, T $\gtrsim$ 1450K where the decrease of $\langle |S_\ell(\vec{g}_\alpha)|^2 \rangle$ with increase in temperature is more rapid for layers in the surface region of the material. In the harmonic approximation of solids (the Debye-Waller theory) $\ell n \langle |S(\vec{g})|^2 \rangle$ is proportional to T. For crystals deviations from linear dependence on temperature can be explained in terms of anharmonic vibrational effects. Considerations of these effects yield expressions [102-104] for $\ell n \langle |S(\vec{g})|^2 \rangle$ which contain in addition to the term linear in T, terms proportional to T^2 and T^3 and which can be used to fit the data. The coefficients of the quadratic and cubic terms depend on derivatives of the potential beyond the harmonic approximation and can be expressed within certain models in terms of characteristic material parameters, such as the Grüneisen and volume expansion coefficients [104]. Indeed the calculated data given in Figs. 7(a-c) in the low-temperature regime can be fit by such expressions [105]. For the surface region the behavior of the structure factors in the high-temperature regime cannot be fit by the above mentioned expressions, reflecting the onset of defects and disorder (note also that the first appearance of vacancies and adatoms (ℓ = 0) occur at 1450K).

Our results show that structural and dynamical (transport) properties of the system may be discussed in terms of two temperature regimes:

At low-temperatures (T $\lesssim$ 1450K) the response of the system to increase in temperature can be explained in terms of anharmonic vibrational effects, and for these temperatures there is no indication

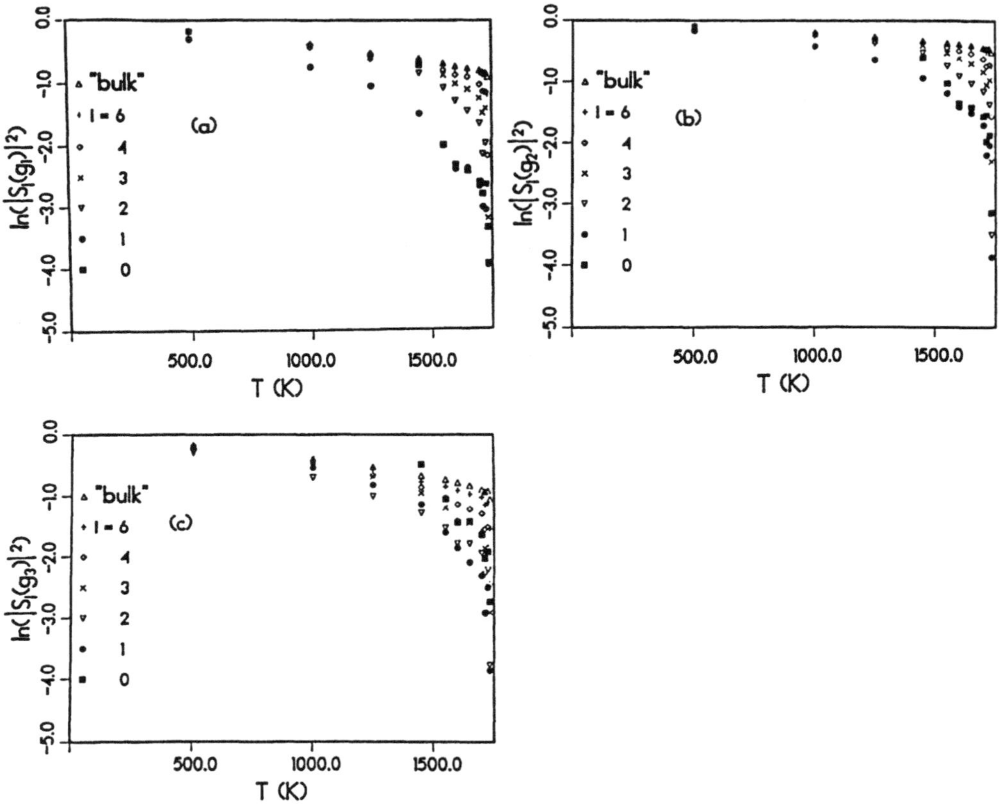

Figure 7: Natural logarithm of the squared magnitude of the equilibrium averaged layer-structure factors, $|S_\ell(\vec{g}_\alpha)|^2$, for $\vec{g}_1$ (in a), $\vec{g}_2$ (in b) and $\vec{g}_3$ (in c), versus temperature. The symbols corresponding to the various layers are defined in the figure. Temperature in degrees Kelvin. $\vec{g}_1 = \frac{2\pi}{a}(2,\bar{2},0)$, along the atomic rows of the surface; $\vec{g}_2 = \frac{2\pi}{a}(0,0,2)$, across the rows; $\vec{g}_3 = \frac{2\pi}{a}(2,2,0)$, normal to the surface plane.

for loss of long-range order due to defect formation and/or significant atomic diffusion. Atoms in the surface region of the material vibrate with larger mean-square vibrational amplitudes than those in deeper

layers in the material, and their displacements are largest in the $(1\bar{1}0)$ direction (along the atomic rows) reflecting the crystal-lographical anisotropy of the surface. Furthermore analysis [105] of the temperature behavior of the structure factors versus temperature indicates that the anharmonicity of the vibrations in the (001)

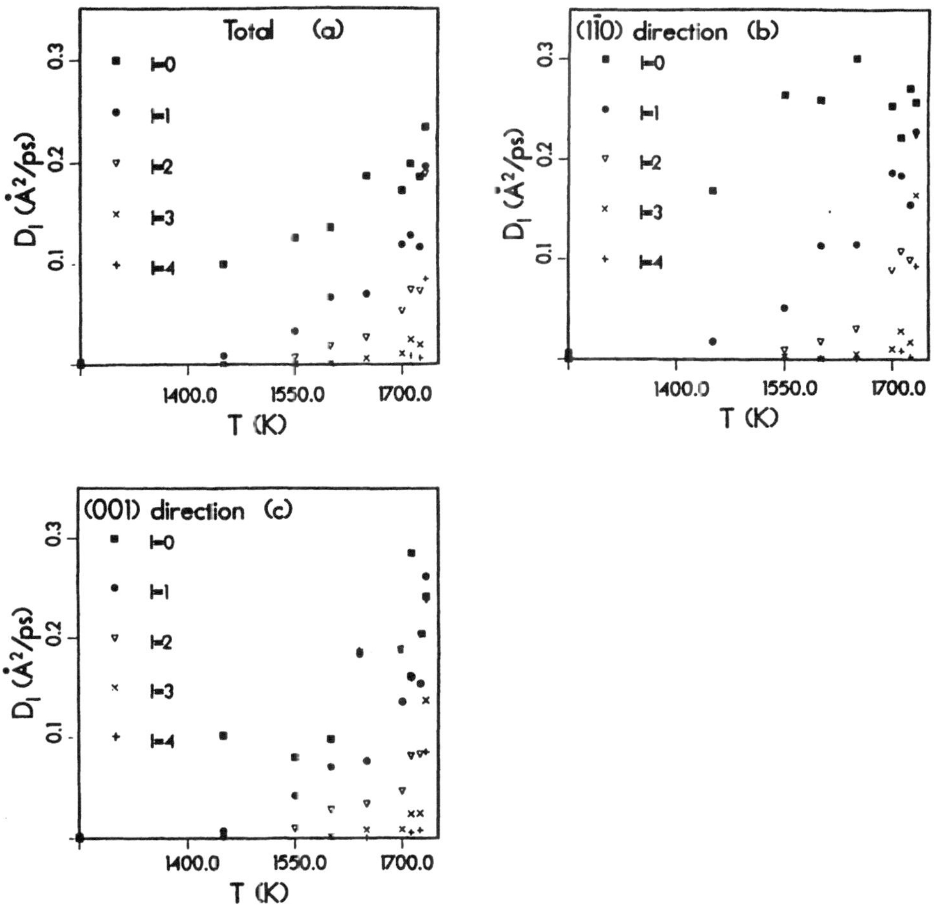

Figure 8: Coefficients of diffusion in layers, versus temperature, for the Ni(110) system. Filled squares correspond to the adlayer, $\ell = 0$; filled circles to the first layer; inverted triangles, $\ell = 2$; X's denote the diffusion coefficients in layer 3 and +'s correspond to those in the fourth layer. (a) Total diffusion coefficients; (b) and (c) Diffusion coefficients along the atomic rows ((1$\bar{1}$0) direction) and across the rows ((001) direction), respectively. Note the marked increase in the rate of diffusion at $T \gtrsim 1700$K (particularly for $\ell \leq 3$). Diffusion coefficients in units of (Å^2/ps) and temperature in degrees Kelvin.

direction (across the atomic rows) is larger than that along the (1$\bar{1}$0) and the (110) directions.

The onset of disorder occurs at $T \approx 1450$K via the generation of vacancies in the topmost layer of the solid accompanied by the formation of an adlayer, a sharp drop in the structure factors (see Fig. 7) and an increase in diffusion at the surface (see Fig. 8). We observed that the

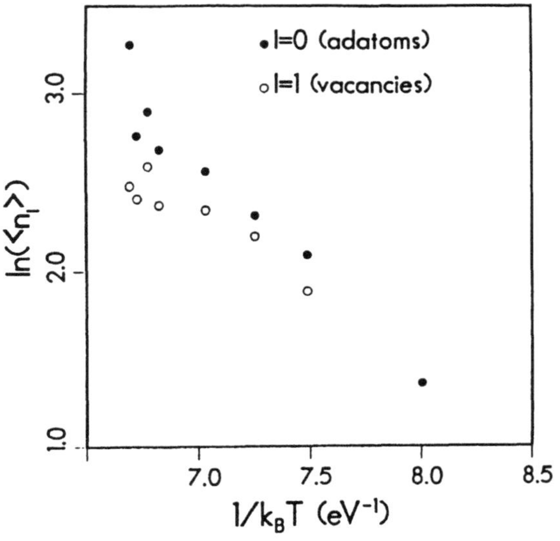

Figure 9. Natural logarithm of the equilibrium average number of adatoms (ℓ = 0, filled circles) and vacancies in the first layer (ℓ = 1, empty circles) versus $(k_BT)^{-1}$ in units of eV. Note the linear relation for the lower temperatures T $\lesssim$ 1700K. The fact that the data for the adatoms and vacancies do not coincide is due to the fact that for T > 1450K not all the adatoms originate from the first layer.

mechanism underlying this process involves initially the formation of divacancy-diadatom pairs. Further development of disorder and it's propagation upon increasing the temperature to the second and third layers of the solid occur via generation of vacancies adjacent to previously formed ones (in the same layer or in the layers above), accompanied by the formation of an adlayer (see Fig. 9), destabilization of the lattice structure, gradual loss of long-range order and higher diffusion rates (of atoms and vacancies). The process culminates at T ~ 1700K in the formation of a quasi-liquid surface region (extending over the adlayer and 2-3 layers of the original solid) characterized by interfacial liquid-like structural and atomic transport properties. In this context it is of interest to note that the approach to the quasi-liquid state is characterized by a crystallographic anisotropy (enhanced loss of long-range order and larger diffusion rates along the $(1\bar{1}0)$ direction, see Fig. 8) and that at all temperatures prior to the melting of the material (at ~ 1733K) the adlayer exhibits a larger degree of order than the adjacent layer underneath it.

From these observations we conclude that disordering of the surface region occurs first via anharmonic effects (T < 1450K), the generation of top layer vacancies accompanied by the formation of an adlayer at T ≃ 1450K, the <u>gradual thickening</u> of the defective surface region (i.e., vacancy formation in deeper layers) at T > 1450K, and the eventual formation of a quasi-liquid surface region (premelting) at T ≃ 1700K. Above 1700K the rate of loss of long-range order in the crystal and the increase in diffusion evolve (both in magnitude and spatially) in a "catastrophic" manner.

It is of interest to remark that while the vacancy driven mechanism of disordering found in this study is similar to that found in a previous study [32] of the (110) surface of Aluminum, the temperature at which the Ni(110) surface premelts (formation of the quasi-liquid layer) found in the present study is much closer to the bulk melting temperature of the material, then that found for Al(110). This may be regarded as a reduced tendency of Nickel to premelt. It is of interest to note that this finding is consistent with that reached by van der Veen et al. [74] on the basis of thermodynamic considerations. A systematic investigation of the microscopic energetic and dynamic origins underlying the propensity of materials to premelt is underway.

Tip-Substrate Interactions in Atomic Force Microscopy, Adhesion and Microindentation

The developments of scanning tunneling microscopy [106] (STM) and of the related atomic force microscopy (AFM) [107] revolutionized our perspectives and abilities to probe the morphological and electronic structure and the nature of interatomic forces in materials, as well as opened new avenues [108] for microscopic investigations and manipulations of technological systems and phenomena such as tribology [109], lithography and in biochemical applications. In both techniques a sharp tip is brought close to the surface and either the tunneling current (STM) or deflection of the cantilever holding the tip (AFM) are monitored as the surface is scanned. Of particular interest for the development of these techniques are questions related to the dynamical response of the substrate and tip which due to their proximity may result in temporary or permanent modifications of the local properties, and be reflected in the recorded data.

To investigate the consequences of the dynamical interactions between the tip and the substrate we have performed MD simulations, employing the SW potentials for silicon [23]. Since both the substrate and tip consist of the same material, these simulations correspond to the case of a reactive tip-substrate system. Our simulations [9,110,111], in both the constant-tip-height and constant-force scan modes, reveal that the <u>local</u> structure of the surface can be stressed and modified as a consequence of the tip-substrate dynamical interaction, even at tip-substrate separations which correspond to weak interaction. For large separations these perturbations anneal upon advancement of the tip while permanent damage can occur for smaller separations. For the material that we simulated (Si), we do not find long-range elastic deformations, which may occur in other circumstances [112] depending upon the elastic properties of the material and the nature of interactions.

We illustrate the method via results for constant-force simulations employing a tip consisting of 102 atoms, arranged in 4 layers and exposing a 16 atom (111) facet. The dynamic tip is supported by a static Si holder consisting of two Si (111) layers and scans in the $[1\bar{2}1]$ direction over a 6 layer dynamical Si (111) substrate with 100 atoms/layer. In the constant-force simulations in addition to the particle equations of motion the center of mass of the tip-holder assembly, Z, is required to obey $M\ddot{Z} = (\vec{F}(t)-\vec{F}_{ext})\cdot\hat{Z} - \gamma\dot{Z}$ where $\vec{F}(t)$ is the total force exerted by the tip-atoms on the static holder at time t, which corresponds to the force acting on the tip atoms due to their interaction with the substrate, $\vec{F}_{ext}$ is the desired (prescribed) force for a given scan, γ is a damping factor and M is the mass of the tip and holder. In these simulations the system is brought to equilibrium for a prescribed value of F_{ext}, and latteral scans proceed in a manner which maintains fully relaxed tip and substrate configurations throughout the scan, while the height of the tip-holder assembly adjusts dynamically according to the above feed-back mechanism.

In Fig. 10 we show results for a scan, for a constant force value $F_{z,ext} = -13.0$ (i.e., 2.15×10^{-8} N). Side views of the system trajectories at the beginning and end stages of the scan are shown in Fig. (10a,b) and (10c), respectively. As seen the tip-substrate interactions

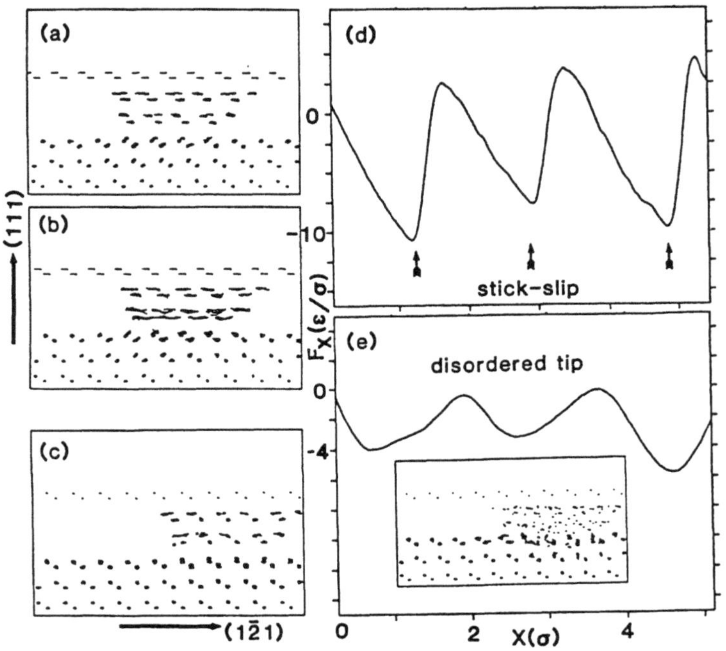

Figure 10. (a-c) Particle trajectories in a constant-force simulation viewed along the [$1\bar{0}1$] direction just before (a) and after (b) a stick-slip event and towards the end of the scan (c), for a large, initially ordered, dynamic tip. (d) The recorded F_x, exhibiting stick-slip behavior. (e) The F_x force in a constant-force scan ($F_{z,ext}$ = 1.0) employing a glassy static tip, exhibiting the periodicity of the substrate. Shown in the inset are the real-space trajectories towards the end of the scan, demonstrating tip-induced substrate local modifications.

induce local modifications of the substrate and tip structure, which are transient (compare the surface structure under the tip at the beginning of the scan (Fig. 10a), exhibiting outward atomic displacements of the top layer atoms, to that at the end of the scan (Fig. 10c), where that region relaxed to the unperturbed configuration). The recorded force on the tip-holder along the scan direction (X) is shown in Fig. (10d), exhibiting a periodic modulation, portraying the periodicity of the substrate. Most significant is the stick-slip behavior signified by the asymmetry in F_x (observed also in the real-space atomic trajectories in Figs. 10(a,b)). Here, the tip atoms closest to the substrate attempt to remain in a favorable bonding environment as the tip-holder assembly proceeds to scan. When the forces on these atoms due to the other tip-atoms exceed the forces from the substrate, they move rapidly by breaking their current bonds to the surface and forming new bonds in a region translated by one unit cell along the scan direction. We note that our constant-force simulation method corresponds to the experiments in Ref. [109] in the limit of a stiff wire (lever) and that the stick-slip phenomena which we observed are a direct consequence of the interplay between the surface forces and interactions among the tip atoms. The F_x force which we record corresponds to the frictional force. From the extrema in F_x (Fig. 10d) and the load ($F_{z,ext}$) used we obtain a coeffi- cient of friction $\mu = |F_x|/|F_{z,ext}| = 0.77$, in the range of typical values obtained from tribological measurements in vacuum (although this comparison should not be taken rigorously).

Finally, we show in Fig. 10e the frictional force obtained in simulations employing a disordered static 102-atom tip, prepared by quenching of a molten droplet, scanning under a load $F_{z,ext} = 1.0$. The significance of this result lies in the periodic variation of the force reflecting the atomic structure of the substrate. This demonstrates that microscopic investigations of structural characteristics and tribological properties of crystalline substrates are not limited to ordered tips.

Among the most innovative applications of atomic force microscopy are studies in the areas of microindentation and adhesion [113]. The use of AFM in these areas opens new avenues in studies of the elasticity, plasticity and hardness of materials and the energetics and atomistic mechanisms which underly the response of solids to applied local perturbations. In addition elucidation of the microscopic energetics and dynamics governing the adhesion between solid bodies [114,115] in contact is of importance for understanding friction, wear, lubrication and tribological interactions and in fracture mechanics and crack propagation where it has been recently suggested [116] that surface forces operating across the crack opening play an important role, in addition to the elastic properties of the material. Furthermore, combined experimental and theoretical efforts in this area provide the opportunity to assess the range of validity of traditional contact mechanics theories [117] and to suggest future modifications and formulations which will allow predictions in the nanoscale regime.

Motivated by the above considerations we have studied [35] the interaction between metal nickel tips and a gold surface, chosen to the difference in their hardness. In our investigations the materials are represented by the EAM many-body potentials and the system consists of a

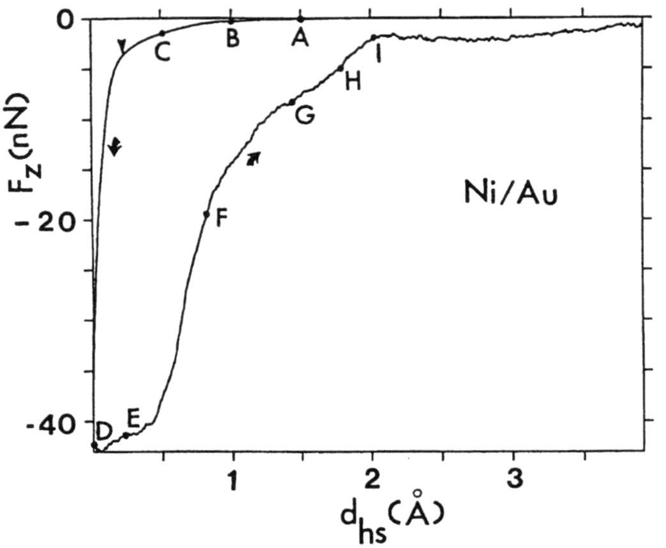

Figure 11. Force on the Ni tip in the Z direction (in nN) due to its interaction with a Au substrate, versus distance. The curve marked by the downward arrow corresponds to the lowering process and that marked by an upward pointing arrow corresponds to the lift-off process. The sharp drop after the point marked by an arrow during the lowering process corresponds to the instability leading to jump-contact. The lettered points indicate distance between the bottom layer of the Ni tip and the topmost layer of the Au surface away from the tip (the corresponding values are given in the text). The distance d_{hs} on the abscissa (in Å) is the displacement of the holder during the lift-off process after contact. Note the marked hysteresis exhibited by the force on the tip.

total of 7526 atoms. The Au substate contains 3 static layers and 8 dynamic ones each consisting of 450 atoms per layer and exposing the (001) surface (the normal to the surface is denoted as the Z axis). The dynamic Ni tip consists of 1400 atoms arranged as a tapered tip of 8 atomic layers with the bottom layer exposing the (001) surface consisting of 72 atoms, the second layer of 128 atoms, and the rest containing 200 atoms per layer. In addition, the tip interacts with a static Ni holder consisting of 1176 atoms.

Following equilibration of the system at 300K, with the Ni tip outside the range of interaction with the Au surface, the tip is lowered very slowly towards the surface, allowing for full relaxation of the system along its path, and controlling the kinetic temperature of the system at 300K via scaling of particle velocities in the bottom layer of the dynamic Au substrate. The calculated force in the Z direction on the tip upon lowering (downward arrow) is shown in Fig. 11, where the distance from the bottom layer of the tip to atoms in the topmost layer of the Au surface, away from the area under the tip, at several stages of the lowering process are denoted (A = 5.7Å, B = 5.2Å, C = 4.7Å and D

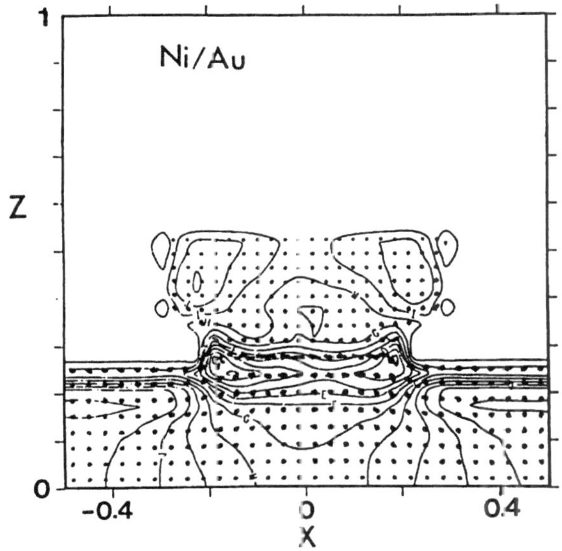

Figure 12. A side view of a slice of the system corresponding to the relaxed configuration of the system at adhesive contact (point D in Fig. 11). Length along the X and Z axes in units of 61.2Å. Note the displacement toward the tip of the topmost layer of the Au substrate under the Ni tip. The contours correspond to the distribution of hydrostatic pressure in the system. The maximum magnitude of the pressure is for atoms at the periphery of the contact (X = ± 0.19 and Z = 0.27) and is equal to -10×10^4 atm. The contours are equally spaced with an increment of 1×10^4 atm. The contours marked E, F and G correspond to -6×10^4, -5×10^4 and -4×10^4 atm, respectively, and those marked H, I and J correspond to -3×10^4, -2×10^4 and -1×10^4 atm, respectively.

= 4.2Å). We note first the monotonic increase of the force on the tip as it approaches the surface, followed by a sharp increase in the force. In fact at a certain distance from the surface (~ 4.18Å, marked by an arrow in Fig. 11), the system exhibits a mechanical instability which leads to a sudden increase of the force between the tip and the substrate and the formation of an adhesive contact. After the contact formation the distance between the bottom layer of the tip and the Au surface (away from the tip) decreases to 4.0Å. Inspection of the relaxed configuration after contact reveals that the contact formation is accompanied by a plastic deformation of the Au substrate with atoms in the topmost layer of the surface displacing toward the Ni tip, followed by upward displacements of the second and deeper layers of the substrate (see Fig. 12). The contours of the local hydrostatic pressure in the materials after contact formation (calculated from the trace of the atomic stress tensors [118]) given in Fig. 12, demonstrate that the atoms at the periphery of the contact (X ~ ± 0.19 and Z ~ 0.27 in Fig. 12, where X = 1 and Z = 1 correspond to 61.2Å, each) are under extreme tensile conditions (-1.0×10^4 atm).

Starting from the materials in contact the tip was then raised. The force versus the distance that the holder of the tip traversed during lifting of the tip (d_{hs}) is shown in Fig. 11 (upward arrow) with the distance between the bottom of the tip to the topmost substrate layer denoted (E = 4.4Å, F = 4.85Å, G = 5.5Å, H = 5.85Å and I = 6.2Å). As seen the force versus distance exhibits a marked hysteresis which is a consequence of the adhesive contact formation and major inelastic deformations in the system (in this context we remark that starting the tip lift-off process prior to jump contact, i.e., at a distance prior to the arrow in Figure 11, results in no hysteresis).

The behavior of the system along the lift-off path is characterized by adherence of a monolayer of Au atoms to the bottom of the tip followed by a formation of a 'neck' made of gold atoms (see Fig. 13 where the atomic configuration during lift-off is shown). The atomic dynamic processes which underly the observed behavior are akin to those which have been postulated to occur in fracture mechanics. In this context it is interesting to note that the tear of the Au layer by the tip initiates at the periphery of the contact (i.e., the region of maximum tensile stress) propagating toward the center of the contact zone as the tip moves away from the surface, which is similar to one of the common postulated mechanisms of crack propagation.

Finally we remark that the hysteresis in the force versus distance observed in our simulations is of similar character to that measured

Figure 13: Atomic configuration of the Ni/Au system during the lift-off process. The nickel tip atoms are at the top of the figure (dark small balls). Atoms of the topmost layer of the gold substrate are in white. Note that a monolayer of Au atoms adhered to the Ni tip, leaving an extended defect in the substrate under the tip.

recently for a tungsten tip interacting with a golc surface [113]. Furthermore collaborative theoretical and experimental work on the nickel-gold system is underway, as well as systematic studies aimed at incorporating atomistic properties and mechanisms into formulations of adhesive contact formation, fracture and microindentation.

Acknowledgement: These studies were supported by the U.S. Department of Energy (DOE) Grant No. FG05-86ER-45234, by the National Science Foundation (NSF) Tribology Program Grant No. MSM-8719669 and by the United States Air Force Office of Scientific Research (AFOSR). Most calculatioms were performed on the CRAY Research Inc. X-MP at the National Magnetic Fusion Energy Computer Center, Livermore, California through a DOE grant. The help of Ms. Audrey Ralston in the preparation of the manuscript is gratefully acknowledged.

References

[1] U. Landman et al., Mat. Res. Soc. Symp. Proc. Vol. 63 (MRS, Boston, 1985), p. 273.
[2] F. F. Abraham, Adv. Phys. 35, 1 (1986); J. Vac. Sci. Technol. B2, 534 (1984).
[3] MRS Bull. Volume XIII (2), February 1988, p. 14-39.
[4] U. Landman, R. N. Barnett, C. L. Cleveland, J. Luo, D. Scharf and J. Jortner, in Few-Body Systems and Multiparticle Dynamics, Ed. D. A. Micha, AIP Conf. Proc. 162 (AIP, New York, 1987), p. 200.
[5] D. W. Heerman, Computer Simulation Methods (Springer, Berlin, 1986).
[6] Computer Simulations of Solids, Eds. C. R. A. Catlow and W. C. Machord (Springer, Berlin, 1982).
[7] U. Landman, in: Computer Simulation Studies in Condensed Matter Physics: Recent Developments, Eds., D. P. Landau, K. K. Mon and H.-B. Schuttler (Springer, Berlin, 1988).
[8] F. F. Abraham, Adv. Phys. 35, 1 (1986) and J. Vac. Sci. Technol. B2, 534 (1984).
[9] U. Landman in New Materials Approaches to Tribology: Theory and Applications, Mat. Res. Soc. Sym. Proc. Vol. 140 (MRS, Pittsburgh, 1989), p. 101.
[10] U. Landman and W. D. Luedtke in: Atomic Modeling in Materials: Beyond Pair-Potentials, Ed. V. Vitek (Plenum, New York, 1988).
[11] U. Landman in Computer Simulation Studies in Condensed Matter Physics: Recent Developments, Eds. D. P. Landau, K. K. Mon and H. -B. Schuttler (Springer, Berlin, 1988), p. 144.
[12] Monte Carlo Methods in Quantum Problems, Ed. M. H. Kalos (Reidel, Dordrecht, 1984).
[13] B. J. Alder, D. M. Ceperley and E. L. Pollock, Acc. Chem. Res. 18, 268 (1985).
[14] B. J. Berne and D. Thirumalai, Ann. Rev. Phys. Chem. 37, 401 (1986).
[15] M. Sprik and M. Klein, Comp. Physics Reports 7, 147 (1988).
[16] R. Kosloff, J. Phys. Chem. 92, 2087 (1988).
[17] J. Jortner, D. Scharf and U. Landman in "Elemental and Molecular Clusters", Eds. G. Benedek, T. P. Martin and G. Pacchioni (Springer-Verlag, Berlin, 1988), p. 148.

[18] R. N. Barnett, U. Landman and A. Nitzan, J. Chem. Phys. 89, 2422 (1988); ibid. 90, 4413 (1989); Phys. Rev. Lett. 62, 106 (1989).
[19] R. N. Barnett, U. Landman, Gunaretnan Rajagopal and A. Nitzan, Israel J. Chem. (in press, 1989).
[20] G. C. Maitland, M. Rigby, E. B. Smith and W. A. Wakeham, Intermolecular Forces, (Clarendon, Oxford, 1981).
[21] M. J. Sangster and M. Dixson, Adv. Phys. 25, 247 (1976).
[22] P. N. Keating, Phys. Rev. 145, 637 (1966).
[23] F. H. Stillinger and T. A. Weber, Phys. Rev. B31, 5262 (1985).
[24] R. Biswas and D. R. Hamann, Phys. Rev. Lett. 55, 2001 (1985).
[25] T. Tersoff, Phys. Rev. B37, 6991 (1988).
[26] See, e.g., W. A. Harrison, Pseudopotentials in the Theory of Metals (Benjamin, Reading, Mass., 1966).
[27] R. N. Barnett, C. L. Cleveland and U. Landman, Phys. Rev. Lett. 54, 1679 (1985).
[28] R. N. Barnett, C. L. Cleveland and U. Landman, Phys. Rev. Lett. 55, 2035 (1985).
[29] P. Hohenberg and W. Kohn, Phys. Rev. B 136, 864 (1964).
[30] K. W. Jacobson, J. K. Norskov and M. J. Puska, Phys. Rev. B 35, 7423 (1987); J. K. Norskov, Phys. Rev. B 26, 2875 (1982).
[31] (a) M. S. Daw and M. I. Baskes, Phys. Rev. B 29, 6443 (1984); (b) S. M. Foiles, M. I. Baskes and M. S. Daw, Phys. Rev. B 33, 7983 (1986).
[32] (a) P. Stolze, J. K. Norskov, and U. Landman, Phys. Rev. Lett. 61, 440 (1988); (b) Surf. Sci. (to be published).
[33] E. T. Chen, R. N. Barnett and U. Landman, Phys. Rev. B 40, 924 (1989).
[34] E. T. Chen, R. N. Barnett and U. Landman, Phys. Rev. B (to be published).
[35] W. D. Luedtke and U. Landman, Phys. Rev. Lett. (to be published).
[36] D. E. Polk, J. Non-Cryst. Solids 5, 365 (1971).
[37] D. Henderson and F. Herman, J. Non-Cryst. Solids 8-10, 359 (1972).
[38] D. Henderson, J. Non-Cryst. Solids 16, 317 (1974).
[39] L. Guttman, Phys. Rev. B 23, 1866 (1981).
[40] F. Wooten, K. Winer, and D. Weaire, Phys. Rev. Lett. 54, 1392 (1985).
[41] Q. Broughton and X. P. Li, Phys. Rev. B 35, 9120 (1987).
[42] M. D. Kluge, J. R. Ray, and A. Rahman, Phys. Rev. B 36, 4234 (1987).
[43] R. Biswas, G. S. Grest and C. M. Soukoulis, Phys. Rev. B 36, 7437 (1987).
[44] W. D. Luedtke and U. Landman, Phys. Rev. B 37, 4656 (1988).
[45] K. Ding and H. C. Andersen, Phys. Rev. B 34, 6987 (1986).
[46] F. F. Abraham and I. P. Batra, Surf. Sci. 163, L752 (1985).
[47] I. P. Batra, F. F. Abraham and S. Ciraci, Phys. Rev. B 35, 9552 (1987).
[48] U. Landman, W. D. Luedtke, R. N. Barnett, C. L. Cleveland, M. W. Ribarsky, E. Arnold, S. Ramesh, H. Baumgart, A. Martinez, and B. Khan, Phys. Rev. Lett. 56, 155 (1986).
[49] F. F. Abraham and J. Q. Broughton, Phys. Rev. Lett. 56, 734 (1986).
[50] U. Landman, W. D. Luedtke, M. W. Ribarsky, R. N. Barnett, and C. L. Cleveland, Phys. Rev. B 37, 4637, 4647 (1988).
[51] W. D. Luedtke and U. Landman, Phys. Rev. B 40, 1164 (1989).

[52] M. Parrinello and A. Rahman, J. Appl. Phys. 52, 7182 (1981); see also M. W. Ribarsky and U. Landman, Phys. Rev. B. (in press, 1988).
[53] T. A. Postal, C. M. Falco, R. T. Kampwirk, I. K. Schuller and W. B. Yelon, Phys. Rev. Lett. 45, 648 (1980).
[54] W. A. Kamitakahara, C. M. Soukoulis, H. R. Shanks, U. Buchenau and G. S. Grest, Phys. Rev. B 36, 6539 (1987).
[55] See citations in reference 48.
[56] See also earlier MD studies of liquid-phase epitaxy: U. Landman, R. N. Barnett, C. L. Cleveland and R. H. Rast, J. Vac. Sci. Technol. A3, 1574 (1985); U. Landman, C. L. Cleveland and C. S. Brown, Phys. Rev. Lett. 45, 2032 (1980) and in Nonlinear Phenomena of Phase Transitions and Instabilities, Ed. T. Riste (Plenum, New York, 1982), p. 379.
[57] M. Faraday, Phil. Trans. 147, 145 (1857).
[58] For a recent review see H. J. Leamy, G. H. Gilmer, and A. G. Dirks, in Current Topics in Materials Science, Ed. E. Kaldis (North-Holland, Amsterdam, 1980), Vol. 6, Chap. 4.
[59] See reviews in MRS Bulletin 13, 18-50, 29-67 (1988).
[60] S. S. Lau and J. W. Mayer in Treatise of Material Science and Technology (Academic, New York, 1982), Vol. 24, Chap. 3; C. E. C. Wood in Physics of Thin Films (Academic, New York, 1980), Vol. 2, p. 35; E. Kasper, Appl. Phys. A 28, 129 (1982).
[61] (a) J. A. Thornton, J. Vac. Sci, Technol. 4, 3059 (1986).
(b) M. Sikkens, J. J. Hodgkinson, F. Horowitz, H. A. Macleod and J. J. Wharton, Opt. Eng. 25, 142 (1986).
[62] See articles in Vapor Deposition, Eds. C. F. Powell, J. H. Oxley and J. M. Blocher Jr. (Wiley, New York, 1966).
[63] G. Honjo and K. Yagi in Current Topics in Materials Science, Ed. E. Kaldis (North-Holland, Amsterdam, 1980), Vol. 6, p. 195.
[64] Epitaxial Growth, Ed. J. W. Matthews (Academic, New York, 1975).
[65] H. J. Leamy, G. H. Gilmer and A. C. Dirks in Ref. 63, p. 309.
[66] See J. A. Venables, Vacuum 33, 701 (1983), and references therein.
[67] M. Schneider, J. K. Schuller and A. Rahman, Phys. Rev. B 36, 1340 (1987).
[68] E. T. Gawlinski and J. D. Gunton, Phys. Rev. B 36, 4774 (1987).
[69] R. Biswas, G. S. Grest and C. M. Soukoulis, Phys. Rev. B 38, 8154 (1988).
[70] M. Schneider, A. Rahman and J. K. Schuller, Phys. Rev. Lett. 55, 604 (1985).
[71] M. Schneider, A. Rahman and J. K. Schuller, Phys. Rev. B 34, 1643 (1986).
[72] W. D. Luedtke and U. Landman, Phys. Rev. B (in press, 1989).
[73] W. E. Lukens and C. W. J. Wagner, J. Appl. Cryst. 9, 159 (1976).
[74] See review by J. van der Veen, B. Pluis, and A. W. Denier van der Gon, in Chemistry and Physics of Solid Surfaces, Vol. VII (Springer-Verlag, Berlin, 1988).
[75] J. G. Dash, in Proceedings of the Solvay Conference on Surface Science: Invited Lectures and Discussions, Austin, Texas, 1987, edited by F. W. de Wette, Springer Series in Surface Science, Vol. 14 (Springer-Verlag, New York, 1988), p. 142.
[76] M. Faraday, Proc. Roy. Soc. Lond. 10, 440 (1860); in Faraday Diary, Vol. IV (Bell and Sons, London, 1933), pp. 79-81; see also W. A. Weyl, J. Colloid Sci. 6, 389 (1951).

[77] G. Tammann, Z. Phys. Chem. 68, 205 (1910); Z. Phys. 11, 609 (1910).
[78] J. F. van der Veen and J. W. M. Frenken, Surf. Sci. 178, 382 (1986) and references therein.
[79] D. Nenow, Prog. Cryst. Growth and Charact. 9, 1893 (1984), and references therein.
[80] W. K. Burton, N. Cabrera and F. C. Frank, Philos. Trans. Roy. Soc. London 243, 299 (1951).
[81] F. A. Lindemann, Z. Phys. 14, 609 (1910).
[82] J. N. Stranski, Z. Physik 119, 22 (1942).
[83] J. Frenkel, Kinetic Theory of Liquids, (Clarendon Press, London, 1946), pp. 425-6.
[84] J. K. Kristensen and R. M. J. Cotterill, Phil. Mag. 36, 437 (1977).
[85] T. Kuroda and R. Lacmann, J. Cryst. Growth 56, 189 (1982), D. Nenow and A. Trayanov, J. Cryst. Growth 79, 801 (1986).
[86] L. Pietronero and E. Tosatti, Solid State Commun. 32, 255 (1979).
[87] C. S. Jayanthi, E. Tosatti, A. Fasolino and L. Pietronero, Surf. Sci. 152/153, 155 (1985).
[88] C. S. Jayanthi, E. Tosatti and L. Pietronero, Phys. Rev. B31, 3456 (1985).
[89] A. Trayanov and E. Tosatti, Phys. Rev. Lett. 59, 2207 (1987); Phys. Rev. B38, 6961 (1988).
[90] G. An and M. Schick, Phys. Rev. B 39, 9722 (1989).
[91] R. Lipowsky, Phys. Rev. Lett. 49, 1575 (1982).
[92] R. Lipowsky and W. Speth, Phys. Rev. B 28, 3983 (1983).
[93] R. Lipowsky, U. Breuer, K. C. Prince and H. P. Bonzel, Phys. Rev. Lett. 62, 913 (1989).
[94] J. Q. Broughton and G. H. Gilmer, Acta Metall. 31, 845 (1983).
[95] J. Q. Broughton and G. H. Gilmer, J. Chem. Phys. 79, 5095, 5105, 5119 (1983).
[96] V. Pontikis and P. Sindzingre, Physica Scripta T19, 375 (1987).
[97] V. Rosato, G. Cicotti and V. Pontikis, Phys. Rev. B 33, 1860 (1986).
[98] S. Valkealahti and R. M. Nieminen, Phys. Scripta 36, 646 (1987).
[99] S. M. Foiles, Phys. Rev. B 32, 3409 (1985).
[100] Handbook of Chemistry and Physics (CRC Press, Cleveland, 1974).
[101] The two dimensional unit cell in the surface plane is defined by the unit vectors $\vec{a}_1 = \frac{a}{2}(1,-1,0)$ and $\vec{a}_2 = a(0,0,1)$ which lie along and across the atomic rows, respectively. The third direction is defined by the vector $\vec{a}_3 = \frac{a}{2}(1,0,1)$. The spacing between atomic layers is $d = a/2\sqrt{2}$, and a is the FCC cube edge length which for nickel is equal to 3.52 Å at 0K and 3.59 Å at 1700K. The dimensions of the calculational cell in the surface plane are 10 $\vec{a}_1$ and 7 $\vec{a}_2$, with 70 atoms/layer.
[102] A. A. Maradudin and P. A. Flinn, Phys. Rev. 129, 2529 (1963).
[103] B. T. M. Willis, Acta. Cryst. A25, 277 (1969).
[104] J. Prakash and M. P. Hemkar, J. Phys. Soc. Japan 36, 1608 (1974).
[105] R. N. Barnett and U. Landman (to be published).
[106] G. Binnig and H. Rohrer, IBM J. Res. Develop. 30, 355 (1986).
[107] G. Binnig, C. F. Quate and Ch. Gerber, Phys. Rev. Lett. 56, 930 (1986).

[108] P. H. Hansma and J. Tersoff, J. Appl. Phys. 61, R1 (1986).
[109] C. M. Mate, G. M. McClelland, R. Erlandsson and S. Chiang, Phys. Rev. Lett. 59, 1942 (1987).
[110] U. Landman, W. D. Luedtke and A. Nitzan (submitted to Phys. Rev. Lett., Aug., 1987); see also ref. 7.
[111] U. Landman, W. D. Luedtke and M. W. Ribarsky, J. Vac. Sci. Technol. (in press, July/August, 1989).
[112] J. M. Soler, A. M. Baro, N. Garcia and H. Rohrer, Phys. Rev. Lett. 57, 444 (1986); see comment by J. B. Pethica, ibid 57, 3235 (1986).
[113] N. A. Burnham and R. J. Colton, J. Vac. Sci. Technol. (in press, July/August, 1989).
[114] D. Tabor, J. Coll. Int. Sci. 58, 2 (1977).
[115] R. G. Horn, J. N. Israelachvili and F. Pribac, J. Coll. Int. Sci. 115, 480 (1987).
[116] B. R. Lawn, Appl. Phys. Lett. 47, 809 (1985).
[117] K. L. Johnson, Contact Mechanics, (Cambridge, Univ. Press, Cambridge, 1985); see also references 2-7 cited in ref. [115].
[118] T. Egami and D. Srolovitz, J. Phys. F: Met. Phys. 12, 2141 (1982).

Atoms Embedded in Electron Gas

M.J. Puska

Laboratory of Physics, Helsinki University of Technology,
SF-02150 Espoo, Finland

Abstract. The model problem of an atom embedded in electron gas is discussed. The characteristic features of the solutions are described and connections to real physical systems are drawn. Parameters, which are needed in applying the Effective Medium Theory, and which can be calculated using the embedded atom model, are defined and tabulated for several elements.

1. Introduction

An atom embedded in homogeneous electron gas constitutes a model problem, which can be solved accurately within the density functional theory [1,2]. The only parameters of this model are the nuclear charge Z of the atom and the density $\bar{n}$ of the homogeneous electron gas. The great value of this model is that it can, in spite of its simplicity, mimic various aspects of the interaction of a constituent atom with its electronic environment in real physical systems and thereby give transparent physical insight into many problems. Especially this is true when the actual host is a simple metal, but also in the case of more complicated electronic environments like transition metals or semiconductors the solution for an atom in jellium can be used as an important contribution when creating the entire model. There are several examples about different aspects of atom-environment interactions which have been studied by the help of this model problem: potential energy surfaces of atoms on surfaces and in bulk metals [3-5], cohesive properties of metals [6], kinetic energy shifts of photoemission and Auger electrons [7-9], damping rate of an vibrating atom on a surface [10], electronic stopping power for slow ions [11] *etc.*

In the present context, the model of an atom in homogeneous electron gas is important, because it is the starting point of the Effective Medium Theory (EMT) [6,12] and provides the parameters needed in the applications. The purpose of this contribution is to enlighten some aspects of the model problem but first of all to give a uniform data basis for the EMT parameters.

2. Solution of the model problem

The electronic structure for an atom embedded in homogeneous electron gas can be solved in the density functional theory [1]. Usually, the Local Density Approximation (LDA) for the exchange and correlation is made and the Kohn-Sham equations are

Springer Proceedings in Physics, Vol. 48 Many-Atom Interactions in Solids
Editors: R.M. Nieminen · M.J. Puska · M.J. Manninen

written in the non-relativistic form as

$$[-\frac{1}{2}\nabla^2 + v_{eff}(r)]\psi_i = \epsilon_i \psi_i, \tag{1}$$

$$v_{eff}(r) = v_{ext}(r) + \int \frac{n(r')}{|\vec{r} - \vec{r}\,'|} d\vec{r}\,' + v_{xc}(n(r)), \tag{2}$$

$$n(r) = \sum_i^{occ} |\psi_i(r)|^2 = \sum_i^{bound} |\psi_i(r)|^2 + \int_0^{\epsilon_F} n(\epsilon, r) d\epsilon. \tag{3}$$

This set of equations is solved self-consistently for the electron density $n(r)$ responding to the external potential v_{ext} formed by the rigid positive background charge $\bar{n}$ and the point charge Z. Above, $v_{xc}(n(r))$ is the exchange-correlation potential in LDA [13]. The screening of the point charge Z and the positive background consists of bound states and delocalized states (combined spatial and energy-axis density $n(\epsilon, r)$) occupied from zero-energy (bottom of the conduction band) up to the Fermi level ϵ_F. Eqs. (1)-(3) are written in the spin-compensated formalism, which means that the partly filled electron nl-shells of free atoms either become completed or they emerge in the continuum when the free atom is brought from vacuum into the electron gas.

In the numerical solution of the problem, the impurity induced perturbation in the effective potential has to be cut off at a given 'large' distance R. At R the scattering states integrated numerically outwards from origin are joined to spherical waves. This matching determines the scattering phase shifts $\delta_l(\epsilon)$ for the partial waves with angular momentum l at energy ϵ. Furthermore, the continuum part of the impurity induced change in the density of states $\Delta D(\epsilon)$ is obtained from the energy derivatives of the phase shifts as

$$\Delta D(\epsilon) = \frac{2}{\pi} \sum_l \frac{\partial \delta_l}{\partial \epsilon}(2l+1). \tag{4}$$

The total energy E^{hom} of an atom embedded in homogeneous electron gas is calculated from the energy functional of the density functional theory. A practical form is the following

$$\begin{aligned} E^{hom} &= \sum_i^{bound} \epsilon_i + Z\epsilon_F - \frac{2}{\pi}\sum_l \int_o^{\epsilon_F} d\epsilon (2l+1)\delta_l(\epsilon) - \int_0^\infty dr 4\pi r^2 v_{eff}(r) n(r) \\ &+ \frac{1}{2}\int_0^\infty dr 4\pi r^2 (n(r) - \bar{n})(\phi(r) - Z/r) \\ &+ \int_0^\infty dr 4\pi r^2 [n(r)\epsilon_{xc}(n(r)) - \bar{n}\epsilon_{xc}(\bar{n})], \end{aligned} \tag{5}$$

where $\phi(r)$ is the Coulomb potential in the system and $\epsilon_{xc}(n)$ is the exchange-correlation energy per electron in homogeneous electron gas with density n. In practice, the integrals in Eq.(5) are carried out numerically from the origin to the cut-off radius R and thereafter analytically using the asymptotic forms of the induced densities and potentials.

From the practical point of view, the program solving the electronic structure of an atom in homogeneous electron gas is, due to the existence of the scattering states, somewhat more complicated than a standard free atom program using the LDA. The long range Friedel oscillations raise some aspects which have to be handled very carefully. The calculation of the integrals in the total energy functional is one of these aspects [14]. The other one is the solution of the Poisson equation for the Coulomb potential. In order that the self-consistency iterations do not diverge, this has to be done *e.g.* by the special technique called the screened Green function method [15].

3. General properties of the solutions

As an example of atom-induced densities $\Delta n(r) = n(r) - \bar{n}$ in jellium we show in Fig. 1 that for Al in electron gas with the density parameter $r_s = (3/4\pi\bar{n})^{1/3} = 3\ a_o$. The induced density is compared with the free atom density. The induced electron density is enhanced over the free atom density near the Al nucleus resulting in a local overscreening. At large r-values there are strong Friedel oscillations. Fig. 2 shows as examples the continuum parts of the change in density of states $\Delta D(\epsilon)$ for atoms C through F for $r_s = 2\ a_o$. When energy approaches zero there is a weak divergence due to s-wave scattering [1]. The most prominent feature seen in Fig. 2 is the resonance peak developing gradually when going from C to O. This resonance is due to the p-wave scattering and it contains the $2p$-electrons which rise to continuum states when the free atom enters the electron gas. For example there are four electrons in the resonance for O. F binds a full $2p$ shell so that there is no resonance but $\Delta D(\epsilon)$ integrated up to the Fermi level shows that one electron is removed from the conduction band.

As an example of the use of the scattering phase shifts Fig. 3 shows the stopping power $\frac{dW}{dR}$ for slow ions in electron gas as a function of the atomic number of the ion[11]. The stopping power is calculated from the phase shifts as

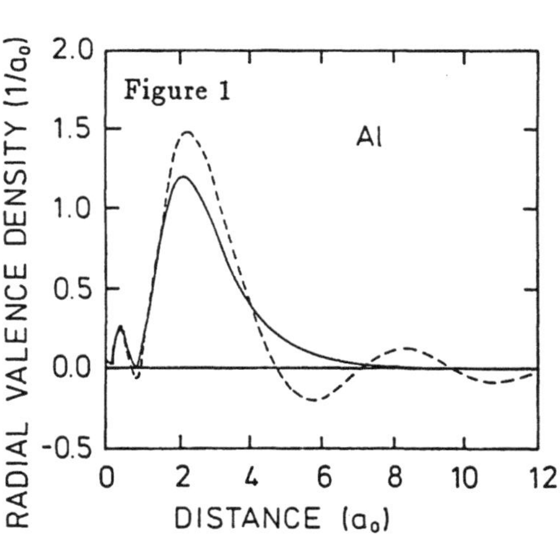

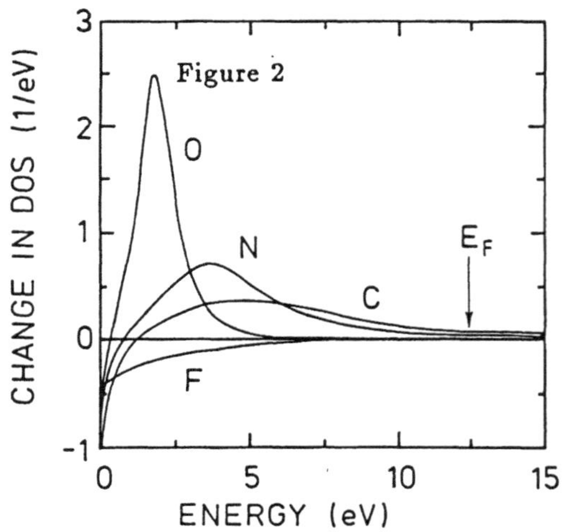

Figure 1. Induced valence electron density of an Al atom in homogeneous electron gas with $r_s = 3\,a_o$ (dashed) compared with the free-atom density (solid line).

Figure 2. Induced density of states for C, N, O, and F atoms in electron gas with $r_s = 2\,a_o$. The position of the Fermi level is denoted by an arrow.

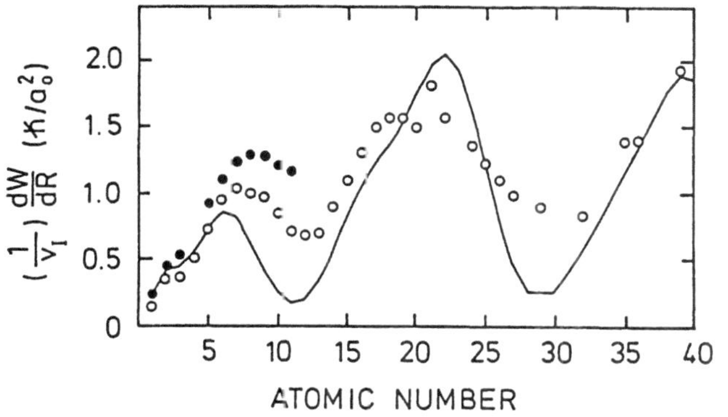

Figure 3. Stopping power for well channeled ions [11]. The solid line gives the theoretical results obtained via Eq. (6) using phase shifts for $r_s = 2\,a_0$. Open circles and full dots denote (scaled) experimental values for carbon and aluminium, respectively.

$$\frac{dW}{dR} = \frac{3\hbar v_I}{k_F r_s^3} \sum_{l=0}^{\infty} (l+1) \sin^2[\delta_l(\epsilon_F) - \delta_{l+1}(\epsilon_F)], \tag{6}$$

where v_I is the velocity of the ion, and k_F is the Fermi wave vector. The theoretical curve in Fig. 3 corresponds to the phase shifts for $r_s = 2$. The experimental values are for carbon (open circles) and aluminium (full dots) as the target materials. The theoretical and experimental results show the same oscillations when the atomic number of the moving ion increases. These oscillations reflect the filling of the different atomic shells.

The immersion energy $\Delta E^{hom}(\bar{n})$ of an atom to the electron gas of density $\bar{n}$ is defined as

$$\Delta E^{hom}(\bar{n}) = E^{hom}(\bar{n}) - E^{atom}, \tag{7}$$

where E^{atom} is the total energy for a free atom calculated in the same LDA as $E^{hom}(\bar{n})$. Fig. 4 shows some examples of the immersion energies as a function of the electron gas

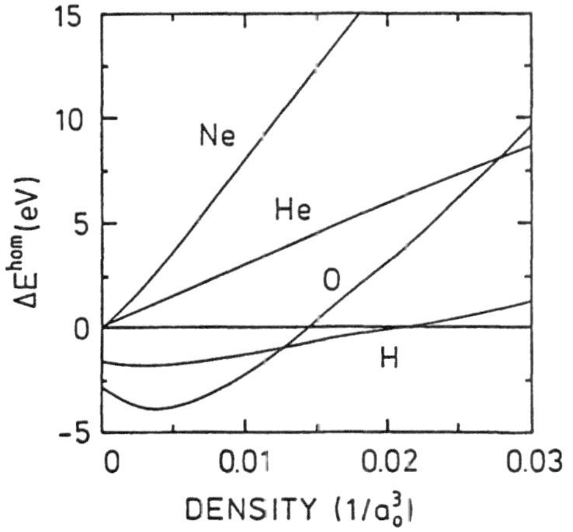

Figure 4. Immersion energy ΔE^{hom} as a function of electron gas density for H, He, O, and Ne.

TABLE I. Fourth order polynomial fits for the $\Delta E^{hom}(\bar{n})$-functions. The coefficients of the polynomial $C_0 + C_1\bar{n} + C_2\bar{n}^2 + C_3\bar{n}^3 + C_4\bar{n}^4$, in which $\bar{n}$ is in a_o^{-3} and the resulting $\Delta E^{hom}(\bar{n})$ in eV, are given. In the case of atoms H through K the fits are based on the $\Delta E^{hom}(n)$-data with $r_s = 2...6$ and for the other atoms the range is $r_s = 2...4$. At a given density the errors due to fitting are typically of the order of 0.05 eV.

Atom	C_0	C_1	C_2	C_3	C_4
H	-1.6818	-87.577	1.8681×10^4	-6.9150×10^5	9.1509×10^6
He	0	317.94	-1720.9	3.9798×10^4	-6.0245×10^5
Li	-0.79428	249.82	2.6050×10^4	-1.4086×10^6	2.3878×10^7
Be	-0.38583	129.35	3.3575×10^4	-1.4188×10^6	2.1363×10^7
B	-1.0594	-186.17	5.5972×10^4	-2.0429×10^6	2.7581×10^7
C	-1.1461	-391.75	6.5598×10^4	-2.2075×10^6	2.8437×10^7
N	-0.33060	-548.39	7.7180×10^4	-2.5634×10^6	3.2838×10^7
O	-2.8274	-584.25	9.1474×10^4	-3.2424×10^6	4.3332×10^7
F	-4.9285	-37.088	6.6923×10^4	-2.7663×10^6	4.2167×10^7
Ne	0	602.36	3.1600×10^4	-1.4976×10^6	2.4672×10^7
Na	-0.95286	-871.49	2.5766×10^4	-1.2667×10^6	2.0369×10^7
Mg	-0.53091	971.99	2.3818×10^4	-8.8033×10^5	1.0465×10^7
Al	-1.4206	733.32	5.3373×10^4	-1.9678×10^6	2.4637×10^7
Si	-4.5171	1285.9	4405.8	-1.3260×10^5	8.0513×10^5
P	-2.0335	315.09	1.1235×10^5	-4.4773×10^6	6.2006×10^7
S	-4.0987	563.55	1.0304×10^5	-4.2344×10^6	5.9608×10^7
Cl	-4.6631	1077.0	8.8932×10^4	-4.6874×10^6	7.8046×10^7
Ar	0	1718.5	4.0282×10^4	-2.7028×10^6	4.8247×10^7
K	-0.85952	2007.7	2.0963×10^4	-1.8902×10^6	3.6706×10^7
Ca	-0.70742	2033.2	-1400.0	-4.6175×10^5	1.2788×10^7
Sc	-0.72820	1582.4	1.9456×10^4	-9.6608×10^5	1.5841×10^7
Ti	-0.47598	1422.5	1.0054×10^4	-2.2485×10^5	1.6372×10^6
V	0.97604	1273.4	8894.8	-242.70	-2.8503×10^6
Cr	3.1760	1032.5	2.4709×10^4	-6.8899×10^5	8.0295×10^6
Mn	2.6230	966.93	2.4802×10^4	-6.5923×10^5	7.6476×10^6
Fe	0.33384	932.32	2.2879×10^4	-5.1903×10^5	5.3503×10^6
Co	-1.0239	899.41	2.4678×10^4	-5.4992×10^5	5.5764×10^6
Ni	-1.9744	944.85	2.6567×10^4	-6.1546×10^5	6.2613×10^6
Cu	-2.5572	1431.0	-1.5150×10^4	1.4424×10^6	-2.8217×10^7
Pd	-1.9814	1826.7	3.8287×10^4	-8.6813×10^5	8.7007×10^6
Ag	-1.8160	2112.9	3.2850×10^4	-8.1826×10^5	9.0745×10^6
Pt	-2.3439	2590.2	3.7881×10^4	-6.0090×10^5	6.8512×10^5
Au	-2.0063	2879.9	3.3559×10^4	-5.3844×10^5	7.3195×10^5

density. In the case of inert gas atoms $\Delta E^{hom}(\bar{n})$ is always positive and rises rather linearly from zero as the electron density increases. This means that the interaction of the inert gas atom with the electron gas is always repulsive. The linear rise of $\Delta E^{hom}(\bar{n})$ is due to the increase of kinetic energy when more and more electron gas states have to be orthogonal against the bound states induced by the atom. For other atoms than the inert gases the immersion energy curves exhibit a clear minimum. This means that these reactive atoms have a tendency to dissolve into the electron gas, if its density is small enough. The $\Delta E^{hom}(\bar{n})$ curves for these atoms end at $\bar{n} = 0$ to the electron affinity because the zero-density limit is a negative ion. For the practical purposes, we have fitted fourth order polynomials to the calculated $\Delta E^{hom}(\bar{n})$ -data. The coefficients are given in Table I. The fits give faithfully the original data and they can be used for interpolations, but it is important that they are not used for extrapolation far outside the r_s-ranges given in the caption of Table I (especially, when the highest order coefficient C_4 is negative, the values calculated from the expansion diverge at very high densities towards negative infinity).

4. Parameters for the effective medium theory

Parameters needed in the applications of the effective medium theory can be calculated from the $\Delta E^{hom}(\bar{n})$-functions described above and from the induced electron densities $\Delta n(r)$ [6]. The basic ansatz made in EMT is that the electron density of a system of interacting atoms can be approximated by superimposing densities $\Delta n(r)$ induced by the atoms in homogeneous electron gas. Thereafter the total energy of the system can be calculated approximatively by the help of the $\Delta E^{hom}(\bar{n})$-functions.

The central EMT-parameter describing the induced density $\Delta n(r)$ is the neutral sphere radius s defined so that the total electron density $n(r)$ integrated from origin to s neutralizes the nuclear charge Z. The energy function used in EMT is not $\Delta E^{hom}(\bar{n})$ but the so-called cohesive function $E_c(\bar{n})$, which is defined as

$$E_c(\bar{n}) = E^{hom}(\bar{n}) - \alpha(\bar{n})\bar{n}. \tag{8}$$

Above, the last term cancels the Coulomb interaction of the positive background charge with the electron density, and the function α is defined as

$$\alpha(\bar{n}) = \int_0^{s(\bar{n})} dr 4\pi r^2 \phi(r). \tag{9}$$

The cohesive function $E_c(\bar{n})$ is shown in Fig. 5 for some simple metals. The minima of these curves occur at a certain density n_o. According to EMT [6], the corresponding value $E_c(n_o)$ is the cohesive energy of the simple metal. Furthermore, n_o gives the neutral sphere radius s_o which is recognized as the Wigner-Seitz radius. The second derivative of the $E_c(\bar{n})$ curve is directly proportional to the bulk modulus of the metal. It is useful for applications to make a third order polynomial fit to the $E_c(\bar{n})$-functions around the minima as

$$E_c(\bar{n}) = E_o + E_2(\bar{n}/n_o - 1)^2 + E_3(\bar{n}/n_o - 1)^3. \tag{10}$$

This introduces three new parameters E_o, E_2, and E_3, which have the dimension of energy.

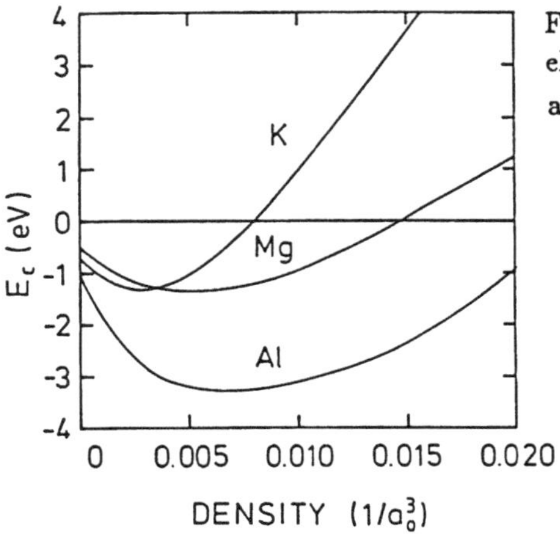

Figure 5. Cohesive function E_c as a function of electron gas density for simple metals Mg, Al, and K.

Parameters n_o, s_o, E_o, E_2, and E_3 suffice for describing a *perfect* simple metal. Normally one is interested in cases with a lower symmetry. Then one needs three more parameters in order to describe the Coulomb interaction when the neutral spheres overlap strongly with each other [6]. The total energy of the system is then given as

$$E_{tot} = \sum_i E_c(\bar{n}_i) + E_{AS}, \tag{11}$$

where $\bar{n}_i$ is the average electron density at the site of atom i due to the surrounding atoms. E_{AS} is the so-called atomic spheres correction, which can be calculated from the approximative expression [6]

$$\begin{aligned} E_{AS} &= \alpha \sum_i \{ [\frac{1}{12} \sum_{j \neq i} \exp[-\eta_2 (r_{ij} - \beta s_o)]]^{\eta/(\eta+\eta_1)} \\ &- \frac{1}{12} \sum_{j \neq i} \exp[-\eta (r_{ij}/\beta - s_o)]\}. \end{aligned} \tag{12}$$

The new parameters are: $\alpha = \alpha(n_o)$ (Eq. (9)), η, and η_1. η gives the dependence of the neutral sphere radius s on the background density $\bar{n}$ in an exponential fit

$$\bar{n}(s) = n_o e^{-\eta(s-s_o)}. \tag{13}$$

η_1 describes the decay of the electron density used in the superimposing ansatz. In practice, it cannot accurately be calculated from first principles, but is used as an adjustable parameter to fit the experimental elastic constant c_{44} [12]. In Eq. (12) β is a geometrical factor: $\beta = (16\pi/3)^{(1/3)} 2^{-(1/2)}$ and $\eta_2 = (\eta + \eta_1)/\beta$.

The EMT-parameters calculated for several elements are given Table II. In the case of the rare gas atoms the $E_c(\bar{n})$-function rises monotonically, so that the parameters n_o and s_o cannot be defined. However, the background density still depends exponentially on the neutral sphere radius (Eq. (13)). The parameters needed are given in Table III for He, Ne, and Ar. Moreover, in the case of rare gase atoms $\alpha(\bar{n})$ can also be calculated and it approaches at the important region of low electron densities the value

TABLE II. Parameters needed in the effective medium theory. For the definitions, see text.

Atom	s_o (a_o)	n_o (a_o^{-3})	η (a_o^{-1})	E_0 (eV)	E_2 (eV)	E_3 (eV)	α (eVa_o^3)
H	1.716	0.0074	5.013	-2.149	0.336	-0.175	82.9
Li	2.964	0.0050	1.465	-1.691	0.538	-0.281	528.
Be	2.492	0.0099	1.943	-2.290	1.045	-0.350	533.
B	2.216	0.0139	2.440	-4.500	1.618	-0.862	516.
C	2.029	0.0164	3.058	-5.788	2.372	-1.213	494.
N	1.921	0.0164	4.052	-5.423	2.735	-1.600	478.
O	1.881	0.0138	5.416	-6.978	1.867	-0.875	472.
F	1.973	0.0070	9.006	-6.058	0.665	-0.183	473.
Na	3.430	0.0034	1.553	-1.388	0.646	-0.533	1068.
Mg	3.233	0.0052	1.729	-1.378	0.707	-0.289	1236.
Al	3.013	0.0069	2.001	-3.252	1.087	-0.449	1293.
Si	2.852	0.0078	2.411	-4.554	1.452	-0.955	1313.
P	2.749	0.0076	2.977	-4.826	1.624	-0.775	1316.
S	2.741	0.0057	4.227	-5.765	1.068	-1.334	1320.
Cl	2.869	0.0029	7.505	-4.828	0.388	-0.167	1352.
K	4.096	0.0025	1.311	-1.338	0.490	-0.177	2236.
Ca	3.776	0.0042	1.305	-1.823	0.859	-0.191	2285.
Sc	3.269	0.0074	1.450	-3.192	1.661	-0.856	2015.
Ti	2.904	0.0109	1.629	-3.629	2.164	-0.617	1796.
V	2.715	0.0134	1.756	-2.832	2.739	-0.918	1671.
Cr	2.561	0.0157	1.909	-1.411	2.940	-1.031	1572.
Mn	2.487	0.0168	2.060	-2.048	2.878	-1.032	1511.
Fe	2.450	0.0163	2.173	-4.207	2.893	-1.035	1468.
Co	2.386	0.0174	2.376	-5.149	2.706	-0.912	1428.
Ni	2.437	0.0150	2.524	-5.123	2.159	-0.577	1435.
Cu	2.586	0.0113	2.487	-3.562	1.287	-0.295	1495.
Pd	2.698	0.0123	2.791	-4.672	2.198	-0.587	2400.
Ag	2.901	0.0086	2.489	-3.160	1.083	-0.237	2487.
Pt	2.808	0.0116	2.777	-4.915	2.183	-0.604	3172.
Au	3.002	0.0083	2.530	-3.405	1.267	-0.289	3290.

α_{at} for free atoms. The α_{at} -values and the third order expansion coefficients of the $E_c(\bar{n})$-functions are given also in Table III.

In order to describe other systems than simple metals one needs a method to approximate the so-called one-electron energy term. For example, in the case of transition metals this term takes into account the effects due to the narrow d-bands, which are not included via the $E_c(\bar{n})$ -functions based on the interaction of the transition metal atom with the 'sp-like' homogeneous electron gas [6]. Also in the case of semiconductors and insulators the effects due to the band gap should be treated separately. Some models exist for the one-electron energy term, and some of the parameters needed can

TABLE III. Parameters needed in the effective medium theory for the rare gases He, Ne, and Ar. The neutral sphere radius s and the background density $\bar{n}$ are related as $\bar{n}(s) = n_o \exp(-\eta s)$. The cohesive function E_c is given by the expansion $E_c(\bar{n}) = E_1\bar{n} + E_2\bar{n}^2 + E_3\bar{n}^3$. α_{at} is the integral (Eq. (9)) over the Coulombic potential of a free atom.

Atom	n_o (a_o^{-3})	η (a_o^{-1})	E_1 (eVa_o^3)	E_2 (eVa_o^6)	E_3 (eVa_o^9)	α_{at} (eVa_o^3)
He	2.397	2.876	98.893	1412.4	-22747.	146.7
Ne	24.17	3.436	33.912	9153.2	-1.2787×10^5	561.6
Ar	7.674	2.415	44.633	20375.	-3.3231×10^5	1499.

be calculated from the solutions for atoms in electron gas [6,16]. The models are not very general and we will not try to give here any parameters for them.

Also in the case of systems which contain different types of atoms one cannot at the present stage of the theory give a general description. These kinds of systems are for example alloys and atoms chemisorbed on solid surfaces [12]. However, some trends can be seen directly from the $E_c(\bar{n})$ -functions. For example, the minimum of the $E_c(\bar{n})$ -curve (E_o) for a given gas atom gives the main contribution to its chemisorption energy on a metal surface. It is interesting to note that the experimental chemisorption energies for hydrogen are rather close to the EMT E_o -parameter but those for oxygen are much smaller than the EMT estimation [17]. The discrepancy is mainly due to LDA [18]. We have calculated the EMT parameters for oxygen also by using the Self-Interaction-Correction formalism (SIC) [19]. It raises the $E_c(\bar{n})$ -curve upwards about 2 eV [20]. The other EMT-parameters remain in practice unchanged. For comparison, in the case of Al the effect of using SIC instead of LDA is of the order of 0.1 eV.

5. Conclusions

In this article we first described the solution and the general features of the electron structure for an atom embedded in homogeneous electron gas. The model can in many cases mimic the essential phenomena when an atom is interacting with a real electronic environment. The model forms the basis for the Effective Medium Theory. We believe the tabulations of the EMT-parameters given in this article will turn out to be very useful in the future applications of EMT.

References

[1] M. J. Puska, R. M. Nieminen, and M. Manninen, Phys. Rev. B **24**, 3037 (1981)

[2] M. J. Stott and E. Zaremba, Can. J. Phys. **60**, 1145 (1982)

[3] M. J. Stott and E. Zaremba, Phys. Rev. B **22**, 1564 (1980)

[4] J. K. Nørskov and N. D. Lang, Phys. Rev. B **21**, 2131 (1980)

[5] P. Nordlander, S. Holloway, and J. K. Nørskov, Surf. Sci. **136**, 59 (1984)

[6] K. W. Jacobsen, J. K. Nørskov, and M. J. Puska, Phys. Rev. B **35**, 7423 (1987)

[7] C. O. Almbladh and U. von Barth, Phys. Rev. B **13**, 3307 (1976)

[8] G. W. Bryant and G. D. Mahan, Phys. Rev. B **17**, 1744 (1978)

[9] R. M. Nieminen and M. J. Puska, Phys. Rev. B **25**, 67 (1982)

[10] M. Persson and B. Hellsing, Phys. Rev. Lett. **49**, 662 (1982)

[11] M. J. Puska and R. M. Nieminen, Phys. Rev. B **27**, 6121 (1983)

[12] K. W. Jacobsen, in these proceedings and in Comm. Cond. Matter Phys. **14**, 129 (1988)

[13] The results given in the tables of this article are calculated by using the local exchange-correlation functional by D. M. Ceperley and B. J. Alder, Phys. Rev. Lett. **45**, 566 (1980), as parametrized by J. P. Perdew and A. Zunger, Phys. Rev. B **23**, 5048 (1981)

[14] The most recent computer code, which uses analytic extensions of the energy integrals and corrects the errors in the phase shifts due to the finite integration radius, is made by G. Eliasson.

[15] M. Manninen, R. Nieminen, P. Hautojärvi, and J. Arponen, Phys. Rev. B **12**, 4012 (1975)

[16] S. Myllyrinne, M. Manninen, and J. K. Nørskov, in these proceedings.

[17] U. Yxklinten, M. Limbäck, and B. I. Lundqvist, in these proceedings.

[18] R. O. Jones, J. Chem. Phys. **82**, 325 (1985)

[19] J. P. Perdew and A. Zunger, Phys. Rev. B **23**, 5048 (1981)

[20] M. J. Puska, to be published

The Structural and Electronic Properties of Molten Metals, Semimetals, and Semiconductors

J. Hafner

Cavendish Laboratory, University of Cambridge, Madingley Road, Cambridge, CB3 0HE, UK and
Institut für Theoretische Physik, Technische Universität Wien, Wiedner Hauptstr. 8/10, A-1040 Wien, Austria

Abstract. The structural and electronic properties of molten metals, semimetals and semiconductors (including liquid III-V and II-VII compounds) are studied using a combination of nearly-free-electron perturbation theory and ab-initio methods. The atomic structure is calculated using molecular dynamics simulations based on effective interatomic forces derived from second-order pseudopotential theory. The electronic structure is calculated self-consistently using linear-muffin-tin-orbital (LMTO) supercell calculations. It is found that the trend from close-packed metallic structures in groups I to III to open covalent structures in groups IV to VI of the Periodic Table is well understood in terms of the interplay of volume and pair forces. The investigation of the molten III-V and II-VI compounds shows that the local atomic arrangement in the melt is dominated by the strong anion-anion potential and a trend to preferred heterocoordination in the liquid II-VI alloys. The relevance of these results for the construction of empirical interatomic forces is discussed.

1.Introduction

Over the last years a lot of effort has gone into the construction of interatomic potentials for semimetallic and semiconducting elements. The procedure is usually the following: a plausible analytic form is assumed for the two- and three-body interactions and the open parameters of this ansatz are determined by fitting to a large data base, involving both experimental data and data from ab-initio total energy calculations [1-4]. The main aim of the exercise is to create the basis for computer simulation studies of the properties of disordered materials, e.g. of the atomic structure of the molten and amorphous phases or the atomic arrangement at interfaces or grain boundaries [5-8]. It is probably not unfair to say that these attempts have not been entirely successful, for more than one reason. One reason is that the data base used to determine the parameters of the interatomic potentials is never really complete, inclusion of new data alters the potentials sometimes in a quite significant way. Hence the range of problems that can be dealt with a given set of potentials is determined by the data base originally used to fit the parameters. A more serious reason for the relative failure of these attempts to decompose the total energy of a solid or liquid into a set of classical two- and three-body forces is that they deliberately ignore some important lessons from modern condensed matter theory. Both

Springer Proceedings in Physics, Vol. 48 Many-Atom Interactions in Solids
Editors: R.M. Nieminen · M.J. Puska · M.J. Manninen

tight-binding (TB) and nearly-free-electron (NFE) theories show that an expansion of the total energy into a series of two- and many-body forces will be poorly convergent [9]. Even in its simplest form a TB-calculation leads to a square-root dependence of the total energy on the coordination number. NFE-perturbation theory shows that the binding energy is given by a large volume-contribution and much smaller contributions expressible in terms of two-, three- and many-body potentials. These interatomic potentials are state-dependent via the variation of the dielectric response function with electron density and electronic structure. Thus electron theory predicts the interatomic interaction in say Si to be different in the high-density metallic phase and in the low-density semiconducting phase. The TB and NFE pictures are consistent which each other and lead to a description of the physical mechansim determining the atomic structures of solids and liquids that is quite different from one based on classical pair- and many-body forces.

In a system where the atoms interact by classical pair potentials only the most stable structure is determined by packing requirements leading to a nearest-neighbour distance close to the main minimum in the pair potential. Except for very special forms of the pair potential, the most stable crystalline phase will be cubic- or hexagonal-close-packed with coordination number N_C=12. The crystal melts into a hard-sphere-like liquid with N_C~10-11. Open structures with low coordination numbers can be stabilized only by strong three-body interactions. The three-body forces put a strong penalty on the formation of close-packed atomic arrangements, usually by assuming an angular-dependence with a sharp minimum at the characteristic bond-angle of the stable phase [1-4]. This energetically favourable bond angle can be determined only if the stable configuration is known.

The picture following from electron theory is a rather different one. The nearest-neighbour distance is determined to lowest order by the volume- and pair forces. If this distance falls close to a minimum in the pair potential, a close-packed structure will be stable. If it falls on a repulsive hump of the pair interaction a distorted structure with a lower coordination number will be energetically more favourable by putting the atoms into the attractive minima of the potential. This allows us to understand the occurrence of open structures without the introduction of three-body forces [10-12].

In my paper I shall show that the trend in the liquid structures of the s,p-bonded elements from groups I to VI of the Periodic Table (i.e. from hard- or soft-sphere like melts such as liquid Al or Na to highly viscous melts such as Te where a polymeric picture is sought to be adequate) is quite well understood in terms of the interplay of volume- and pair forces alone. The study deals mainly with the liquid elements because, although the trends in the crystalline and in the liquid structures is largely parallel [11], the melt is usually more metallic [13] and a NFE-perturbation approach to the interatomic bonding forces is more appropriate. A further point is that for crystals any investigation of phase stability is restricted to a comparison of the total energy of a relatively small number of structural alternatives and it is impossible to make sure that this includes the absolute minimum in the total energy. On the other hand a computer simulation of the liquid explores the entire configuration space and leads unambiguously to the structure with the lowest free energy. In the last section the investigation is extended to molten III-V and II-VI alloys.

2. Interatomic Forces and the Atomic and Electronic Structures of Liquid Metals, Semimetals and Semiconductors

2.1 Trends in the Interatomic Forces

To second order in the electron-ion potential the total energy may be written as [10-12]

$$E_{tot} = E_o(V) + N^{-1} \sum_{i<j} \Phi(R_{ij};V) \tag{1}$$

where $E_o(V)$ is a volume energy and the second term is given by a sum over volume-dependent interatomic forces. The volume energy is given by

$$E_o(V) = Z\,(E_{eg} + 1/2\, V\, B_{eg}) + 1/2\ \Phi_{ind}(R=0;V) \tag{2}$$

where Z is the valence, V the atomic volume , and E_{eg} and B_{eg} are the total energy and bulk modulus of the electron gas. The third term in (2) represents the self-energy of the pseudoatom, i.e. the non-Coulomb part of the interaction of an ion core with its own screening electron cloud. The effective pair potential $\Phi(R)$ is given by (we use atomic units)

$$\Phi(R;V) = 2\,Z^2/R + \iint d\mathbf{r}d\mathbf{r}'\ w(\mathbf{r}-\mathbf{R})\ \Gamma(\mathbf{r}-\mathbf{r}';V)\ w(\mathbf{r}'-\mathbf{R}') \tag{3}$$

where $\Gamma(\mathbf{r};V)$ is the density response function of the interacting electron gas. For a local pseudopotential Fourier transformation of (3) leads to

$$\Phi(q;V) = 8\pi Z^2/Vq^2 + V\, w(q)^2\, \chi(q)\,/\,\varepsilon(q) \tag{4}$$

where χ and ε are the susceptibility and the dielectric function of the electron gas. The reciprocal space formulation is interesting because one sees that using the generalized Poisson equation for the screening electron density one can decompose the effective pair interaction according to [14] ($\tilde{w}=w/\varepsilon$ is the screened pseudopotential)

$$\begin{aligned}\Phi(q) &= 8\pi Z^2/Vq^2 - V\,(1-G(q))\,8\pi\,\rho_{sc}(q)^2\,/\,q^2 + V\,\chi(q)\,\tilde{w}(q)^2 \\ &= \Phi_{no}(q) + \Phi_{1e}(q)\end{aligned} \tag{5}$$

into a pairwise interaction Φ_{no} between neutral pseudoatoms (consisting of the ionic cores and their screening charges ρ_{sc} of conduction electrons, G(q) is the local-field factor for exchange and correlation corrections), and a one-electron term Φ_{1e} representing just the sum over the one-electron energies calculated to second order in the pseudopotential.

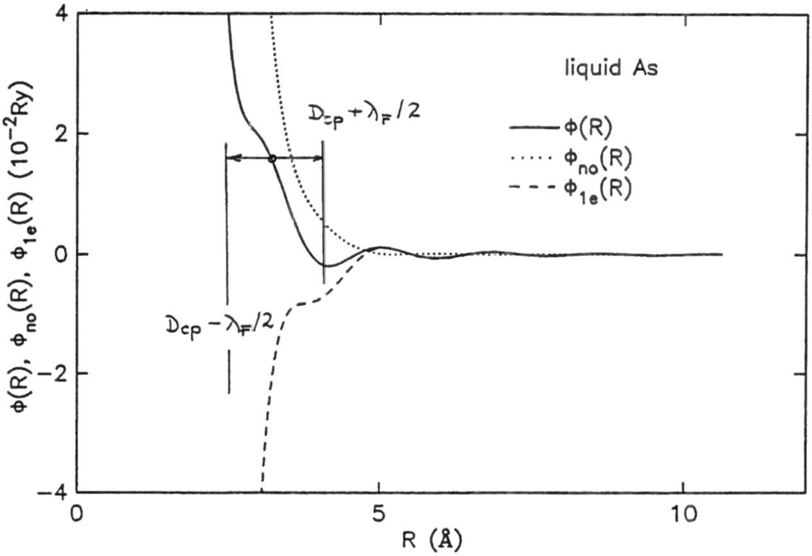

Figure 1. Interatomic pair potential $\Phi(R)$ for liquid As, decomposed into a neutral pseudo-atom interaction $\Phi_{no}(R)$ and a one-electron (or band-structure) term $\Phi_{1e}(R)$. The re-arrangement of the nearest-neighbour shell induced by the electronic bonding effects is indicated. See text.

The characteristic form of the pair interactions is well known. At short interatomic distances, $\Phi(R)$ is essentially of a screened Coulomb form, this part of the pair interaction is dominated by the pseudo-atom term Φ_{no} (Fig.1). If the electron-ion interaction is modelled by a simple empty-core pseudopotential, the diameter of the repulsive core is just given by [10,12]

$$\begin{aligned} D_{rep} &= 2\,R_c + 2\,\lambda_{TF} \\ &= 2\,R_c + 2\text{x}\ 0.456\,R_s^{1/2} \end{aligned} \tag{6}$$

where λ_{TF} is the Thomas-Fermi screening length. Note that λ_{TF} is proportional to the square-root of the electron-density parameter R_s (R_s is the diameter of a sphere containing just one conduction electron). The form of the pair potential at intermediate and large distances is determined by the Friedel oscillations arising from the singularity in the susceptibility at $q=2\,k_F$. The wavelength of the oscillations is given by the Friedel wavelength

$$\lambda_F = 2\pi/2\,k_F = 1.637\,R_s. \tag{7}$$

The amplitude of the oscillations is set by the on-Fermi-sphere matrix element $w(2k_F)$ of the pseudopotential. Note that the oscillations appear only in the one-electron part of the pair interaction (Fig.1).

It was shown by Hafner and Heine [12] that the trend from close-packed metallic to open covalent structures is determined by the variation of the interatomic potentials with electron density, and that it is well described even by the simplest pseudopotential. The core radius R_c is chosen to fit the known electronic properties of the element, the atomic volume is taken from experiment (this is in some way equivalent to fitting the non-locality of the potential by a second independent parameter). At increasing electron density (decreasing R_s) the repulsive core of $\Phi(R)$ moves over the first attractive minimum because λ_F (equ. 7) changes more rapidly than D_{rep} (equ. 6). At high electron density the interatomic distance D_{cp} at close-packing falls on a repulsive hump of the potential, hence a close-packed structure is unstable. The variation of the pair potentials with density explains the structural phase transitions under pressure [9,12]. For details of the argument, see refs. [11,12]. For the crystalline structures the simple pair potential argument explains only the static stability of say the diamond structure of Ge relative to a fcc lattice. In the open structures there are always some transverse acoustic vibrational modes which cannot be stabilized by pair forces alone. In the liquid these modes are overdamped and we expect the volume- and pair forces to give a complete description of the structural properties of the melt.

2.2 Molecular Dynamics Simulations of the Liquid Structure

We have carried out systematic molecular dynamics studies of all the molten s,p-bonded elements form group I to group VI (for details see refs. [15-20]). Figs. 2(a,b) show the variation of the pair potentials and the pair correlation functions in the series Al - Ge - As - Te. For Al the close packing distance D_{cp} falls into a minimum in $\Phi(R)$, and the Friedel wavelength is compatible with the requirements of sphere packing, hence a hard-sphere-like structure will be stable. For Ge, As, and Te D_{cp} falls on a repulsive hump in $\Phi(R)$. In this case it is energetically favourable to split the first shell into two subshells centred at

$$D_{1,2} = D_{rep} \pm \lambda_F/2 \ , \qquad (8)$$

the positions of the second and more distant coordination shells are determined by the positions of the minima in $\Phi(R)$. The coordination numbers are determined by packing requirements: the smaller D_1 is compared to D_{cp}, the smaller is the nearest neighbour coordination number N_c. This explains the oberserved trend in N_c: Al - N_c=10.9 (MD), N_c=10.5 (exp.); Ge - N_c=7.6 (MD), N_c=6.5 (exp.); As - N_c=3.0 (MD), N_c=3.0 (exp.); Te - N_c=2.63 (MD), N_c=2.56 (exp.). A similarly good agreement is found for the bond distances [10-15]. The ratio of the first- and second neighbour distances determines the bond angles (Fig.3). In liquid Al we find two pronounced peaks at $\theta=60^o$ and $\theta=115^o$. This shows that the local order in liquid Al is not too different from the cubo-cotahedral arrangement in fcc Al or in a dense distorted polytetrahedral packing. In l-Ge the bond angle distribution is very diffuse, the bond angles of all the crystalline polymorphs (diamond structure (α-Ge), white-tin str. (β-Ge), simple hexagonal (sh) structure) appear in the local configurations in the melt. In l-As we find a main peak at the bond-angle of crystalline rhombohedral As ($\theta=97.2^o$) and a smaller peak close to $\theta=60^o$. This shows

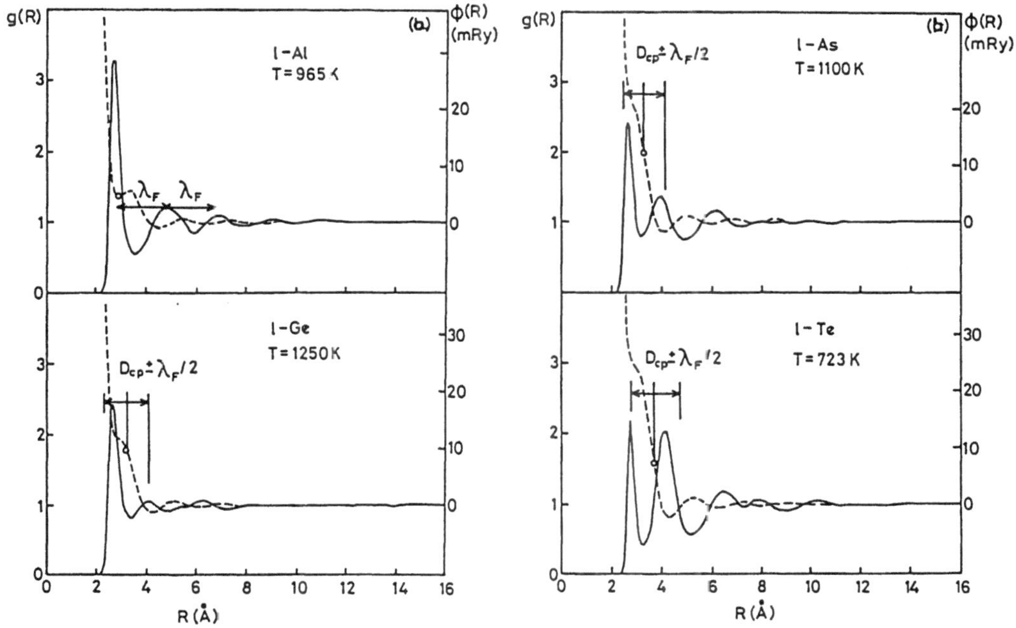

Figure 2. Interatomic pair potentials Φ(R) and pair correlation function g(R) determined by a molecular dynamics simulation for Al, Ge (a) and As, Te (b).

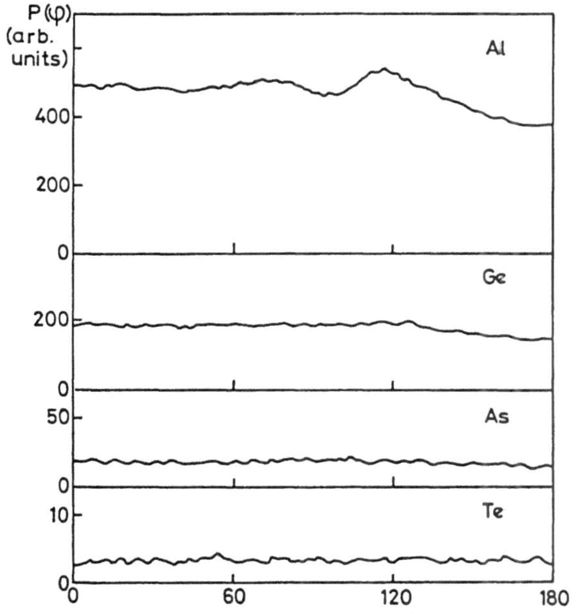

Figure 3. Bond angle distribution functions for liquid Al, Ge, As and Te. The vertical bars mark the bond angles in the crystalline phases (see text).

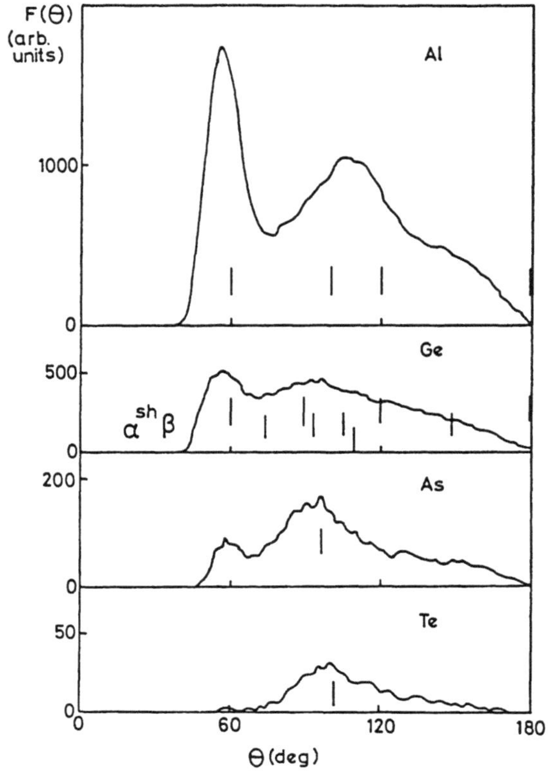

Figure 4. Distribution of the dihedral angles in liquid Al, Ge, As and Te.

that although around most atomic sites the local order is close to that in the crystal, a certain number of As-atoms forms isocele triplets or distorted tetrahedra. Note that gaseous As is a molecular gas consisting of tetrahedral As_4 molecules. With increasing temperature one finds a gradual transition for a crystal-like to a gas-like local order. In l-Te the bond angle distribution shows a single peak close to the bond angle in trigonal Te ($\theta=102^o$). Note that these preferred bond angles result from the modulation of the random packing of the atoms by the Friedel oscillations in the potential. Finally Fig.4 shows that the distribution of the dihedral angles is nearly random in all four elements.

The most convincing argument however is certainly a rapid glimpse into the three-dimensional arrangement of the atoms. Fig.5 shows an instantaneous atomic configuration. A slice two nearest-neighbour bonds thick has been cut out of a large MD-model and projected on the xy-plane. The atoms and the nearest-neighbour bonds have been drawn, the size of the symbols representing the atoms has been scaled with the z-coordinate. L-Al forms a dense, distorted polytetrahedral packing. In l-Ge we find tetrahedral arrangement of atoms like in α-Ge, but also more dense hard-sphere-like local arrangements. In l-As the most frequent local unit is a flattened trigonal prims like in the crystal, but the the connectivity of these units is random. Liquid Te finally consists of broken, entangled chains.

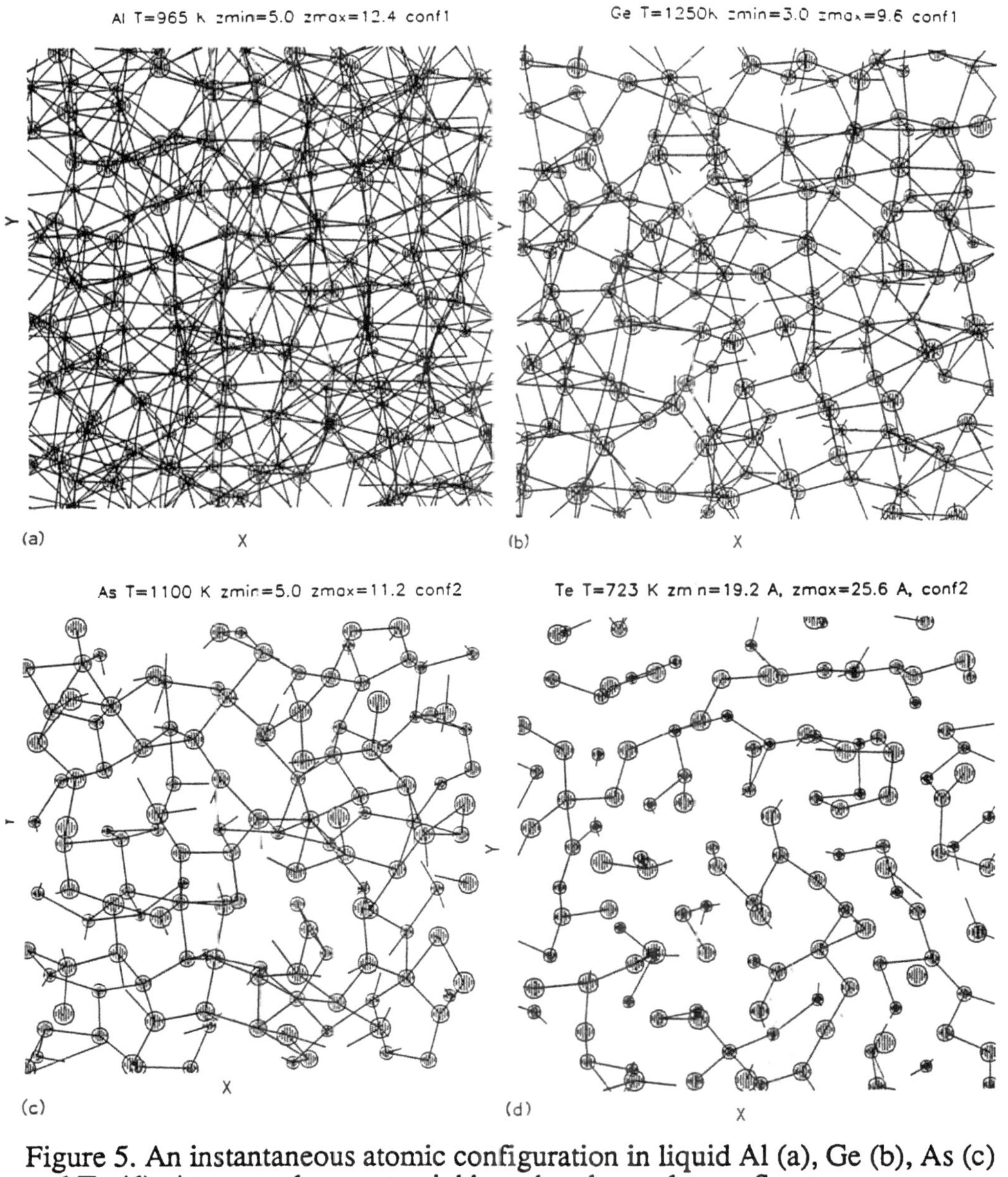

Figure 5. An instantaneous atomic configuration in liquid Al (a), Ge (b), As (c) and Te (d). Atoms and nearest neighbour bonds are drawn. See text.

2.3 Electronic Structure of Liquid Metals

Parallel to the simulation of large clusters runs for small 64-atom cells have been used to create atomic configurations suitable for self-consistent electronic structure calculations using a linear-muffin-tin-orbital supercell method [19-21]. Only liquid Al and Si show the nearly-free-electron-like electronic density of states usually expected to hold for all liquid simple metals (Fig.6). In the elements with higher valence and/or higher atomic number a pseudogap appears in the middle of the valence band, separating s- and p-like subbands. The theoretical results are fully confirmed by photoemission experiments [19,20]. The opening of a pseudogap is clearly related to the decrease in the coordination number. At low N_C the overlap between atomic orbitals centred at neighbouring sites is not strong enough to promote the strong s,p-hybridization necessary for the formation of a free-electron band. In the heavier elements the separation of s- and p-states is further enhanced by relativistic effects [20], in Ge the partial penetration of the 4s-electrons into the 3d-core creates an additional attractive potential lowering the s-states relative to the p-states [19].

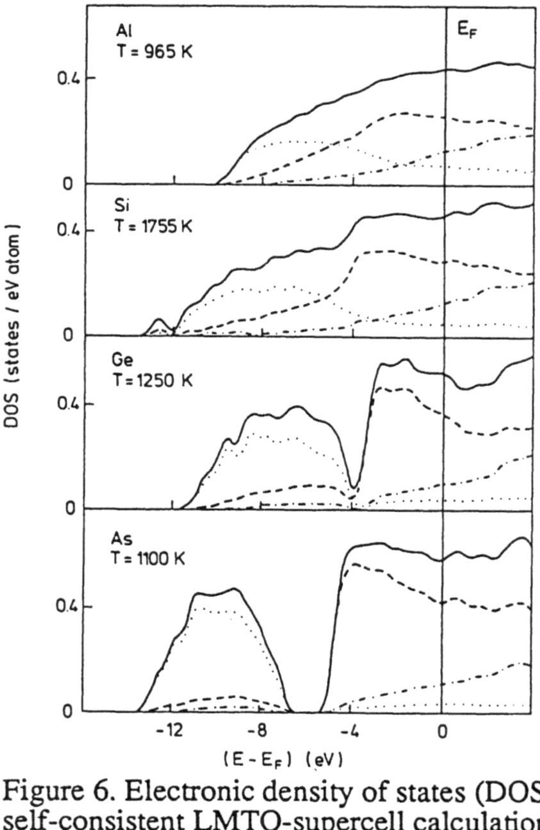

Figure 6. Electronic density of states (DOS) in liquid Al, Si, Ge and As from a self-consistent LMTO-supercell calculation. Full line - total DOS, dotted line - s-partial, dashed line - p-partial, and dot-dashed line d-partial DOS.

2.4 Friedel oscillations and Peierls Distortion

We have shown that the local order in the liquid structures is similar to that in the crystalline structures and that the liquid structures are stabilized by the interplay of volume- and pair forces. Hence we should be able to relate this argument to the conventional interpretation of the crystal structures of the elements from groups IV to VI. The electronic mechanism stabilizing the structures of the lighter group V and VI elements is usually described as a Peierls distortion, within either a tight-binding or a pseudopotential formalism [22]. In the group V (VI) elements there are three (four) electrons and hence the one-dimensional p-bands of a simple cubic lattice (which are assumed to be non-interacting) are 1/2 (2/3) filled. Thus a Peierls distortion will lead to a lattice periodicity of 3a (2a) where a is the lattice parameter of the sc lattice and to a decrease of the coordination number from six to three (two). Recently Gaspard et al [23] have tried to generalize this argument to disordered materials. The problem is that if the Peierls argument correctly defines the conditions for a lattice instability to occur, it does not predict the actual magnitude of the distortion. This requires further assumptions on the nature and strength of the repulsive interaction which held the atoms apart.

Note that the argument we have used here is precisely a generalized real-space formulation of the Peierls argument, although it does not refer to Brillouin zones and electron bands. However, if we formulate the argument in real space, it is more appropriate to chose a close-packed structure as a starting point (it is well known that the As- and Te- structures can equally well be described as arising from distortion of close-packed layers). Equ. (8) tells us that if the interatomic distance falls on a repulsive maximum in the pair potential, the structure will distort and lower the symmetry of the local atomic arrangement. The splitting of the pair potential into a neutral pseudo-atom-part and a band-structure term according to (5) is helpful to stress the analogy with the original Peierls argument. Fig.1 shows that the one-electron band-structure term drives the distortion of the liquid structure, while the repulsive forces are compatible with a close-packed structure. The real-space argument is even more general as it allows for a calculation of the actual magnitude of the distortion and applies also to the group IV-elements. We will elaborate this argument in more detail in a forthcoming publication.

3. Structure and Properties of Molten III-V and II-VI Compounds

The same methods can also be applied to the molten semiconductor compounds. Here the problem is to understand why III-V compounds become metallic on melting (albeit with a smaller increase in the coordination number as the semiconducting elements), while the II-VI compounds remain semiconducting and show no increase in the coordination number on melting [24]. Results of preliminary studies [25] show that the structure of the molten compounds is dominated by the strong anion-potentials, the local atomic environment being similar to that in the pure molten elements. For liquid GaAs

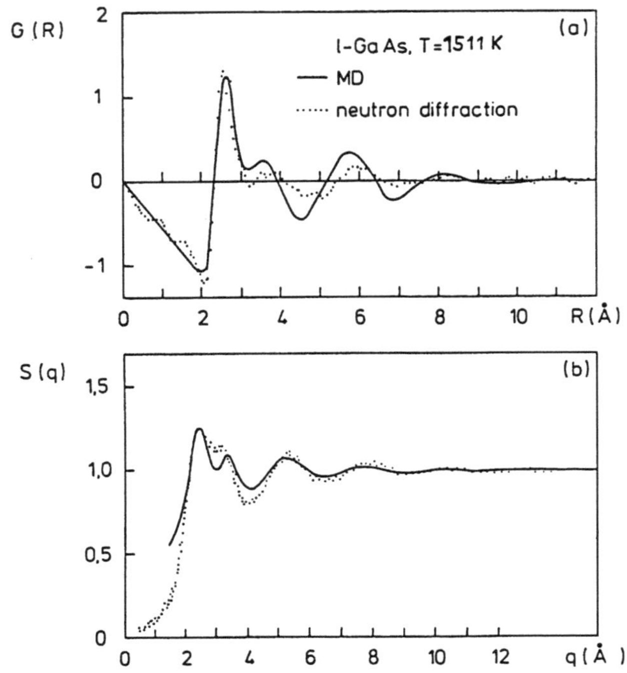

Figure 7. Neutron-scattering reduced pair correlation function G(R) and static structure factor for liquid GaAs.

calculations similar to those described for the elements lead to a good description of the liquid structure (Fig.7). It is mainly the strong As-As interaction that leads to a smaller increase of the coordination number than in liquid Ge, and a similar argument holds for liquid CdTe. Whereas the MD simulations predict only a weak chemical order in GaAs, a strong Cd-Te interaction makes molten CdTe much more ionic [26].

4. Conclusions and Outlook

The structures of the molten elements and III-V and II-VI compounds are well described in terms of the interplay of volume- and pair forces. This approach accounts for covalent bonding effects only as far as they are describable in terms of overlapping spherical pseudoatoms. Nevertheless it leads to a correct description in the characteristic trends in the coordination numbers, bond distances and bond angles. This suggests that it is a better starting point for the modelling of interatomic interactions than the conventional decomposition of the total energy into classical pair-, three-, and many-body forces.

Acknowledgement

It is a pleasure to acknowledge the hospitality of Clare College and the Cavendish Laboratory, Cambridge, where this paper was written with the financial support of the Royal Society and the Science and Engineering Research Council.

REFERENCES

[1] F.H.Stillinger and T.A.Weber, Phys.Rev. **B31**,5262(1985).
[2] E.Pearson, T.Takai, T.Halicioglu, and W.A.Tiller , J.Cryst.Growth **70**,33(1984).
[3] J.Tersoff, Phys.Rev.Lett. **56**,632(1986); P.C.Kelires and J.Tersoff, Phys.Rev.Lett. **61**,562(1988).
[4] R.Biswas and D.R.Hamann, Phys.Rev.Lett. **55**,2201(1985)
[5] R.Biswas, G.S.Grest, and C.Soukoulis, Phys.Rev. **B36**, 7437 (1987).
[6] M.D.Kluge, J.R.Ray, and A.Rahman, Phys.Rev. **B36**4234(1987).
[7] K.Ding and H.C.Andersen, Phys.Rev. **B34**,6987(1986).
[8] W.D.Luedtke and U.Landmann, Phys.Rev. **B40**,1164(1989).
[9] See the paper by V.Heine and J.Hafner in these proceedings.
[10] V.Heine and D.Weaire, Solid State Physics **24**,247(1970).
[11] J.Hafner, *From Hamiltonians to Phase Diagrams* (Springer, Berlin 1987).
[12] J.Hafner and V.Heine, J.Phys. F (Metal Physics) **13**,2479(1983).
[13] V.M.Glazov, S.N.Chizevskaya, and N.N.Glagoleva, *Liquid Semiconductors* (Plenum, New York 1969).
[14] R.Evands, in *Electrons in Disordered Metals and at Metallic Surfaces*, ed. by P.Phariseau et al. (Plenum, New York 1979), p.417.
[15] A.Arnold, N.Mauser, and J.Hafner, J.Phys.-Cond.Matt.**1**,965(1989).
[16] J.Hafner, Phys.Rev.Lett. **62**,748(1989).
[17] A.Arnold and N.Mauser, Comp.Phys.Comm. (submitted).
[18] J.Hafner, J.Phys.-Cond.Matt. (submitted).
[19] W.Jank and J.Hafner, Europhys.Lett. **7**,283(1988).
[20] W.Jank and J.Hafner, Phys.Rev. B (submitted) and to be published.
[21] S.S.Jaswal and J.Hafner, Phys.Rev. **B38**,7311(1988); J.Hafner and S.S.Jaswal, ibid. p.7320.
[22] P.B.Littlewood, CRC Critical Reviews in Solid State and Materials Science **11**,243(1983).
[23] J.P.Gaspard, F.Marianelli, and A.Pellegatti, Europhys.Lett. **3**,1095 (1987).
[24] J.P.Gaspard, C.Bergman, C.Bichara, R.Bellissent, P.Chieux, and J.Goffart, J.Non-Cryst.Solids. **97+98**,1283(1987).
[25] J.Hafner and W.Jank, J.Phys.-Cond.Matt. **1**,4235(1989).
[26] J.Hafner, to be published.

Part II

Metals

First-Principles Interatomic Potentials in Transition Metals: Multi-ion Interactions and Their Analytic Representation*

J.A. Moriarty

Lawrence Livermore National Laboratory, University of California, Livermore, CA 94550, USA

Abstract. The first-principles, density-functional version of the generalized pseudopotential theory (GPT), previously developed for empty- and filled-*d*-band metals, has now been extended to pure transition metals with partially-filled *d* bands. Within this formalism, a rigorous real-space expansion of the bulk total energy has been obtained in terms of widely transferable, *structure-independent* interatomic potentials, including both central-force pair interactions and angular-force triplet and quadruplet interactions. In the central transition metals, the three-ion and four-ion potentials are essential to a proper description of materials properties, but are necessarily multi-dimensional functions which cannot be easily tabulated. We have, therefore, also developed a simplified version of the theory in which these potentials can be expressed *analytically* while retaining the most important physics of the full first-principles treatment. The analytic potentials are suitable for wide-scaled applications, including molecular-dynamics and Monte-Carlo simulation. The first-principles and analytic theories are here reviewed together with selected applications to (*i*) the prediction of high-pressure structural phase transitions; (*ii*) phonon anomalies and their close correlation with the present multi-ion potentials; and (*iii*) the vacancy-formation energy.

1. Introduction

Working within the framework of generalized pseudopotential theory (GPT) [1,2] and local-density quantum mechanics [3], we have recently succeeded in developing a first-principles interatomic-potential expansion of the total energy suitable for transition metals [4]. In the bulk elemental metal with atomic volume Ω, the total-energy functional takes the form

$$E_{tot}(R_1 \ldots R_N) = E_0(\Omega) + \frac{1}{2}\sum_{i,j}{}' v_2(ij) + \frac{1}{6}\sum_{i,j,k}{}' v_3(ijk) + \frac{1}{24}\sum_{i,j,k,l}{}' v_4(ijkl) . \quad (1)$$

The leading volume term E_0 in this expansion includes all one-ion *intra-atomic* contributions to the total energy and already leads to a good description of transition-metal cohesion in lowest order. The interatomic potentials v_2, v_3, and v_4 are given in terms of weak pseudopotential and *d*-state tight-binding and hybridization matrix elements coupling different sites, and the series is rapidly convergent beyond three-ion interactions. As in the corresponding theory of nontransition metals [2,5], the interatomic potentials are volume dependent but *structure independent* and thus rigorously transferable at a given volume to all bulk structures, either ordered or disordered. At constant volume Ω, the central-force pair potential v_2 is a one-dimensional function of the ion-ion separation distance $R_{ij} = |\mathbf{R}_i - \mathbf{R}_j|$:

$$v_2(ij) \equiv v_2(R_{ij};\Omega) , \quad (2)$$

while the angular-force triplet potential v_3 and quadruplet potential v_4 are, respectively, the three- and six-dimensional functions

$$v_3(ijk) \equiv v_3(R_{ij},R_{jk},R_{ki};\Omega) \quad (3)$$

*Work performed under the auspices of the U. S. Department of Energy by Lawrence Livermore National Laboratory under contract No. W-7405-Eng-48.

and

$$v_4(ijkl) \equiv v_4(R_{ij},R_{jk},R_{kl},R_{li},R_{ki},R_{lj};\Omega) \ . \tag{4}$$

In the full first–principles GPT, the potentials v_2, v_3, and v_4 are all nonanalytic functions with long–range oscillatory tails. Unlike the simple radial function v_2, however, the multi-dimensional functions v_3 and v_4 cannot be readily tabulated for application purposes. This has greatly inhibited the widespread application of this theory to the central transition metals, where the multi–ion potentials are essential to an accurate description of materials properties. To improve this situation, we have now developed a simplified model transition–metal GPT [6] in which v_3 and v_4 are approximated by analytic short–range forms that retain the dominant physics of the full theory. Salient features of the first–principles GPT are reviewed in Sec. 2, while the model GPT is discussed in Sec. 3. Representative applications of both theories are presented in Sec. 4.

2. First–Principles Generalized Pseudopotential Theory (GPT)

The first–principles GPT has been developed completely within the modern framework of local–density quantum mechanics [2,4]. Briefly, both the valence and inner–core electrons in the metal are govern by a one–electron Schrödinger equation

$$(T + V)|\psi_\alpha\rangle = E_\alpha|\psi_\alpha\rangle \ , \tag{5}$$

where T is the kinetic–energy operator and V is the self–consistent potential. The inner–core electrons are amenable to a purely atomiclike treatment and, invoking the usual small–core approximation, one may transform Eq. (5) to an exactly equivalent pseudo–Schrödinger equation

$$(T + W)|\phi_\alpha\rangle = E_\alpha|\phi_\alpha\rangle \tag{6}$$

for the remaining *s*, *p*, and *d* valence electrons, where W is an optimized nonlocal pseudopotential operator. For transition metals, W is effectively weak for the nearly–free–electron *s* and *p* electrons, just as in simple metals, but is strong for the more localized *d* electrons. An appropriate basis set with which to represent the valence states in Eq. (6) then is one consisting of both plane waves $|\mathbf{k}\rangle$ and five localized *d* states $|\phi_d\rangle$ centered on each ion site. The latter are chosen to be exact eigenstates of a suitable atomiclike reference Hamiltonian:

$$(T + v_{at})|\phi_d\rangle = E_d^0\ |\phi_d\rangle \ . \tag{7}$$

A second effectively weak potential is then

$$\Delta \equiv \delta V - \langle\phi_d|\delta V|\phi_d\rangle \ , \tag{8}$$

where $\delta V \equiv v_{at} - V$. The nearly–free–electron *s* and *p* bands are characterized by free–electron energies $\epsilon_{\mathbf{k}}$ and small plane–wave matrix elements $W_{\mathbf{k}\mathbf{k}'} \equiv \langle\mathbf{k}|W|\mathbf{k}'\rangle$, while the tight–binding–like *d* bands are characterized by a mean energy

$$E_d \equiv \langle\phi_d|T+V|\phi_d\rangle = E_d^0 - \langle\phi_d|\delta V|\phi_d\rangle \tag{9}$$

and small *d*–state overlap matrix elements

$$S_{dd'}(R_{ij}) \equiv \langle\phi_d(r-R_i)|\phi_{d'}(r-R_j)\rangle \tag{10}$$

and

$$\Delta_{dd'}(R_{ij}) \equiv \langle\phi_d(r-R_i)|\Delta|\phi_{d'}(r-R_j)\rangle \ . \tag{11}$$

The remaining *sp–d* hybridization between the bands is then described by small plane–wave–*d*–state matrix elements

$$S_{\mathbf{k}d} \equiv \langle \mathbf{k} | \phi_d \rangle \tag{12}$$

and

$$\Delta_{\mathbf{k}d} \equiv \langle \mathbf{k} | \Delta | \phi_d \rangle \,. \tag{13}$$

The reference system which defines the localized *d* states $|\phi_d\rangle$ through Eq. (7) and upon which the total–energy expansion (1) is based consists of N isolated transition–metal ions placed in a compensating uniform electron density Z/Ω. Each ion exhibits a strong *d* resonance in the presence of the free–electron gas such that the number of valence *d* electrons retained by the ion is

$$Z_d = (10/\pi)\, \delta_2(\epsilon_F) \,, \tag{14}$$

where δ_2 is the $l = 2$ phase shift of the ion potential and ϵ_f is the Fermi energy of the free–electron gas:

$$\epsilon_F = \frac{\hbar^2}{2m} \left[\frac{3\pi^2 Z}{\Omega} \right]^{2/3} . \tag{15}$$

For a pure transition metal, Z effectively represents the number of non–*d* valence electrons per atom and one must additionally require that

$$Z + Z_d = \text{constant} \,, \tag{16}$$

where the constant is determined by the Periodic Table for the element in question. Equations (14)–(16) then constitute three equations in three unknowns and must be iterated to yield self–consistent values of Z, Z_d, and ϵ_f. The resulting entity, self–consistent ion plus compensating free–electron gas, we call a "zero–order pseudoatom" and the reference system of N neutral zero–order pseudoatoms represents an excellent first approximation to a real transition metal. (See, for example, Figs. 3–7 of Ref. 4.) In particular, the zero–order pseudoatom values of Z and Z_d are already in good accord with estimates of these quantities from self–consistent band–structure calculations. For the 3*d* and 4*d* transition metals at normal density, Z takes on values in the narrow range $1.1 \lesssim Z \lesssim 1.7$, with a typical value of 1.4. For a given element as a function of volume, Z and Z_d also properly reflect any transfer of electrons in or out of the *d* states, so that for the early and central transition metals (e.g., Mo) Z decreases with compression representing an $s \to d$ electron transfer, while for late transition metals (e.g., Cu) Z increases with compression representing a $d \to s$ electron transfer. This latter behavior is very important in the context of high–pressure structural phase transitions, as discussed in Sec. 4.1. In addition, all elementary aspects of the electronic structure, including the position and width of the *d* bands, can be accurately estimated from the properties of the pseudoatom. The fact that the *d*–band width is proportional to the *d*–resonance width associated with the phase shift δ_2 is also sufficient to ensure that the zero–order pseudoatom cohesive energy, $E_{coh}{}^{pa}$, which is the leading contribution to E_0 in Eq. (1), already well approximates the cohesive energy of the actual metal. In all of the 3*d* and 4*d* transition metals, $E_{coh}{}^{pa}$ agrees to within 20% with self–consistent band–structure calculations of the cohesive energy.

The remaining task is then to develop the series (1) in terms of the weak matrix elements $W_{\mathbf{kk}'}$, $S_{dd'}$, $\Delta_{dd'}$, $S_{\mathbf{k}d}$, and $\Delta_{\mathbf{k}d}$ and related quantities. This requires the introduction of Green's functions and an intricate analysis in which the volume and structural components of all terms are carefully separated. This analysis will not be reviewed here, but the interested reader is referred to Ref. 4 for the details.

3. Model GPT for Central Transition Metals

The model GPT is developed within exactly the same framework but with the introduction of additional simplifying approximations. The simplifications begin with an ansatz: we neglect the troublesome long–range oscillatory tails of the potentials v_2, v_3, and v_4. In transition metals, these tails arise primarily through indirect hybridization interactions, such as $\Delta_{dk}\Delta_{kd'}$, which couple *d* states on different sites. It has been found from the first–principles GPT, however, that these hybridization tails destructively interfer in the calculation of materials properties when the *d* bands are partially occupied. Physically, this destructive interference is a consequence of the nonspherical nature of the Fermi surface in such transition metals which inhibits any net long–range hybridization interaction between ions. Clearly, the destructive interference is most complete for the central transition metals (e.g., V, Nb, Cr, and Mo), so that the model GPT is focused primarily on this subset of materials.

Next, we simplify the description of the remaining direct *d*–state couplings between ions. As in the first–principles GPT, the two–ion *d*–state nonorthogonality contributions arising from $S_{dd'}$ matrix elements are absorbed into a repulsive hard–core–like potential v_2^{hc}. The three–ion and four–ion contributions from $S_{dd'}$ are now dropped, however. We then introduce canonical *d* bands to express the remaining $\Delta_{dd'}$ matrix elements analytically and retain direct couplings to only fourth order in the dominant volume component $\Delta_{dd'}^{vol}$. The *d*–state band–structure contributions to the interatomic potentials can then be expressed diagramatically, and the relevant diagrams contributing to v_2, v_3, and v_4 are illustrated in Fig. 1(a). Each line in a diagram connecting two ion sites represents a 5 x 5 matrix made up of elements $\Delta_{dd'}^{vol}$, while the energy contribution corresponding to a complete diagram is proportional to the trace of the resulting matrix product. In a pure canonical *d*–band description,

$$\Delta_{dd'}^{vol}(R_{ij}) = \alpha_m (R_{WS}/R_{ij})^5 \,, \tag{17}$$

where $5 = 2l+1$ for $l = 2$ and R_{ws} is the Wigner–Seitz radius. Here the α_m are constants and there are fixed ratios between the *m* components such that $dd\sigma{:}dd\pi{:}dd\delta$ is 6:–4:1. Utilizing such a description of $\Delta_{dd'}^{vol}$, permits all of the diagrams to be evaluated analytically. In practice, we actually generalize the R_{ij} dependence of Eq. (17) to

$$\Delta_{dd'}^{vol}(R_{ij}) = \alpha_m (R_{WS}/R_{ij})^p \,, \tag{18}$$

where p is a material–dependent constant, and we also include a smooth Gaussian cutoff at large distance.

The two–ion *d*–state potential v_2^d in the model GPT consists of one second–order and one fourth–order contribution. From Eq. (18) and Fig. 1(a), one is led to a general form

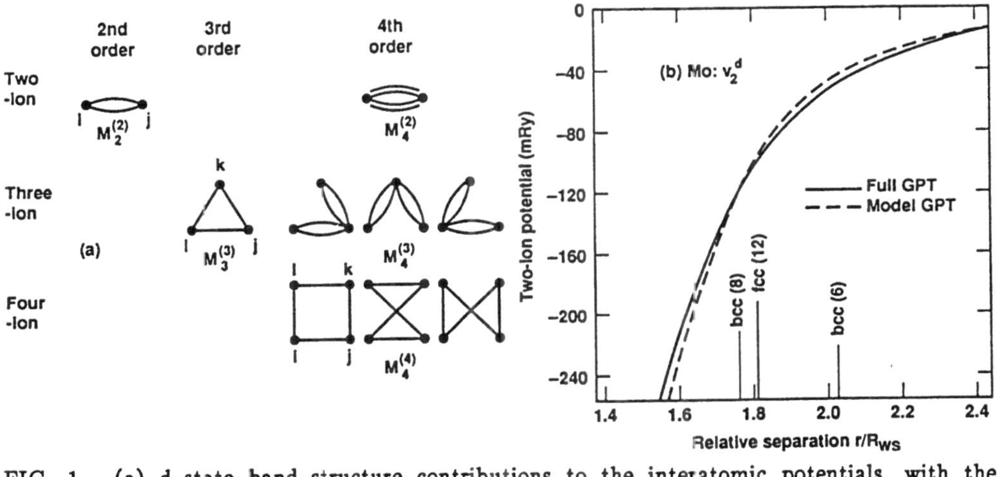

FIG. 1. (a) d–state band–structure contributions to the interatomic potentials, with the notation as in Ref. 4; (b) two–ion d–state potential v_2^d for Mo, as calculated from the model GPT with p = 4 in Eq. (19) and from the full first–principles GPT neglecting *sp–d* hybridization.

$$v_2^d(r) = v_a[f(r)]^4 - v_b[f(r)]^2 = v_a(r_0/r)^{4p} - v_b(r_0/r)^{2p} , \tag{19}$$

where

$$f(r) \equiv (r_0/r)^p \tag{20}$$

is a characteristic radial function, with $r_0 \equiv 1.8R_{ws}$. The quantities v_a and v_b are material-related constants which depend primarily on *d*-band filling. From general theoretical considerations, one expects $v_a > 0$ and $v_b > 0$ with $v_b >> v_a$, so that $v_2{}^d$ is an entirely attractive potential, as shown in Fig. 1(b) for Mo. In this case, the choice p = 4 leads to a good representation of the corresponding first-principles GPT calculation of $v_2{}^d$. To the *d*-state potential must be added the hard-core contribution $v_2{}^{hc}$ and a simple-metal-like contribution for the *s* and *p* electrons, $v_2{}^{sp}$, so that the total two-ion pair potential is

$$v_2(r) = v_2^{sp}(r) + v_2^{hc}(r) + v_2^d(r) . \tag{21}$$

Both $v_2{}^{sp}$ and $v_2{}^{hc}$ are components of the first-principles theory as well and these quantities are retained without change in the model GPT.

The three-ion triplet potential is dominated by *d*-state contributions, so that we always take $v_3 = v_3{}^d$, as is done in the first-principles GPT. From Fig. 1(a) it can be seen that there are one third-order and three fourth-order contributions to $v_3{}^d$, leading to a form

$$v_3(r_1,r_2,r_3) = v_c f(r_1)f(r_2)f(r_3)L(\theta_1,\theta_2,\theta_3) + v_d\{[f(r_1)f(r_2)]^2P(\theta_3)$$
$$+ [f(r_2)f(r_3)]^2P(\theta_1) + [f(r_3)f(r_1)]^2P(\theta_2)\} , \tag{22}$$

where θ_1, θ_2, and θ_3 are the angles subtended by r_1, r_2, and r_3, respectively. Here L is the angular function

$$L(\theta_1,\theta_2,\theta_3) \equiv L(x_1,x_2,x_3) = \frac{1}{144}[54 + 330\, x_1x_2x_3 - 105(x_1^2x_2^2 + x_2^2x_3^2 + x_3^2x_1^2)$$
$$+ 735(x_1x_2x_3)^2] , \tag{23}$$

with $x_1 \equiv \cos\theta_1$, $x_2 \equiv \cos\theta_2$, and $x_3 \equiv \cos\theta_3$, and P is the additional angular function

$$P(\theta) \equiv P(x) = \frac{1}{1448}(533 + 510\, x^2 + 405\, x^4) , \tag{24}$$

with $x \equiv \cos\theta$. Both L and P are universal functions which depend only on *d* symmetry and apply to all transition metals. The coefficients v_c and v_d are additional material-dependent constants that again mostly reflect *d*-band filling. For almost half-filled *d* bands, v_c is near zero while v_d is positive, so that $v_d >> |v_c|$ and v_3 is an entirely repulsive potential, as is shown in Fig. 2(a) for Mo. Again it is seen that the model GPT is a rather faithful representation of the first-principles theory.

Similarly, for the four-ion quadruplet potential we take $v_4 = v_4{}^d$. From Fig. 1(a) there are three fourth-order contributions to $v_4{}^d$, so that one has a form

$$v_4(r_1,r_2,r_3,r_4,r_5,r_6) = v_e[f(r_1)f(r_2)f(r_4)f(r_5)M(\theta_1,\theta_2,\theta_3,\theta_4,\theta_5,\theta_6)$$
$$+ f(r_2)f(r_3)f(r_5)f(r_6)M(\theta_5,\theta_7,\theta_8,\theta_9,\theta_{10},\theta_{11})$$
$$+ f(r_3)f(r_1)f(r_6)f(r_4)M(\theta_3,\theta_4,\theta_5,\theta_6,\theta_{11},\theta_{12})] , \tag{25}$$

where M is a third universal angular function which depends only upon *d* symmetry and v_e is an

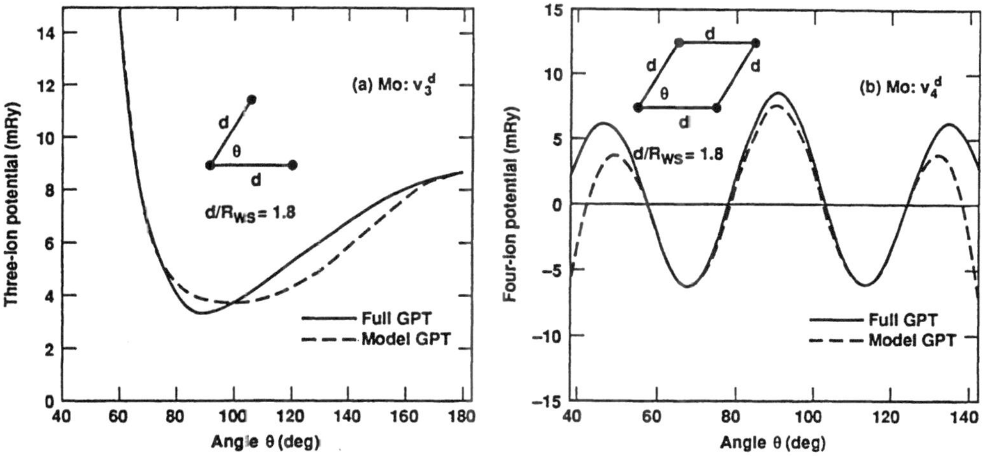

FIG. 2. Multi–ion potentials for Mo, as calculated from the model GPT with p = 4 in Eqs. (22) and (25) and from the full first–principles GPT neglecting *sp–d* hybridization. (a)Three–ion potential $v_3 = v_3{}^d$ for the special case $r_1 = r_2 = 1.8R_{ws}$; (b) four–ion potential $v_4 = v_4{}^d$ for the special case of in–plane interactions with $r_1 = r_2 = r_4 = r_5 = 1.8R_{ws}$.

additional material–dependent parameter. The function M has also been obtained analytically, but with six independent variables its form is rather more complicated than L or P. The full analytic result is given in Ref. 6, together with precise definitions of the six distances $r_1 \ldots r_6$ and twelve angles $\theta_1 \ldots \theta_{12}$ associated with four–ion geometry. Unlike v_2 and v_3, the nature of v_4 is purely oscillatory, but again the model GPT well reproduces its structure, as shown in Fig. 2(b) for Mo. While v_4 consequently leads to a much smaller cohesive–energy contribution than either v_2 or v_3, it is important to many structural properties connected with bcc stability, as can be inferred from Fig. 2(b).

The material–dependent parameters v_a, v_b, v_c, v_d, and v_e in the model GPT can be determined in various ways by imposing theoretical constraints and fitting to the first–principles potentials, to calculated first–principles physical properties, and/or to experimental data. We have developed a number of different schemes to do this, and all give very similar results. The applications discussed below on Nb and Mo represent preliminary results obtained from a representative scheme. A more complete analysis is given in Ref. 6.

4. Applications

We now discuss a few selected applications of both the first–principles and model GPT which illustrate the current capabilities of these theories to calculate the physical properties of real materials.

4.1 Prediction of High–Pressure Structural Phase Transitions

A revealing test of any theory of the total energy of a metal is its ability to calculate structural energy differences between different phases, both at normal density ($\Omega=\Omega_0$) and also at high pressure ($\Omega<\Omega_0$) where structural phase transitions are possible. Under high pressure, the normal boundaries of the Periodic Table separating insulators, simple metals, and transition metals are altered due to changes in the electronic structure, as band broadening closes band gaps and turns insulators into simple metals and *s–d* electron transfer turns simple metals into transition metals, with empty *d* bands filling in the pre– and early transition metals and filled *d* bands emptying in the late and post– transition metals. Some years ago we made a systematic study [7] of high–pressure structural phase stability in the third–period simple metals Na, Mg, Al, and Si, where the lowering and partial filling of the initially empty 3*d* bands under compression drives sequences of transitions among the fcc, bcc, and hcp structures. Parallel calculations were done with the first–principles GPT and the linear–muffin–tin–orbital (LMTO) band–structure method [8], yielding predictions in generally good qualitative agreement with each other. A total of nine new high–pressure phases were predicted, all but one of which were beyond the limits of experimental accessiblity at the time of the calculations. Since that time, however, advances in high–pressure technology have extended the accessible pressure range to more than 100 GPa (1 Mbar), with the result that three of our predictions have now been confirmed in diamond–anvil experiments: a new bcc phase in Mg at 50 GPa [9], a new hcp phase in Si at 40 GPa [10], and a new fcc phase in Si at 80 GPa [11].

In the context of the GPT, most nontransition metals do *not* require multi–ion potentials to obtain a good total energy via Eq. (1) and consequently may be treated with the full first–principles theory using v_2 only. In the central transition metals, on the other hand, including explicit multi–ion contributions is essential in the calculation of structural properties. In the case of Mo, for example, we find with the model GPT that more than half of the large calculated bcc–fcc energy difference at normal density arises from v_4. Moreover, under sufficiently high pressure the bcc structure in Mo can actually be destablized due to $s \rightarrow d$ electron transfer. With the model GPT including v_3 and v_4, we predict a bcc $\rightarrow$ hcp transition at about 220 GPa (2.2 Mbar), as shown in Fig. 3(a). Parallel LMTO calculations [12] give a similar result with a predicted transition pressure of 320 GPa and a correspondingly larger bcc–fcc energy difference at normal density. Simultaneous with these calculations, the destabilization of bcc Mo was indeed observed at 210 GPa and 4100 K in a very elegant shock–wave experiment [12]. Similar high–pressure phase transitions are expected in Cr and W.

In the late transition metals, the s–d electron transfer is reversed into a weak $d \rightarrow s$ transition which has the effect of increasing rather than decreasing the stability of the normal–density fcc structure. In the case of Cu we have applied the first–principles GPT to a wide range of high–pressure properties [14–16], including structural phase stability. We predict that this metal should remain fcc to beyond three–fold compression, as shown in Fig. 3(b). Experimentally [18], the stability of the fcc structure has now been confirmed to 180 GPa [1.8 Mbar, corresponding to $\Omega/\Omega_0 = 0.65$ in Fig. 3(b)]. At high pressure the $d \rightarrow s$ transition also extends into the post–transition metals, but with somewhat different consequences. At normal density these metals are typically characterized by distorted structures. Surprisingly, we find that such structures can remain favored over more close–packed ones even at multi–megabar pressures. In the case of Hg, in fact, we predict from first–principles GPT calculations [19] that the fcc, bcc, and ideal hcp structures will be mechanically unstable to beyond 1000 GPa (10 Mbar). In addition, we

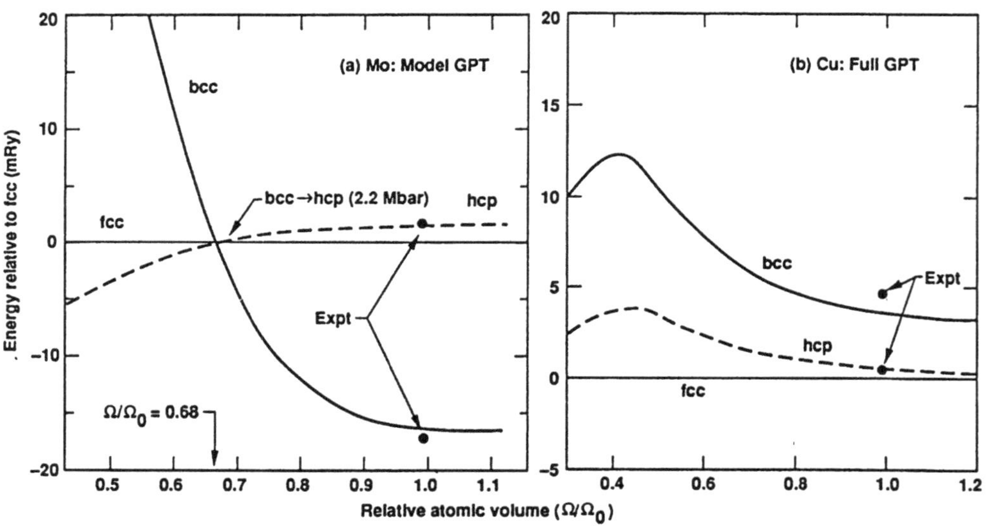

FIG. 3. Calculated high–pressure structural phase stability in transition metals. (a) Mo, as obtained from the model GPT including v_3 and v_4; (b) Cu, as obtained from the first–principles GPT with v_2 only. Experimental points are estimates based on extrapolations of alloy data: Ref. 13 for Mo and Ref. 17 for Cu.

predict a high-pressure phase transition near 100 GPa from the observed β-Hg structure to a distorted hcp structure with a very large c/a axial ratio. Independently, this latter phase has been observed in a recent diamond-anvil experiment [20].

4.2 Transition-Metal bcc Phonons: Correlations with v_3 and v_4

Another revealing application for transition metals is the bcc phonon spectrum, which shows a number of anomalous features driven by the angular forces associated with our multi-ion potentials. In the [100] direction, the most interesting trend is the sensitive behavior of the longitudinal (L) branch with *d*-band filling. For simple metals with empty *d* bands above the Fermi level, such as the alkali metals K, Rb and Cs, the L branch is nearly degenerate with the corresponding transverse (T) branch, but always lies slightly higher in frequency. With small *d*-band filling, as occurs for example in bcc Ba, the L branch is pushed below the T branch, however. As we have shown in first-principles GPT calculations on Ba [21], this behavior is quite naturally driven by the attractive three-ion potential v_3 present under such conditions. For the central bcc transition metals, on the other hand, the L branch is pushed well above the T branch and proportionally so as one increases *d*-band occupation in going from the VB metals (V, Nb, and Ta) to the VIB metals (Cr, Mo, and W). In turn, this latter behavior is driven by the repulsive v_3 present under these conditions, as we have demonstrated in the context of the first-principles GPT for V and Cr [4,14].

For the case of [110] bcc phonons, the interesting trends occur in the transverse T_1 and T_2 branches. In the alkali and alkaline-earth metals, the T_1 branch is characteristically soft ($T_1 << T_2$), while in the central transition metals this branch is relatively stiff ($T_1 \approx T_2$). Furthermore, in the latter case, these branches show a systematic reversal in their ordering at the Brillouin-zone boundary, with $T_1 < T_2$ for the group VB metals while $T_1 > T_2$ for the group VIB metals. As shown in Fig. 4 for Nb and Mo, the stiffening of T_1 is driven by v_4 rather than

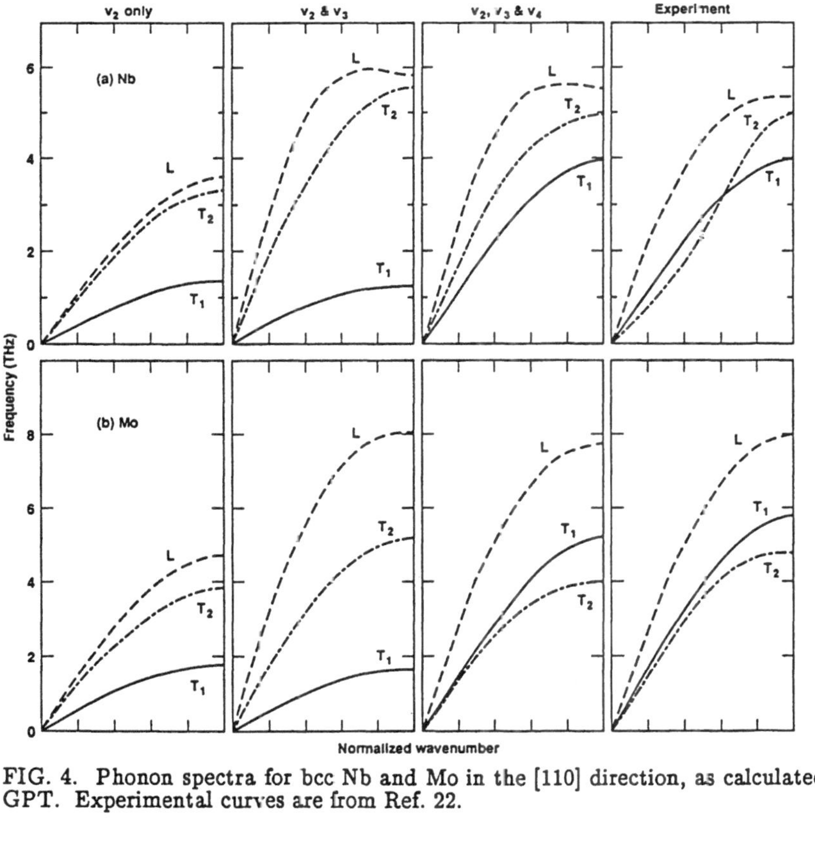

FIG. 4. Phonon spectra for bcc Nb and Mo in the [110] direction, as calculated from the model GPT. Experimental curves are from Ref. 22.

v_3 in this instance, with both the proper ordering and magnitude of the phonon frequencies being obtained when and only when both multi-ion potentials are included. Note, however, in the case of Nb that experiment reveals a pronounced softening of the T_2 branch at small wavenumber which the model GPT result does not contain. This effect is driven by higher-order, long-range d-state couplings that have been neglected in the model GPT, but are included in the full GPT.

4.3 Unrelaxed Vacancy-Formation Energy

Conceptually, the unrelaxed vacancy-formation energy at constant volume, E_{vac}, may be thought of as arising from a two-step process: (i) a bulk atom is removed from the interior of the metal and placed on the surface; and (ii) the metal is then uniformly compressed to maintain constant volume. This gives rise to both cohesive-energy and virial-pressure contributions to E_{vac}, so that in the present context one may write

$$E_{vac} = E_{vac}^{(2)} + E_{vac}^{(3)} + E_{vac}^{(4)} , \qquad (26)$$

where

$$E_{vac}^{(n)} = -(n-1)E_{coh}^{(n)} + \Omega\, P_{vir}^{(n)} , \qquad (27)$$

with

$$E_{coh}^{(n)} \equiv \frac{1}{Nn!} \sum_{i,j,\ldots}{}' v_n(ij\ldots) \qquad (28)$$

and

$$P_{vir}^{(n)} \equiv -\frac{1}{6N\Omega(n-2)!} \sum_{i,j,\ldots}{}' R_{ij} \frac{\partial}{\partial R_{ij}} [v_n(ij\ldots)] . \qquad (29)$$

Table I lists calculated components of the cohesive and vacancy-formation energies of Mo so obtained from the model GPT. In this case there are very small net contributions to E_{vac} from the two-ion and four-ion potentials, and the result is dominated by three-ion-potential contributions. Good overall agreement with experiment is obtained, and this agreement is *not* sensitive to the details of the model GPT calculation. Similar good results have been obtained for V, Nb, and Cr.

TABLE I. Cohesive and vacancy-formation energies and their components for Mo, as calculated from the model GPT at normal density ($\Omega = \Omega_0 = 105.1$ a.u.). Energies in Ry.

	E_0	v_2	v_3	v_4	total	experiment
E_{coh}	−0.42	−0.31	0.08	−0.01	−0.66	−0.50[a]
E_{vac}	–	−0.00	0.26	−0.02	0.24	0.22[b]

[a] Reference 23. [b] Reference 24.

5. Current and Future Directions

Many additional applications of the model GPT should be possible for the central transition metals. We are currently attempting to combine this approach with molecular-dynamics and Monte-Carlo computer simulation methods to study point and line defects in the bulk, including grain boundaries. We hope to similarly extend our capabilities of studying high-pressure structural phase stability to high temperature and thereby calculate complete pressure–temperature phase diagrams for elemental metals. Beyond elemental metals, it should be possible to extend both the first–principles and model GPT to compounds and alloys, as has been done successfully by Hafner [5] for nontransition metals within the framework of simple–metal pseudopotential theory. Finally, further extension of the transition–metal GPT to surfaces also appears to be possible by making an internal transformation of the total–energy functional to an embedded–atom [25] or effective–medium [26] representation, in which the embedding function is identified as the bulk volume term, E_0.

References

1. J. A. Moriarty, Phys. Rev. B **5**, 2066 (1972).
2. J. A. Moriarty, Phys. Rev. B **16**, 2537 (1977) and B **26**, 1754 (1982).
3. W. Kohn and L. J. Sham, Phys. Rev. **140**, A1133 (1965); L. Hedin and B. I. Lundqvist, J. Phys. C **4**, 2064 (1971).
4. J. A. Moriarty, Phys. Rev. B **38**, 3199 (1988).
5. J. Hafner, *From Hamiltonians to Phase Diagrams* (Springer, Berlin, 1987).
6. J. A. Moriarty, Phys. Rev. B (to be published).
7. J. A. Moriarty and A. K. McMahan, Phys. Rev. Lett. **48**, 809 (1982); A. K. McMahan and J. A. Moriarty, Phys. Rev. B **27**, 3235 (1983).
8. H. L. Skriver, *The LMTO Method* (Springer, Berlin, 1984).
9. H. Olijnyk and W. B. Holzapfel, Phys. Rev. B **31**, 4682 (1985).
10. H. Olijnyk, S. K. Sikka, and W. B. Holzapfel, Phys. Lett. **103A**, 137 (1984).
11. S. J. Duclos, Y. K. Vohra, and A. L. Ruoff, Phys. Rev. Lett. **58**, 775 (1987).
12. R. S. Hixson, D. A. Boness, J. W. Shaner, and J. A. Moriarty, Phys. Rev. Lett. **62**, 637 (1989).
13. A. R. Miedema and A. K. Niessen, CALPHAD **7**, 27 (1983).
14. J. A. Moriarty, Phys. Rev. Lett. **55**, 1502 (1985).
15. J. A. Moriarty, in *Shock Waves in Condensed Matter*, edited by Y. M. Gupta (Plenum, New York, 1986), p. 101.
16. W. J. Nellis, J. A. Moriarty, A. C. Mitchell, M. Ross, R. G. Dandrea, N. W. Ashcroft, N. C. Holmes, and G. R. Gathers, Phys. Rev. Lett. **60**, 1414 (1988).
17. L. Kaufman and H. Bernstein, *Computer Calculation of Phase Diagrams* (Academic, New York, 1970).
18. P. M. Bell, J. A. Xu, and H. K. Mao, in *Shock Waves in Condensed Matter*, edited by Y. M. Gupta (Plenum, New York, 1986), p. 125.
19. J. A. Moriarty, Phys. Lett. **131A**, 41 (1988).
20. O. Schulte and W. B. Holzapfel, Phys. Lett. **131A**, 38 (1988).
21. J. A. Moriarty, Phys. Rev. B **34**, 6738 (1986).
22. B. M. Powell, P. Martel, and A. D. B. Woods, Can. J. Phys. **55**, 1601 (1977).
23. C. Kittel, *Introduction to Solid State Physics*, 5th ed. (Wiley, New York, 1976), p. 74.
24. K. Maier, M. Peo, B. Saile, H. E. Schaefer, and A. Seeger, Philos. Mag. **40**, 701 (1979).
25. M. S. Daw and M. I. Baskes, Phys. Rev. B **29**, 6433 (1984).
26. K. W. Jacobsen, J. K. Nørskov, and M. J. Puska, Phys. Rev. **35**, 7423 (1987).

The Effective-Medium Theory Beyond the Nearest Neighbour Interaction

H. Häkkinen

Department of Physics, University of Jyväskylä,
SF-40100 Jyväskylä, Finland

Abstract. The effective medium theory has been extended to include more than the nearest neighbours in the total energy calculation of a metallic system. The effects of this extension are discussed in the case of aluminium and copper.

1. Introduction

During this decade, a whole new class of potentials for treating many-atom interactions in metals has been developed : the "embedded-atom method" (EAM) [1], the "Finnis-Sinclair potential" (FS) [2], the "effective-medium theory" (EMT) [3, 4], and the "glue model" [5]. In all these models the total energy of a metallic system composed of N atoms can be written as

$$E_{tot} = \sum_{i=1}^{N} F(n_i) + \frac{1}{2}\sum_{i=1}^{N}\sum_{j(\neq i)} V(r_{ij}), \quad (1)$$

where $V(r_{ij})$ is a pair potential, n_i is a quantity describing the local electron density at the atomic site i and $F(n_i)$ is the corresponding energy term. The local density n_i is dependent of the distances from atom i to its neighbours, and in this way F is structure-dependent and implicitely includes many-atom interactions. The derivation and the physical interpretation of the terms in Eq. (1) is different in each of the models mentioned above,although the resulting applicable formulas have many common features.

The work reported in this paper is concerned with the extension of the range of interatomic interactions beyond the nearest neighbours in EMT. This extension was first outlined by Jacobsen *et al.* [3] and was tested in molecular dynamics simulations of bulk copper by Häkkinen and Manninen [6]. This paper shortly summarizes our results obtained so far in the dynamic simulations of aluminium and copper using effective medium theory. The comparision of the simulation data given by the nearest neighbour model and the extended model reveals the effects of the extension of the interatomic interactions.

2. The Effective Medium Theory

The nearest neighbour formalism of the effective medium theory (EMT) has been derived in detail in ref. [3]. It has been proved to be succesful in pre-

 Springer Proceedings in Physics, Vol. 48 Many-Atom Interactions in Solids
Editors: R.M. Nieminen · M.J. Puska · M.J. Manninen

dicting many properties of fcc-metals, like the structure of a pure metal surface or an adsorbate-surface system [3,7], the impurities in metals [8], and the high temperature dynamics of a metal surface [9]. However, the extension of the theory beyond the nearest neighbour interaction makes it possible to study a larger amount of interesting phenomena in metals, like extended dislocations or highly disordered structures. This section reviews the ideas used in the derivation of the applicable formulas in EMT and outlines the extension beyond the nearest neighbour interaction.

The approach of Jacobsen *et al.* [3] is based on the ansatz that the total electron density of a metallic system (composed of N atoms) can be written as a superposition of free-atom like densities $\Delta n_i(\mathbf{r})$:

$$n(\mathbf{r}) = \sum_{i=1}^{N} \Delta n_i(\mathbf{r}), \tag{2}$$

where $\Delta n_i(\mathbf{r})$ is an electron density induced in a homogeneous electron gas due to the presence of atom i. In the calculation of the electron densities EMT uses the so-called atomic sphere approximation, where the system is thought to consist of atomic spheres centered at the fcc-lattice sites. The radius of a sphere is determined by the requirement that the sphere is neutral. In the equilibrium lattice this definition is equal to the definition of a spherical Wigner-Seitz cell associated to each atom. The requirement of neutrality produces an unique relationship between the radius of a given atomic sphere and the embedding density at the corresponding lattice site. The embedding density $\bar{n}_i$ at each lattice site i can be calculated from the positions of neighbouring atoms as a spherical average over a sphere of radius s_i centered at the lattice site i:

$$\begin{aligned} \bar{n}_i &= \left\langle \sum_{j(\neq i)}^{n.n.} \Delta n_j \right\rangle_i \\ &= \frac{1}{\frac{4}{3}\pi s_i^3} \int_{s_i} d\mathbf{r} \sum_{j(\neq i)}^{n.n.} \Delta n(|\mathbf{r} - \mathbf{R}_j|) \\ &\equiv \sum_{j(\neq i)}^{n.n.} \Delta \bar{n}_j(s_i, R_j), \end{aligned} \tag{3}$$

where $\Delta \bar{n}_j(s_i, R_j)$ is the density tail coming from neighbour j at a distance R_j and $n.n.$ means a nearest neighbour sum. For each element there is an optimum embedding density, which in fact determines the equilibrium lattice constant of the solid phase.

The total binding energy of a metallic system is written in EMT as

$$E_{tot} = \sum_{i=1}^{N} E_{C,i}(\bar{n}_i) + \Delta E_{1el} + \Delta E_{AS}. \tag{4}$$

In this expression the main part of the cohesion comes from the first term. The second and the third terms are corrections, which are discussed later on. The density-dependent energy function $E_{C,i}$ is defined as

$$E_{C,i}(\bar{n}_i) = E_i^{hom}(\bar{n}_i) - \alpha(\bar{n}_i)\bar{n}_i, \qquad (5)$$

where $E_i^{hom}(\bar{n}_i)$ is the embedding energy of atom i in a homogeneous electron gas of density $\bar{n}_i$ and $\alpha(\bar{n}_i)$ is an integral over a neutral sphere of radius s_i of the induced electrostatic potential $\Delta\phi$:

$$\alpha(\bar{n}_i) = \int_{s_i} \Delta\phi(\mathbf{r})d\mathbf{r}. \qquad (6)$$

The subscript i has been written explicitely to remind that the local electron density is dependent on the atomic site. The cohesive function $E_C(\bar{n})$ has been calculated by Puska *et al.* [10] for a great number of elements. These functions are unique in the sense that the same function can be used for a given element both in a pure solid or in an alloy. Qualitatively, the shape of the cohesive function is related to the chemical activity of the element in question: the inert gases have a linear repulsive E_C, whence in the other elements E_C has usually a single minimum. This minimum is due to the competition between the kinetic repulsion, which dominates in the embedding function E^{hom}, and the electrostatic attraction $-\alpha(\bar{n})\bar{n}$. In the density region of interest, E_C-functions are parametrised with third order polynomials, where the zero-order coefficient is (by definition) the equilibrium cohesive energy per atom and the second-order coefficient is related to the bulk modulus.

The second term in the Eq. (4) is the correction to the cohesion energy coming from one-electron energy levels. This correction is important in transition metals, which have partially filled d-bands (and many of them have a bcc-structure). For non-transition metals like aluminium and noble metals this one-electron energy correction is assumed to be small. However, it has been found to play a crucial role in determining the equilibrium configurations of small noble metal clusters [11].

The third term in Eq. (4) is the so-called atomic sphere correction. It arises from the fact that the neighbouring Wigner-Seitz spheres begin to overlap when the atomic array is distorted away from the ideal lattice symmetry. The atomic sphere correction for atom i can be approximated to consist of the electrostatic interaction $\alpha\bar{n}_i$ and a sum of pair interactions $V(r_{ij})$. For an ideal fcc-lattice the atomic sphere correction is defined to be zero.

The extension beyond the nearest neighbour interaction makes use of the fact that the density tails $\Delta\bar{n}_j$ in the Eq. (3) are represented with exponential functions, which extend over the nearest neighbour distances. Same functions can then be used in the extension of the interactions with appropriate modification of the parameters. This modification is based on two requirements: (i) the embedding density in the ideal fcc-lattice site must be the same both in the nearest neighbour formalism and in the extended formalism (this guarantees the same value for the zero-temperature lattice constant in both formalisms) and (*ii*) the atomic sphere correction must always vanish in the ideal fcc-lattice for all reasonable lattice constants (this makes it possible to define the pair potential in the atomic sphere correction term). These ideas for going beyond the nearest neighbour interaction were outlined already in the original presentation of EMT by Jacobsen *et al.* [3] and were implemented and tested by Häkkinen and Manninen [6] (for a more detailed documentation see ref. [6]).

3. The Effects of the Extension

3.1 Thermal Expansion and Melting of Cu

The effects of the extended interactions were studied by simulating thermal behaviour of bulk copper up to the melting point with a constant pressure molecular dynamics method, which allows fluctuations both in the shape and volume of the simulation cell [12]. We carried out two independent simulation sets using both the nearest neighbour formalism and the extended formalism (interactions extended to the third neighbours) with simulation cells of 256 and 500 atoms, respectively. Periodic boundary conditions were applied in three dimensions producing bulk geometry. At each temperature the system was equilibrated during 2000 time steps (6 ps) and the next 2000-4000 time steps (6-12 ps) produced the data used in the calculation of thermodynamical quantities.

The calculated zero-pressure lattice constant at elevated temperatures is shown in the Fig. 1 together with the experimental thermal expansion curve [13]. It can be seen that in the solid phase there are no significant differences between the nearest neighbour and the extended formalisms, but the nearest neighbour formalism fails to describe the volume expansion of the liquid copper, as well as the density change at the melting point. The thermodynamical quantities obtained from simulations have been collected in Table 1, from which one can see that the nearest neighbour formalism also overestimates the latent heat of melting by a factor of three. The "bulk melting point" obtained in our simulations is the first temperature where the sample was seen to melt within the used waiting time. It is not directly comparable to the experimental melting point of copper because of the significant overheating inherent with the molecular dynamics calculations of bulk samples. Simulation of a thick film with the solid-melt equilibrium interface would probably be the best method to determine the bulk melting point in the molecular dynamics[14].

Table 1 shows also the defect formation energies (vacancy, stacking fault) for Cu predicted by the nearest neighbour and the extended EMT. It is interesting to see that the extension of the atomic interactions in EMT brings the calculated vacancy formation energy E_V^F nearer to the experimentally measured value in Cu[15]. We have found the same trend also in the case of Al, although the value obtained for E_V^F with the extended model 1.09 eV is still far from the measured

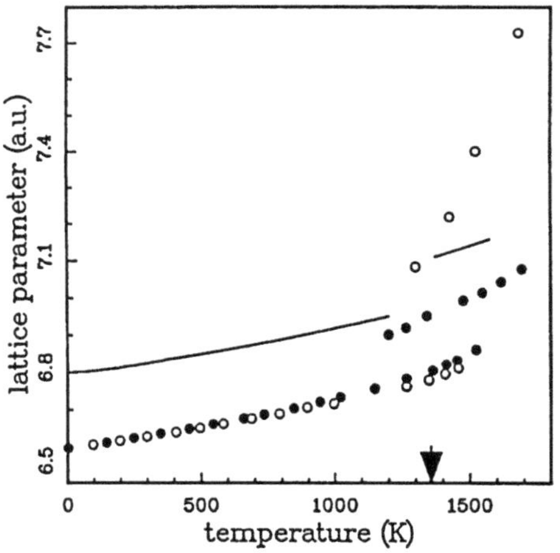

Figure 1. Lattice parameter vs temperature for copper. The calculated values with the extended potential are shown by solid circles. The open circles show the results of the nearest neighbour formalism. The solid lines represent the experimentally measured thermal expansion in solid and liquid phase. The arrow indicates the experimental melting point of 1356 K. The experimental data is taken from ref. [13].

Table 1. Thermodynamical quantities for copper obtained from molecular dynamics simulations with the nearest neighbour (n.n.) EMT and the extended (ext.) EMT. α and α_V are the coefficients for linear expansion at room temperature and volume expansion of liquid, respectively. T_M is the bulk melting point, ΔE_M is the latent heat of melting and D is the coefficient of self-diffusion just above the melting point. The "melting points" 1520 K and 1550 K are, in fact, the upper limits of the observed overheating region, as discussed in the text. Also included are the energies for formation of a vacancy E_V^F and a stacking fault γ. The last column shows the experimental data (references in brackets).

	EMT (n.n.)	EMT (ext.)	expt.
α (10^{-6} K^{-1})	18.1	18.1	16.7 [13]
α_V (10^{-4} K^{-1})	8.4	1.5	1.0 [13]
T_M (K)	<1520	<1550	1356 [13]
ΔE_M (eV/atom)	0.42	0.14	0.14 [13]
D (10^{-4} cm^{2}s^{-1})		0.6	1.0 [19]
E_V^F (eV)	1.59	1.47	1.31 [15]
γ (ergs/cm^2)	0	79	30 - 80 [16]

value of 0.66 eV[15]. The calculations of the stacking fault energies are discussed below.

3.2 Stacking Fault Energy in Al and Cu

One of our motivations to extend the range of the atomic interactions in EMT was the fact that the nearest neighbour formalism is not suitable for studying extended dislocations in fcc-metals because it does not give any energy for the stacking fault formed in the glide plane. The stacking fault energy determines the equilibrium distance between the partials of the extended dislocation. The stacking fault produces a hcp phase inside a fcc-crystal and the atomic interactions must then be extended (at least) to the third neighbours in the fcc-lattice to distinguish these two phases. The calculated energy for the infinite unrelaxed stacking fault as a function of the range of interactions is given in Table 2 for Al an Cu. The extending of interactions beyond the fourth neighbours in fcc-lattice makes the stacking fault energy worse. However, one can argue that the exponential approximation of the density tails is not a good description for the true atom induced densities far out. A similar sensitivity of the stacking fault energy to the range of the interaction has also been found for empirical pair potentials and pseudopotentials [16]. It is worth noticing that the fcc-metals with high stacking fault energies, such as Al and Ni, appear to be problematic also in the other novel density-dependent potential models [17]. These metals are either multivalent sp-bonded (Al) or have partially filled d-bands (Ni), and the approximations made in EMT to calculate electron densities prevent taking account for any directional effects or electronic redistribution in the stacking

Table 2. Calculated energies γ for unrelaxed stacking fault in aluminium and copper with the extended effective medium theory. The cut-off radius is given with numbers 3/4, 4/5 ... to show that the range of interactions has been in the middle of 3rd and 4th, 4th and 5th etc. neighbours in fcc-lattice. The stacking fault energies are in units of ergs/cm^2. The last line shows the range of the experimentally estimated values for γ from ref. [16].

cut-off	γ (Al)	γ (Cu)
3/4	55	79
4/5	-9	-13
5/6	7	7
6/7	-14	-17
——		
expt.	140-200	30-80

fault region. The details of the calculations of the stacking fault energies as well as the results for extended dislocation structures in Al and Cu predicted by the extended EMT will be reported elsewhere [18].

4. Conclusions

The extension of the atomic interactions beyond the nearest neighbours in EMT is of importance in the dynamic simulations of high temperature processes and enlarges the area of possible applications for the theory. It gives a good description for the defect energies (vacancy, stacking fault) in Cu. The theory could be further developed by parametrizing one-electron energy corrections [11] to the form suitable for the molecular dynamics simulations of transition metals or small metal clusters.

Acknowledgements

The author is indebted to M. Manninen, M. J. Puska, P. Stoltze and J. K. Nørskov for useful discussions. The Emil Aaltonen Foundation has supported this work financially.

References

1. M.S.Daw and M.I.Baskes: Phys. Rev. Lett. **50**, 1285 (1983); Phys. Rev. B **29**, 6443 (1984).
2. M.W.Finnis and J.R.Sinclair: Phil. Mag. A **50**, 45 (1984).
3. K.W.Jacobsen, J.K.Nørskov and M.J.Puska: Phys. Rev. B **35**, 7423 (1987).
4. M.Manninen: Phys. Rev. B **34**, 8486 (1986).
5. F.Ercolessi, E.Tosatti and M.Parrinello: Phys. Rev. Lett. **57**, 719 (1980).
6. H.Häkkinen and M. Manninen: J. Phys: Cond. Matter (in press).
7. K.W.Jacobsen and J.K.Nørskov: Phys. Rev. Lett. **59**, 2764 (1987); **60**, 2496 (1988).
8. O.B.Christensen, P.D.Ditlevsen, K.W.Jacobsen, P.Stoltze, O.H.Nielsen and J.K. Nørskov: Phys. Rev. B **40**, 1993 (1989).
9. P.Stoltze, J.K.Nørskov and U. Landman: Phys. Rev. Lett. **60**, 440 (1988).
10. M.J.Puska, R.M.Nieminen and M.Manninen: Phys. Rev. B **24**, 3037 (1981).
11. S.Myllyrinne, M.Manninen and J.K.Nørskov, in this proceedings.
12. M.Parrinello and A.Rahman: Phys. Rev. Lett. **45**, 1196 (1980).
13. D.N.Fischel, H.B.Crawford, D.A.Douglas and W.C.Eisler (eds.): *Handbook of Physics* (American Institute of Physics, New York 1972); R.C.Weast (ed.): *Handbook of Chemistry and Physics*, 66th ed. (CRC Press, Boca Raton 1985).
14. E.T.Chen, R.N.Barnett and U.Landman: Phys. Rev. B **40**, 924 (1989).

15. R.W.Siegel: "Atomic Defects and Diffusion in Metals", in *Point Defects and Defect Interactions in Metals*, ed. by J.I.Takamura, M.Doyama and M.Kiritani (University of Tokyo Press, Tokyo 1982).
16. R.M.J.Cotterill and M.Doyama: Phys. Rev. **145**, 465 (1966).
17. G.J.Ackland, G.Tichy, V.Vitek and M.W.Finnis: Phil. Mag. A **56**, 735 (1987).
18. H.Häkkinen, S.Mäkinen and M.Manninen, to be published.
19. D.B.Butrymowicz, J.R.Manning and M.E. Reed: *Diffusion Rate Data and Mass Transport Phenomena for Copper Systems* (National Bureau of Standards, Washington 1977).

Defect Migration in Perfect Lattices

R. Thetford

Theoretical Physics Division, Harwell Laboratory,
Didcot, Oxon, OX11 0RA, UK

Using semi-empirical Finnis-Sinclair potentials, I have investigated the migration of vacancies and self-interstitials in perfect lattices of the b.c.c. transition metals vanadium, niobium, tantalum, molybdenum and tungsten. DEVIL, a program that uses the method of conjugate gradients to perform static relaxation, is well suited to the calculation of energies of formation and migration. The DEVIL results presented here tend to support the 'one-interstitial' model for recovery of metals after radiation damage, against the 'two-interstitial' model. Calculations on anisotropic migration do give some hope to the two-interstitial school.

1. Introduction

The semi-empirical potentials of Finnis and Sinclair [1984] form a halfway step between the full *ab initio* calculations described by other authors at this workshop (for example Car [1989], Gillan [1989]), and earlier calculations using pair potentials. The Finnis and Sinclair formulation of many-body potentials for the b.c.c. transition metals builds in some of the results of tight-binding theory, and overcomes some of the problems with pair potentials. [Finnis and Sinclair, 1984].

I have been using the static relaxation program DEVIL [Thetford, 1985] to calculate various properties of vanadium, niobium, tantalum, molybdenum and tungsten; concentrating mainly on point defects (vacancies and interstitial atoms), and using the Finnis and Sinclair potentials as modified by Ackland and Thetford [1987].

Considerable controversy surrounds self-diffusion in radiation-damaged metals, and what mechanism is responsible for the so-called 'stage III recovery'. Proponents of the one-interstitial model (OIM) maintain that this stage is due to vacancy migration, with interstitial migration responsible for stage 'I_E'. The advocates of the two-interstitial model (TIM) invoke motion of a metastable interstitial state to explain stage I_E, ascribing the stage III recovery to migration of the ground state interstitial. Balluffi [1978] and Young [1978] describe the controversy well, and Wollenberger [1983] gives a sober review. (For an idea of the debate, see Schüle [1981], Jacques and Robrock [1981] and Frank [1981], particularly the discussion on this last paper. The battle recommenced at the next 'Point Defects' conference in Berlin: see, for example, the flatly contradictory statements in van Veen [1986] and Frank and Seeger [1986].) Any work that throws some light on this problem, although likely to be hailed by one set of protagonists and assailed by the other, must be useful.

This paper considers the migration of self-interstitial atoms and monovacancies (hereafter called simply interstitials and vacancies) in vanadium, niobium,

Springer Proceedings in Physics, Vol. 48 Many-Atom Interactions in Solids
Editors: R.M. Nieminen · M.J. Puska · M.J. Manninen

tantalum, molybdenum and tungsten. Sections 1.1 and 1.2 of this introduction describe the interstitial migrations studied here. Section 2 outlines the calculational procedure used, §3 gives the results of the interstitial migrations, and §4 gives the results of the vacancy migrations. In §5 I consider the implications of these results for the one- and two-interstitial models.

As all the calculations in this paper are performed at constant lattice parameter (constant volume), the calculated quantities are *energies (E)* of migration and formation. Experiments carried out at constant pressure will measure *enthalpies (H)*, and while the difference is not large for these metals (there being little change in the volume of the block as a vacancy migrates, for example), I have made the distinction throughout.

1.1 Migration of ⟨111⟩ crowdions and dumbbell interstitials

Consider for a moment a metal in which the interstitial forms a ⟨111⟩ dumbbell or a ⟨111⟩ crowdion. In order to use a 'diagonal' of the block, consider migration in the [11$\bar{1}$] direction. Let the dumbbell have length $2l_d$ and be centred on (0,0,0). The two atoms forming the dumbbell are then at $(\pm l_d/\sqrt{3})(1,1,-1)$; typically l_d is $0.38a$ in a b.c.c. lattice, giving dumbbell positions of $\pm(0.219,0.219,-0.219)a$.

Suppose now that for a crowdion centred at $(\frac{1}{4},\frac{1}{4},-\frac{1}{4})a$ the distance to the nearest neighbours is $2l_c$. The 'nearest-neighbour' atoms will then be at $(\frac{1}{4},\frac{1}{4},-\frac{1}{4})a \pm 2l_c(1,1,-1)$. In particular, if l_c takes the tungsten value of 0.382, one of these atoms is at $(-0.192,-0.192,0.192)a$.

Figure 1, a view of the (1$\bar{1}$0) plane, shows the two configurations: they are very close together. Moving the atoms from one configuration to the other is unlikely to cost a lot of energy: the distance involved is only $0.05a$, and the relaxed position of the rest of the lattice ought to be similar in the two cases.

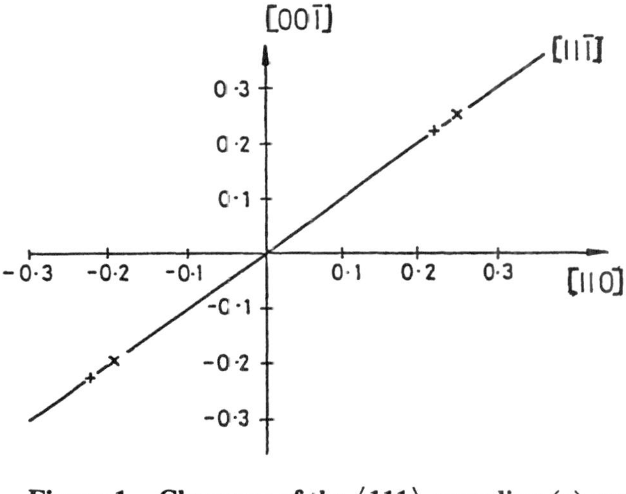

Figure 1. Closeness of the ⟨111⟩ crowdion (x) and the ⟨111⟩ split dumbbell (+).

Hence for metals where both the $\langle 111 \rangle$ dumbbell and the $\langle 111 \rangle$ crowdion are at least metastable, the interstitial formation energies should be similar. Moreover, migration in any of the $\langle 111 \rangle$ directions should be energetically favourable.

1.2 Migration of $\langle 110 \rangle$ dumbbell interstitials

Consider now a metal in which the stable interstitial configuration is a $\langle 110 \rangle$ split dumbbell. Once more, let the dumbbell have length $2l_d$ and be centred on (0,0,0). The two atoms forming the dumbbell are then at $(\pm l_d/\sqrt{2})(1,1,0)$; typically $l_d = 0.377a$ so the dumbbell atoms are at $\pm(0.267,0.267,0.00)a$.

The shortest migration along $[11\bar{1}]$ is to a split dumbbell centred at $(\frac{1}{2},\frac{1}{2},-\frac{1}{2})a$. This dumbbell can be aligned in any of the six $\langle 110 \rangle$ directions, but the ones of interest here are [110], $[01\bar{1}]$ and $[10\bar{1}]$: the others are not favourably oriented for migration from the given start position. By symmetry, migration to only one of the orientations $[01\bar{1}]$ and $[10\bar{1}]$ need be considered. Figure 2 shows the initial position of the dumbbell centred at $(0,0,0)a$, and the two possible final positions studied.

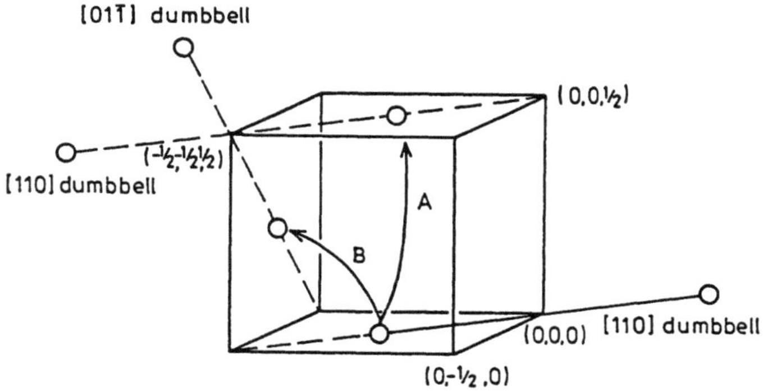

Figure 2. Possible migration paths for the [110] split interstitial.

Step A is the migration to a parallel dumbbell. The 'migrating' atom of the dumbbell pair typically moves from $(0.267,\ 0.267,\ 0.0)a$ to $(0.233,\ 0.233,-0.50)a$: a step that is very close to $[00\bar{1}]$. Jacques and Robrock [1981] have suggested that an interstitial moving in this manner will remain on one $\langle 110 \rangle$ plane for many migration steps: all four such $\langle 111 \rangle$ jumps from the new site are coplanar with the old jump. The length of a step A jump is $0.502a$.

Step B in figure 2 involves a rotation of the dumbbell, and is therefore known as the 'reorientation jump'. The dumbbell centre moves as before, but the dumbbell itself is now pointing along a different $\langle 110 \rangle$ direction. Using once more a typical dumbbell separation of $0.755a$, the migrating atom moves from $(0.267,\ 0.267,-0.0)a$ to $(0.50,\ 0.233,-0.233)a$. This step has length $0.331a$, and its direction is very close to $[10\bar{1}]$.

The nomenclature used here for the migration steps is the same as that used by Harder and Bacon [1986].

2. Calculational Procedure

The static relaxation program DEVIL [Thetford, 1985] is able to freeze the interstitial on various planes along the migration path, permitting energies of migration to be calculated. To step an atom i by a distance δ in the direction of unit vector $\mathbf{n}$, the position $\mathbf{x}_i$ at the end of one relaxation is simply replaced by

$$\mathbf{x}_i' = \mathbf{x}_i + \delta\,\mathbf{n} \tag{1}$$

The migrating atom itself may relax on the plane perpendicular to $\mathbf{n}$. Stepping to the next position involves moving only the migrating atom; the rest of the lattice then relaxes from the position occupied at the end of the previous step. This process should mirror what happens in a real migration.

I performed all of the interstitial migration calculations in a block with edges in the [111], [1$\bar{1}$0] and [11$\bar{2}$] directions. These rather odd boundaries are the same as those required for work on dislocations [Thetford, 1988]. The block has fixed boundaries in the first two directions, and a periodic boundary condition is applied in the third direction. The free region is 24×12×24 lattice planes (giving dimensions $4a\sqrt{3}\times 6a\sqrt{2}\times 4a\sqrt{6}$); including the boundary regions gives a final calculation region of 42×18×24 lattice planes. The free region contains 1152 atoms, the fixed boundaries 1872 and the periodic boundaries contain 2268 images of atoms in the main block.

The distortion round a vacancy is much less than round an interstitial or a dislocation, so a much smaller block can be used to calculate the vacancy formation energy. I used a 10×10×10 block (250 free atoms) with periodic boundary conditions. The block edges are all parallel to the basis cube: i.e. [100], [010] and [001].

Vacancy migration is easily dealt with: define a pair of vacancies, put an interstitial in one and migrate it by the above method towards the other. To allow room for any relaxations associated with the migration, the migration calculations themselves have been performed in a slightly larger (12×12×12) block. This block must have **fixed** boundaries: otherwise, when the interstitial atom is pushed in its migration direction, the whole of the rest of the lattice will translate with it! The block contains 432 free atoms, plus 1026 atoms in the boundary layers.

3. Interstitial Migration Results

Table 1 summarises the results obtained, and the subsections below discuss the calculations in more detail.

Table 1 Calculated interstitial migration energies (eV).

metal	0.25[10$\bar{1}$] jump	$\frac{1}{2}$[00$\bar{1}$] jump	⟨111⟩ migration
V	0.29	0.38	—
Nb	0.22	0.37	—
Ta	0.37	0.36	—
Mo	0.23	0.25	—
W	—	—	0.026

3.1 Niobium: reorientation jump favoured

When I started doing the interstitial migration calculations, I expected the 'reorientation jump', path B in figure 2, to be favoured over path A. The niobium calculation bears out this expectation rather well. The curves of energy against distance along migration direction (figure 3) have similar shapes and are nicely symmetrical. The longer migration path A has a correspondingly higher migration energy. There are no shocks here.

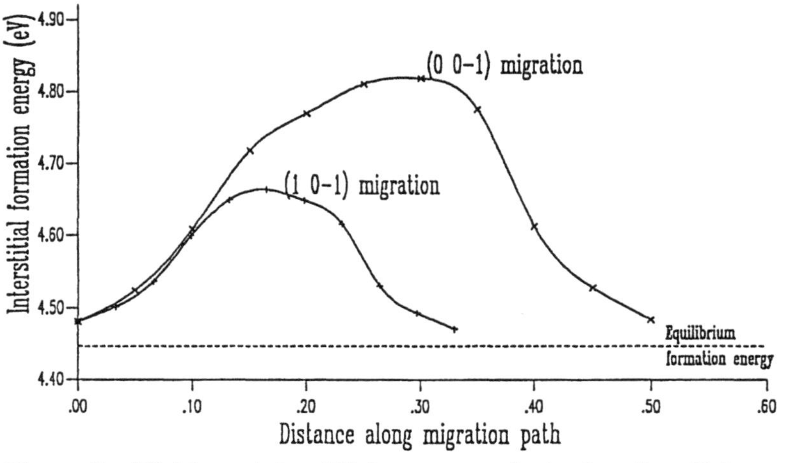

Figure 3. Niobium: interstitial energy against migration distance.

3.2 Vanadium: a second-neighbour jump

The $[10\bar{1}]$ migration in vanadium goes as expected. The migration step has a fairly small energy, the migrating atom follows the obvious path and the curve of energy against distance along the forced migration direction is symmetric. Figure 4 shows this curve.

However, the results for the $\frac{1}{2}[00\bar{1}]$ step were not what I had expected. Instead of the dumbbell going from being centred at $(0,0,0)a$ to $(\frac{1}{2},\frac{1}{2},-\frac{1}{2})a$, the final split interstitial is centred at $(1,1,-1)a$. This migration step has been proposed by Johnson [1985] as a possible mechanism for 2-d migration (the dumbbell always migrates in the $\langle 110 \rangle$ plane that is coplanar with its two atoms). I did not expect to see this step favoured over the shorter jump though, particularly when I was pushing the migrating atom in the correct direction for it to make the short jump.

Figure 5 shows the path of the migrating atom $(X-X')$, starting from the relaxed dumbbell position of $(0.26,\ 0.26,\ 0.0)a$. The positions of the atoms on the perfect lattice are shown as a background reference: these are **not** the relaxed positions. For reasons connected with the shape of the block, the migration is along $[11\bar{1}]$ rather than $[111]$.

The z-coordinate is reduced in steps of $0.1a$, with relaxation allowed in the (001) plane, and for the first five steps, everything goes smoothly. But the relaxed

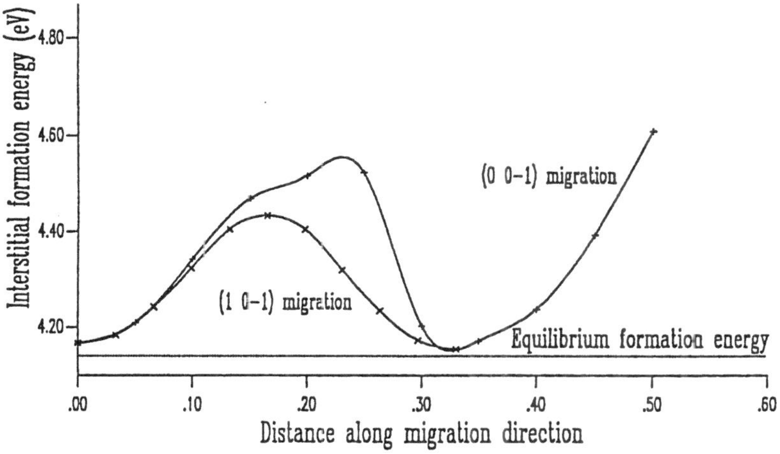

Figure 4. Vanadium: interstitial energy against migration distance.

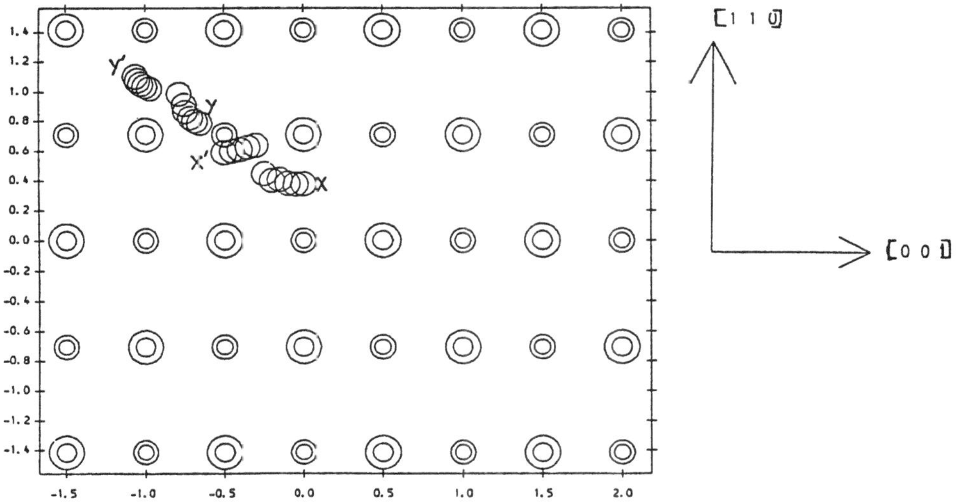

Figure 5. Vanadium: paths of migrating atoms in the 'double jump'.

position for the next step is a long way from the previous path, and the closest pair of atoms in the solid is no longer centred at $(\frac{1}{2},\frac{1}{2},-\frac{1}{2})a$ but at $(1,1,-1)a$. The migration path $(Y-Y')$ of the atom originally at $(\frac{1}{2},\frac{1}{2},-\frac{1}{2})a$ is also plotted, and it too shows the jump: from being part of the dumbbell centred at $(\frac{1}{2},\frac{1}{2},-\frac{1}{2})a$ to being part of the dumbbell centred at$(1,1,-1)a$. Pushing the 'migrating' atom further in the z–direction merely upsets the equilibrium of the newly-formed dumbbell, raising the energy as shown in figure 4.

3.3 Tantalum: 2-d migration possible

In tantalum, the short migration step (path B) once more behaves as expected (figure 6). The path A migration seems to find a flat energy plateau with a wide

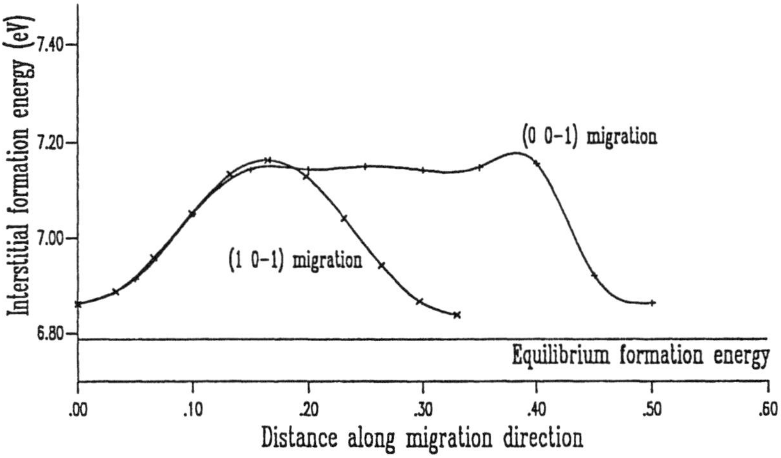

Figure 6. Tantalum: interstitial energy against migration distance.

saddle, before descending into the expected minimum, a jump by $(\frac{1}{2},\frac{1}{2},-\frac{1}{2})a$. This phenomenon appears to be connected with the relatively flat portion of figure 8 of Ackland and Thetford [1987]: the interstitial formation energy changes only slowly as the dumbbell rotates from ⟨110⟩ to ⟨111⟩. There are no unusual features in the plot of the atom paths.

3.4 Molybdenum: complicated by the bent interstitial

Self-interstitial migration in molybdenum is less straightforward, mainly because the ⟨110⟩ dumbbell is not the lowest-energy configuration. The 'bent interstitial', first mentioned by Thetford [1985] and examined in more detail by Ackland and Thetford [1987] is rotated away from ⟨110⟩ by some 15°. Figure 6 in Ackland and Thetford [1987] shows that a rotation very nearly in the (001) plane has the lowest energy, but rotation in any direction will reduce the interstitial formation energy by about 0.04 eV. Starting from a [110] split dumbbell therefore means that the energy will drop slightly to start with, and will rise at the end of the migration paths. Both the $[10\bar{1}]$ (path B) and the $[00\bar{1}]$ (path A) migrations in figure 7 show this nicely.

Both curves are close to being symmetric, upset only by the interstitial striving to attain its minimum-energy bent position. The path B migration has a slightly lower energy than the longer step A, so while 2-d migration is not strongly favoured, neither is it (energetically) excluded.

Again, the plots of the paths of the migrating atoms show nothing unexpected.

3.5 Tungsten: 1-d migration

I have previously reported on calculations of the interstitial migration energy in tungsten [Ackland and Thetford 1987]. The energies of the tungsten ⟨111⟩ crowdion and the ⟨111⟩ dumbbell interstitial were very close. Figure 1 shows that the positions of the atoms involved are also very similar, so the earlier result (a low migration energy along ⟨111⟩) was not a surprise.

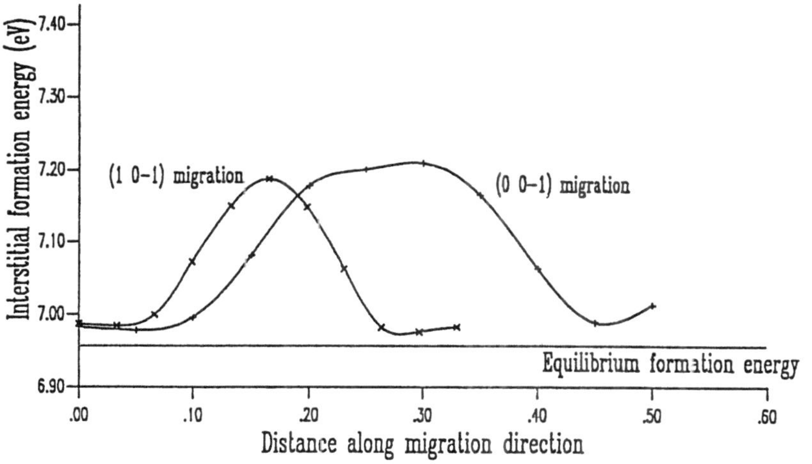

Figure 7. Molybdenum: interstitial energy against migration distance.

The stability of the ⟨111⟩ crowdion and dumbbell in tungsten seems to be an artefact of the potential: as far as I am aware, all the experimental evidence is that the interstitial adopts a ⟨110⟩ split dumbbell configuration, as in all the other b.c.c. transition metals.

4. Vacancy Migration Results

The simulations of vacancy migration do not show the same richness of behaviour as those of interstitial migration. In all cases, the interstitial that is migrating between the two declared vacancies follows a straight path, and the lattice relaxes round it in the expected manner.

Calculated vacancy formation energies for the five b.c.c. metals under consideration are given in table 2, and compared with experimental results. Agreement is good, though the theoretical values tend to be at the low end of the experimental range. The DEVIL calculation used a 10×10×10 block with periodic boundary conditions. The block edges are all parallel to the basis cube: i.e. [100], [010] and [001].

The migration calculations used a slightly larger (12×12×12) block with fixed boundaries, as explained above.

I have calculated migration energies for each of the five metals, and for migration jumps to both first and second neighbours. The jumps to second neighbours have activation energies larger by factors of 3.3 (Mo) to 4.4 (V) times the jumps to nearest neighbours, and are not tabulated. Table 3 compares the DEVIL results for the shortest jump with calculations by Harder and Bacon [1986] and Rebonato *et al.* [1987], and with various experimental values. No experimental result seems to be available for vanadium, due apparently to the difficulty of obtaining pure specimens, but the experimental values agree magnificently for the other metals.

Table 2 Calculated energies (E_v^f) and experimental enthalpies (H_v^f) [compilation of Herzig and Köhler 1981] of vacancy formation (eV).

metal	DEVIL	experiment
V	1.85	2.1
Nb	2.51	2.7 to 3.0
Ta	2.91	2.8 to 3.1
Mo	2.55	3.0 to 3.25
W	3.63	3.6 to 4.0

More recent measurements (for example Ziegler and Schäfer [1986] and Mundy, Ockers and Smedskjaer [1986]) are not significantly different.

Table 3 Calculated and experimental vacancy migration energies

	E_v^m / eV			H_v^m/eV
metal	DEVIL	Harder[a]	Rebonato[b]	experiment
V	0.69	0.76	0.72	0.8[c]
Nb	0.85	0.91	0.87	0.95 ± 0.15[d]
Ta	1.13	1.22	1.12	1.1 ± 0.2[d]
Mo	1.30	1.32	1.32	1.3 ± 0.2[d]
W	1.45			1.5 ± 0.3[e]

[a] Harder and Bacon [1986]
[b] Rebonato, Welch, Hatcher and Bilello [1987]
[c] deduced by subtracting H_v^f=2.1 eV (table 3) from Q_1=2.9 eV [Siegel, 1981].
[d] by electron microscopy [Phillipp, Saile and Urban, 1981]
[e] from positron experiments [Tietze, Takaki, Schwirtlich and Schultz, 1981]

5. Discussion

I had three reasons for doing the perfect-lattice migration calculations described in this paper.

Firstly, I wanted to be sure that the migration algorithm in DEVIL was working properly, before I tried to apply it to the complicated environment around a dislocation.

Secondly, although experimental values for migration enthalpies are clouded with uncertainty, they are important physically. They offer a useful comparison point between experiment and theory, and also between different theories.

Thirdly, details of the migration may influence other properties of the metal lattice. Following the original suggestion of Foreman [1972] (who invoked one-dimensional migration), Evans [1983] has proposed a theory for the formation of void lattices in molybdenum that is based upon 2-d migration of interstitials. Computer simulation [Evans, 1985] shows that in a crystal where interstitials are constrained to move on certain planes, an initially random distribution of voids will turn into a regular void lattice. (Further implications of anisotropic migration of interstitials are discussed by Evans and Foreman [1985].) Had the migration step A from figure 2 been favoured over step B, or even had the double jump seen in vanadium (§3.2) been energetically favoured over the step B single jump, then the migrating interstitial would tend always to have its dumbbell lined up in the same $\langle 110 \rangle$ direction. It would then move in the $\{110\}$ plane defined by this direction, giving the 2-d migration required for the Evans theory.

In all these DEVIL interstitial migration calculations, the energy for the planar step has at least as high an energy as that for the alternative 'reorientation jump'. This fact suggests that the $\langle 111 \rangle$ migration will tend to have a 3-dimensional character. (But note that two-dimensional migration can still arise, even if it is not favoured energetically: work by Lucasson, Maury and Lucasson [1986] on vibrational modes of interstitials suggests that the $\langle 110 \rangle$ dumbbell vibrates on its site with a motion that encourages planar migration.)

An alternative theory of void lattice formation has been proposed by Woo and Frank [1985]. Being advocates of the two-interstitial model (TIM), they suggest that interstitial migration takes place in a two-stage process, the first stage of which is anisotropic. Low-temperature migration is due to a metastable excited $\langle 111 \rangle$ crowdion state, which has a higher formation energy than the 'ground state' $\langle 110 \rangle$ dumbbell. This crowdion migrates along the $\langle 111 \rangle$ direction, via the $\langle 111 \rangle$ dumbbell, and provides the required anisotropy. The tungsten calculation in Ackland and Thetford [1987] shows just how easy this $\langle 111 \rangle$ migration can be. Only at higher temperatures are the ground state $\langle 110 \rangle$ dumbbells able to move.

This theory requires that the $\langle 111 \rangle$ crowdion is more stable than the $\langle 111 \rangle$ dumbbell. Table 3 of Ackland and Thetford [1987] shows that in vanadium and tantalum the two forms of interstitial have essentially equal energy, while the dumbbell is more stable in niobium and the crowdion in molybdenum. As all of the four metals have energy-orientation diagrams similar to figure 8 in Ackland and Thetford [1987] — i.e. with no maximum between $\langle 111 \rangle$ and $\langle 110 \rangle$ — an unstable crowdion scotches the two-interstitial model immediately, for the $\langle 111 \rangle$ dumbbell is unstable with respect to the $\langle 110 \rangle$ dumbbell. Hence the TIM is unlikely to apply to niobium. (The results here do not quite spell doom for the TIM, as the important quantity is not the energy as calculated here (internal energy, at constant volume) but the Gibbs free energy G, which takes into account changes in volume and entropy.)

Consider now the cases where the crowdion *is* stable with respect to the $\langle 111 \rangle$ dumbbell: molybdenum, and perhaps vanadium and tantalum. The instability of the $\langle 111 \rangle$ dumbbell with respect to the $\langle 110 \rangle$ dumbbell interstitial has further consequences for the TIM. As the $\langle 111 \rangle$ dumbbell configuration is the saddle point on the $\langle 111 \rangle$ migration path, the crowdion migration energy is just

$$E_c^m = E_i^{m,1D} = |E_c^f - E_{\langle 111 \rangle}^f| \tag{2}$$

But the energy to convert the crowdion to a ⟨110⟩ dumbbell cannot be any higher (and could be lower), as the ⟨111⟩ dumbbell is a saddle point for a conversion path too! One must therefore invoke entropy and dynamic effects to explain the long-range migration of crowdions without conversion to dumbbells in stage I_E of recovery.

The possibility of migration along ⟨111⟩ directions still remains. Table 1 gives the energies of migration for the three-dimensional reorientation jump (B) and the two-dimensional jump A. But rotation of a ⟨110⟩ dumbbell to ⟨111⟩, followed by motion along ⟨111⟩, is a possible migration route, particularly if the energies for ⟨111⟩ migration are small. Assuming that there are no maxima in energy between the ⟨111⟩ crowdion and the ⟨111⟩ dumbbell, table 3 of Ackland and Thetford [1987] gives enough information to determine the migration energies for this process. The activation energy E_{act} is just the difference between the ⟨110⟩ dumbbell and the greater of E_c^f and $E^f_{\langle 111\rangle}$, while the migration energy is just the crowdion migration energy defined in equation (2). Table 4 lists these results, and reproduces the earlier values for $E_i^{m,3D}$ and $E_i^{m,2D}$ from table 1. A comparison of these values shows that the ⟨111⟩ migration does indeed have lowest activation energy in tantalum, and is very little different in energy from the other two mechanisms in molybdenum.

Table 4 Energies for 3-dimensional, 2-dimensional and 1-dimensional interstitial migration.

metal	$E_i^{m,3D}$ /eV	$E_i^{m,2D}$ /eV	E_{act} /eV	$E_i^{m,1D}$ /eV
V	0.29	0.38	0.45	0.008
Nb	0.22	0.37	0.39	0.062
Ta	0.37	0.36	0.31	0.001
Mo	0.23	0.25	0.24	0.036
W	—	—	—	0.026

The migration along ⟨111⟩ need not be one-dimensional though. Unless one such step, driven by momentum or the excitation of a suitable phonon, favours another in the same direction, the saddle-point ⟨111⟩ interstitial configuration will soon relax to a ⟨110⟩ dumbbell. And as this dumbbell interstitial may well be excited to a different ⟨111⟩ direction next time, long-range one-dimensional migration in unlikely. (In tungsten, of course the ⟨111⟩ crowdion is the DEVIL ground state and ⟨111⟩ migration ought to be very easy.)

Although the migration energies $E_i^{m,1D}$ in table 4 are very low, they are not quite low enough to explain the apparent long-range migration of interstitials seen during the stage I_E recovery. Even down towards liquid helium temperatures, self-interstitial defects produced by radiation damage seem to be able to migrate in vanadium, niobium and tantalum. Table 5 gives the temperatures obtained by Schultz [1981] for the stage I_E recovery of the b.c.c. transition metals, and the

Table 5 Temperatures and corresponding enthalpies for stage I_E recovery [Schultz, 1981].

metal	T_{I_E} /K	H_{mig} /meV
V	<5	<0.4
Nb	<8	<0.7
Ta	<8	<0.7
Mo	35	3.0
W	27	2.3

corresponding enthalpies of migration. Note that these enthalpies are in meV, not eV! Even the apparently tiny migration energies calculated by DEVIL are too high to explain self-diffusion at these extremely low temperatures. So even if the ⟨111⟩ dumbbell interstitial *is* stable against conversion to a ⟨110⟩ dumbbell, the experimental results remain unexplained.

Finally, consider thermally-activated self-diffusion. Vacancies have lower formation energies than self-interstitials, but higher migration energies. Comparing $E_v^f + E_v^{mig}$ with $E_i^f + E_i^{mig}$ should give an indication as to whether self-diffusion is due to interstitials or vacancies. Taking the minimum of $E_i^{m,3D}$, $E_i^{m,2D}$ and E_{act} from table 4 for the value of E_i^m gives table 6. Vacancy migration is favoured every time, and seems in accord with experimental results Q^{SD} for the heat of self-diffusion.

Table 6 Calculated activation energies for migration of monovacancies and ⟨111⟩ crowdions, compared with experimental values of heat of self-diffusion Q^{SD}.

metal	$E_v^f + E_v^{mig}$ /eV	$E_i^f + E_i^{mig}$ /eV	exptl. Q^{SD} /eV
V	2.54	4.45	3.2 [a]
Nb	3.36	4.69	3.62 [b]
Ta	4.04	7.14	4.0±0.2 [c]
Mo	3.85	7.20	4.53 [d]
W	5.08	8.90	5.45 [e]

[a] Maier, Peo, Saile, Schaefer and Seeger [1979]
[b] Einziger, Mundy and Hoff [1978]
[c] Hönes [1979]
[d] Maier, Mehrer and Rein [1979]
[e] Mundy, Rothman, Lam, Hoff and Nowicki [1978]

6. Conclusions

In this paper I have investigated the migration of point defects in perfect lattices. Several interesting phenomena occur.

Tungsten remains an anomaly. I can see no obvious reason why the Finnis-Sinclair potential should predict that the stable form of the interstitial is a $\langle 111 \rangle$ crowdion, but given that it does, DEVIL predicts $\langle 111 \rangle$ migration with a very low activation energy. The tungsten system thus has value as a model for what may happen if the crowdion is metastable in any of these b.c.c. metals.

The most striking result is that self-diffusion in radiation-damaged metals is unlikely to be due the migration of a metastable, frozen-in, $\langle 111 \rangle$ interstitial. Ignoring the effects of configurational entropy and of the different attempt frequencies, table 4 implies that migration of an excited interstitial state is a possibility in tantalum and molybdenum, but this migration will be wholesale: all the interstitials in the material can migrate in this manner. There seems little chance of a small fraction of metastable interstitials being responsible for stage I_E recovery, because the proposed $\langle 111 \rangle$ interstitials are simply unstable! In coming to this conclusion I join most of the rest of the world outside Stuttgart in accepting that the one-interstitial model for resistivity recovery is correct.

I do not suggest that the idea of interstitials in non-equilibrium states migrating should be abandoned. A more rigorous calculation, working with Gibbs free energy rather than internal energy *may* come to a different conclusion. And the self-diffusion seen at extremely low temperatures in metals that have undergone radiation damage below 5K suggests an extremely low migration energy, in a range seen here only for $\langle 111 \rangle$ migration. Perhaps the $\langle 111 \rangle$ crowdion *is* somehow stabilised after all?

Another fundamental result is that the total energy for vacancy migration (i.e. $E_v^f + E_v^{mig}$) is always considerably less than the energy required to excite and migrate an interstitial dumbbell. This suggests that normal thermal self-diffusion is due to vacancies.

Acknowledgement

The work described in this paper is part of the longer term research carried out within the Underlying Programme of the UKAEA.

References

Ackland, G.J. and Thetford, R., 1987, *Phil.Mag.A* **56,** 15-30.

Balluffi, R.W., 1978, *J.Nucl.Mater.* **69/70** 240.

Car, R., 1989, at this workshop.

Einziger, R.E., Mundy, J.N. and Hoff, H.A., 1978, *Phys.Rev. B* **17** 440.

Evans, J.H., 1983, *J.Nucl.Mater.* **119** 180.

Evans, J.H., 1985, *J.Nucl.Mater.* **132** 147.

Evans, J.H. and Foreman, A.J.E., 1985, *J.Nucl.Mater.* **137** 1.

Finnis, M.W. and Sinclair, J.E., 1984, *Phil.Mag.A* **50** 45.

Foreman, A.J.E., 1972, Harwell Report AERE-R.7135, HMSO, London.

Frank, W., 1981, *Proc.Yamada Conf. on Point Defects and Defect Interactions in Metals,* Kyoto, Japan, 1981. Eds. J. Takamura, M. Doyama and M. Kiritani (Univ. of Tokyo Press, 1982) p.203.

Frank, W. and Seeger, A., 1986, Proc.Int.Conf. on Vacancies and Interstitials in Metals and Alloys, Berlin, 1986, published as *Materials Science Forum* **15-18** (1987) 57.

Gillan, M.J., 1989, at this workshop.

Harder, J.M. and Bacon, D.J., 1986, *Phil.Mag.A* **54,** 651.

Herzig, C. and Köhler, U., Proc.Int.Conf. on Vacancies and Interstitials in Metals and Alloys, Berlin, 1986, published as *Materials Science Forum* **15-18** (1987) 451.

Hönes, D., 1979, Diplomarbeit (dissertation) Stuttgart.

Jacques, H. and Robrock, K.-H., *Proc.Yamada Conf. on Point Defects and Defect Interactions in Metals,* Kyoto, Japan, 1981. Eds. J. Takamura, M. Doyama and M. Kiritani (Univ. of Tokyo Press, 1982) p.159.

Johnson, R.A., 1985, *J.Nucl.Mater.* **135** 283.

Lucasson, P., Maury, F. and Lucasson, A., Proc.Int.Conf. on Vacancies and Interstitials in Metals and Alloys, Berlin, 1986, published as *Materials Science Forum* **15-18** (1987) 231.

Maier, K., Mehrer, H. and Rein, G., 1979, *Z.Metallk.* **70** 271.

Maier, K., Peo, M., Saile, B., Schaefer, H.-E. and Seeger, A., 1979, *Phil.Mag. A* **40** 701.

Mundy, J.N., Ockers, S.T and Smedskjaer, L.C., 1986, Proc.Int.Conf. on Vacancies and Interstitials in Metals and Alloys, Berlin, 1986, published as *Materials Science Forum* **15-18** (1987) 199.

Mundy, J.N., Rothman, S.J., Lam, N.Q., Hoff, H.A. and Nowicki, L.J., 1978, *Phys.Rev. B* **18** 6566.

Phillipp, F., Saile, B. and Urban, K., 1981, *Proc.Yamada Conf. on Point Defects and Defect Interactions in Metals,* Kyoto, Japan, 1981. Eds. J. Takamura, M. Doyama and M. Kiritani (Univ. of Tokyo Press, 1982) p.261.

Rebonato, R., Welch, D.O., Hatcher, R.D. and Bilello, J.C., 1987, *Phil.Mag.A* **55** 655.

Schüle, W., 1981, *Proc.Yamada Conf. on Point Defects and Defect Interactions in Metals,* Kyoto, Japan, 1981. Eds. J. Takamura, M. Doyama and M. Kiritani (Univ. of Tokyo Press, 1982) p.209.

Schultz, H., 1981, *Proc.Yamada Conf. on Point Defects and Defect Interactions in Metals,* Kyoto, Japan, 1981. Eds. J. Takamura, M. Doyama and M. Kiritani (Univ. of Tokyo Press, 1982) p.183.

Siegel, R.W., 1981, *Proc.Yamada Conf. on Point Defects and Defect Interactions in Metals,* Kyoto, Japan, 1981. Eds. J. Takamura, M. Doyama and M. Kiritani (Univ. of Tokyo Press, 1982) p.533.

Thetford, R., 1985, Harwell report AERE-M.3507, HMSO, London.

Thetford, R., 1988, Harwell report AERE-M.3679, HMSO, London.

Tietze, M., Takaki, S., Schwirtlich, I.A. and Schultz, H., 1981, *Proc.Yamada Conf. on Point Defects and Defect Interactions in Metals,* Kyoto, Japan, 1981. Eds. J. Takamura, M. Doyama and M. Kiritani (Univ. of Tokyo Press, 1982) p.265.

van Veen, A., 1986, Proc.Int.Conf. on Vacancies and Interstitials in Metals and Alloys, Berlin, 1986, published as *Materials Science Forum* **15-18** (1987) 3.

Woo, C.H. and Frank W., 1985, *J.Nucl.Mater.* **137** 7.

Wollenberger, H.J., 1983, 'Point Defects', in *Physical Metallurgy* eds. R.W. Cahn and P. Haasen (North-Holland, Amsterdam, 1983) pp1139-1221.

Young, F.W., Jr., 1978, *J.Nucl.Mater.* **69/70** 310.

Ziegler, R. and Schaefer, H.E., 1986, Proc.Int.Conf. on Vacancies and Interstitials in Metals and Alloys, Berlin, 1986, published as *Materials Science Forum* **15-18** (1987) 145.

Dynamic Simulations of Dislocation Core Structures in Gold Using Many-Atom Interactions

J. von Boehm[1] *and R.M. Nieminen*[2]

Helsinki University of Technology, SF-02150 Espoo, Finland

The atomic structures of partial dislocations in fcc gold are obtained by using recently suggested many-atom force laws in molecular dynamics simulations. Particular attention is paid to the dislocation core regions.

1. Introduction

Realistic simulation of atomistic properties of complex and low-symmetry materials is emerging as a feasible tool in condensed-matter studies. Key ingredients in such simulations are the interatomic force laws. The traditional recipe of pairwise interactions as applied to metallic substances has serious flaws. For example, two-body forces fail to reproduce the high ratio between the cohesive energy and melting temperature or vacancy formation energy, the low entropy change upon melting, and the breakdown of the Cauchy relation. It is thus obvious that more sophisticated techniques are required in order to obtain reliable structural predictions, not to mention truly dynamic phenomena.

Static and dynamic properties of dislocations in metals are a topic of long-standing interest and important for technological applications. Atomistic simulations of dislocations [3,4] are still few, as the main interest has traditionally focussed on the more macroscopic properties.

The aim of the present work is to incorporate the recent suggestions for describing many-atom interactions in metallic systems into a large-scale simulation of dislocation structure and dynamics. First results for the dissociation of an edge dislocation in gold into two Schockley partials are described below.

2. Many-atom forces

In order to describe correctly the Cauchy discrepancy (the difference between the elastic constants C_{12} and C_{44}), the difference between the vacancy formation energy and cohesive energy, the surface properties *etc.* for metals one has to go beyond the pair-potential approximation. To achieve this goal several recent schemes express the total energy in the following generic form [5-14]

Editors: R.M. Nieminen · M.J. Puska · M.J. Manninen

$$V = \frac{1}{2} \sum_{\substack{i,j=1 \\ (j \neq i)}}^{N} \Phi\left(\left|\mathbf{r}_i - \mathbf{r}_j\right|\right) + \sum_{i=1}^{N} U\left(n_i\right) \tag{1}$$

where

$$n_i = \sum_{\substack{j=1 \\ (j \neq i)}}^{N} \rho\left(\left|\mathbf{r}_i - \mathbf{r}_j\right|\right). \tag{2}$$

In equations (1) and (2) Φ is the (mainly repulsive) pair-potential, $\mathbf{r}_i$ is the position vector of nucleus i and U is the (mainly attractive) many-atom potential. In the embedded atom method (EAM) [5-8] and in the effective medium theory (EMT) [13,14] n_i is the background density at nucleus i formed by the radial electron densities ρ from all other atoms j. In the glue model for gold [9,10] n_i is the effective coordination number of atom i (= 12 for the fcc lattice) and ρ is an empirical short-ranged monotonically decreasing function of distance. In the Finnis-Sinclair scheme [11,12] the many-atom potential assumes the tight-binding form

$$U\left(n_i\right) = -\sqrt{n_i} \tag{3}$$

where n_i is the second moment of the density of states. The short-ranged radial function ρ used to represent n_i is determined empirically.

Since the transformation

$$\begin{aligned} \tilde{\Phi}(r) &= \Phi(r) + 2\lambda\rho(r) \\ \tilde{U}(n) &= U(n) - \lambda n \end{aligned} \tag{4}$$

(λ arbitrary) leaves the total energy V in Equation (1) unchanged the pair-potential and the many-atom potential are not uniquely determined. Therefore a direct comparison of the functions Φ and U is meaningless.

The force $-\nabla_i V$ acting on nucleus i, needed in molecular dynamics (MD) simulations, is obtained from Eqs. (1) and (2):

$$\mathbf{F}_i = -\sum_{j(\neq i)} \left\{\Phi'\left(\left|\mathbf{r}_i - \mathbf{r}_j\right|\right) + \left[U'\left(n_i\right) + U'\left(n_j\right)\right] \rho'\left(\left|\mathbf{r}_i - \mathbf{r}_j\right|\right)\right\} \frac{\mathbf{r}_i - \mathbf{r}_j}{\left|\mathbf{r}_i - \mathbf{r}_j\right|} \tag{5}$$

where ' denotes a derivative with respect to the argument.

3. Initial geometry of the perfect [112] edge dislocation

The face-centered cubic crystal structure of gold is shown in Fig. 1. Energetically the most favourable perfect edge dislocation is the $[11\bar{2}]$ dislocation [15] (z-axis in Fig. 1). The most densely packed plane, the $(11\bar{1})$

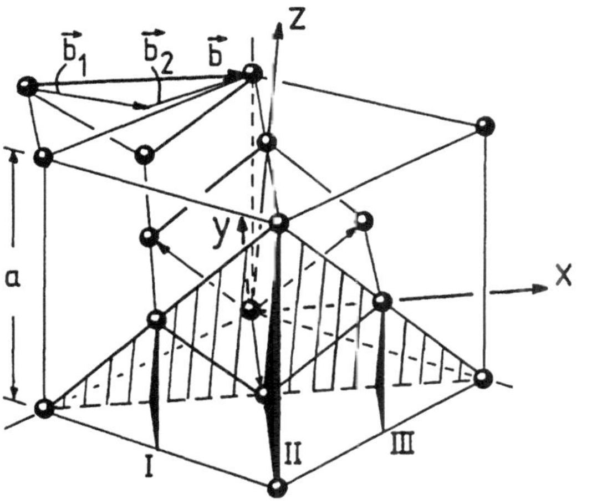

Figure 1. Face-centered cubic crystal structure of gold. a denotes the lattice constant and **b** the Burgers-vector of the perfect [112] edge dislocation. The $(11\bar{1})$-plane acting as the glide-plane is shaded. The atom half-planes I, II and III are indicated with black areas. For further information see the text.

plane (Fig. 1), acts as the glide-plane. The Burgers vector $\mathbf{b} = \frac{a}{2}[\bar{1}10] = \frac{a}{\sqrt{2}}\mathbf{u}_x$ has the lowest possible length and minimizes the stored energy roughly proportional to b^2.

The initial geometry of the perfect [112] edge dislocation was prepared following the Cotterill-Doyama recipe [3]. First, the yz-half-plane II ($y < 0$) of atoms as well as the parallel half-plane I to the left (see Fig. 1) were removed. Second, the parallel half-plane III on the right-hand side was moved by $-\mathbf{b}/2$ to coincide with the yz-plane. Finally the atoms for $x > 0$ were displaced according to the isotropic continuum elasticity theory [15]

$$u_x = \frac{b}{2\pi}\left[\arctan\left(\frac{y}{x}\right) + \frac{1}{2(1-\nu)}\frac{xy}{x^2+y^2}\right]$$

$$u_y = \frac{b}{4\pi(1-\nu)}\left[(1-2\nu)\ln\left(\sqrt{x^2+y^2}\right) + \frac{x^2}{x^2+y^2}\right] \qquad (6)$$

$$u_z = 0$$

(ν is the Poisson ratio = 0.42 for gold [15]) and the atoms for $x < 0$ were displaced symmetrically with respect to the yz-plane. (However, the atoms closest to the z-axis were displaced only by half of the displacements of Eq. (6) to prevent the coincidences with the yz-plane atoms.) This gives a symmetric arrangement with respect to the yz-plane.

4. Splitting of the perfect [112] edge dislocation into two Shockley partials

The Burgers vector may be divided into two partial ones as follows (see Fig. 1):

$$
\begin{aligned}
\mathbf{b} &= \mathbf{b}_1 + \mathbf{b}_2 \\
\mathbf{b} &= \frac{a}{2}[\bar{1}10] = \frac{a}{\sqrt{2}}\mathbf{u}_x \\
\mathbf{b}_1 &= \frac{a}{6}[\bar{2}1\bar{1}] = \frac{a}{2\sqrt{2}}\mathbf{u}_x - \frac{a}{2\sqrt{6}}\mathbf{u}_y \\
\mathbf{b}_2 &= \frac{a}{6}[\bar{1}21] = \frac{a}{2\sqrt{2}}\mathbf{u}_x + \frac{a}{2\sqrt{6}}\mathbf{u}_y .
\end{aligned}
\tag{7}
$$

Since

$$
b^2 = \frac{a^2}{2} > b_1^2 + b_2^2 = \frac{a^2}{3} \tag{8}
$$

the perfect [112] edge dislocation has for energetic reasons a tendency to separate further into two (imperfect) Shockley partial dislocations [15]. An hcp stacking fault ribbon forms between the separated partials. The partial separation r_e may be estimated to be [3]

$$
r_e = \frac{\mu a^2 (2 + \nu)}{48\pi\gamma(1 - \nu)} \tag{9}
$$

where μ is the shear modulus ($= 2.78 \cdot 10^{10}$ $\mathrm{kgm^{-1}s^{-2}}$ for gold [15] and γ the stacking fault energy (= 0.032 $\mathrm{kgs^{-2}}$ for gold [12]). With these values the partial separation for gold becomes 40 Å.

5. Molecular dynamics simulations of the dislocation cores

All the MD simulations presented in this paper were performed with the velocity – Verlet algorithm [16,17]

$$
\begin{aligned}
&1)\quad \mathbf{r}_i(t+\Delta t) = \mathbf{r}_i(t) + \Delta t\, \mathbf{v}_i(t) + \frac{1}{2}\Delta t^2\, \mathbf{a}_i(t) \\
&2)\quad \mathbf{v}_i\left(t + \frac{1}{2}\Delta t\right) = \mathbf{v}_i(t) + \frac{1}{2}\Delta t\, \mathbf{a}_i(t) \\
&3)\quad \mathbf{F}_i(t+\Delta t) \quad \text{and} \quad \mathbf{a}_i(t+\Delta t) \\
&4)\quad \mathbf{v}_i(t+\Delta t) = \mathbf{v}_i\left(t + \frac{1}{2}\Delta t\right) + \frac{1}{2}\Delta t\, \mathbf{a}_i(t+\Delta t) .
\end{aligned}
\tag{10}
$$

The time step used was $\Delta t = 0.5 \bullet 10^{-14}$ s. $\mathbf{v}_i$ and $\mathbf{a}_i$ denote the velocity and acceleration of nucleus i, respectively.

The MD simulations were started with the glue-model [9,10] surrounding the perfect [112] edge dislocation by a cylinder of radius r = 40 Å fixing the atoms inside the surface 30 Å < r < 40 Å and using the periodic boundaries in the [112] direction with the period of $\sqrt{13.5}$ a. However, it turned out quite soon that the picture of the dissociation into two Shockley partials was essentially correct. It was therefore more natural to continue the simulations in the rectangular geometry.

The left-hand side of the initial geometry (symmetric with respect to the yz-plane) is shown in Fig. 2. The filled circles at x = 0 represent the atoms of the two extra planes and those at the left end the fixed atoms. The circles with a bar represent the atoms of the $(11\bar{1})$ glide-plane and the circles with a dot the atoms of the $(11\bar{1})$ plane containing the edge of the dislocation. Figure 2(a) shows the projection of all atoms on the xy-plane and Fig. 2(b) shows the xz-projection of the three atom planes denoted by 1, 2 and 3 in Fig. 2(a). The minimum image convention used requires that the dimension of the MD box in the periodic [112] direction is at least twice the cut-off radius of the interactions [17]. Therefore the periodic boundaries with the period of $\sqrt{6}\,\underline{a}$ containing 12 atom planes were used. Free boundaries were used on the $(11\bar{1})$ surfaces. All simulations started with zero velocities.

The first set of MD simulations were performed both with the glue model [9,10] and with the ATVF-model [12] in a geometry where the x- and y-

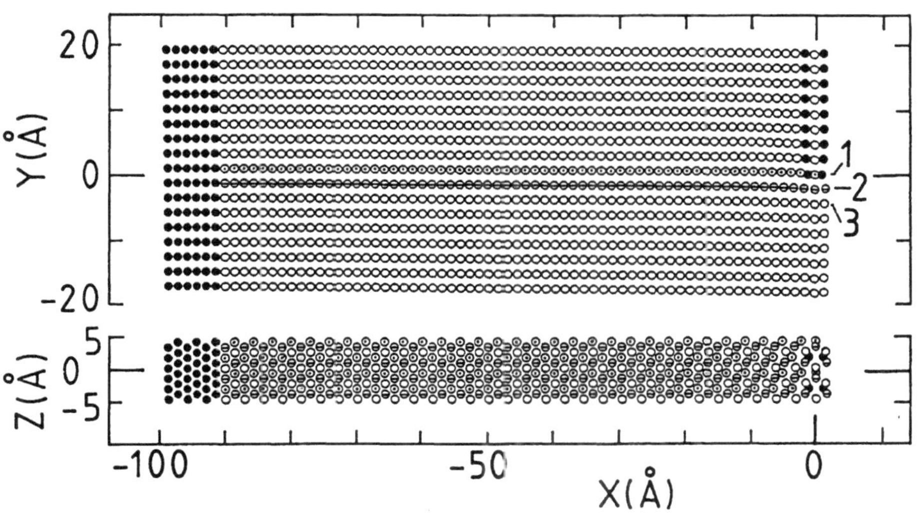

Figure 2. Initial geometry. The circles with a dot, the circles with a bar and the open circles in Fig. 2 (b) represent the atoms in planes 1, 2 (glide plane) and 3, respectively. For further information see the text.

dimensions of the moveable region were 90 and 65 Å respectively (4410 atoms out of which 696 fixed). Since the glue model contains only the nearest neighbour interactions the formation of the hcp stacking fault ribbon does not cost any energy making the separation distance infinite (Eq. (9)). Therefore, it was not surprising to find that the two Shockley partials were pushed by the repulsive forces (after 3380 MD- and 1000 quench-time-steps = 21.9 ps) close (at the distance of 11.5 Å) to the fixed surfaces. This agrees with a recent EMT MD simulation for copper with nearest neighbour interactions where the partials were found to travel until the free $(\bar{1}10)$-edges (see Fig. 1) [18].

The ATVF-model contains interactions up to the third nearest neighbours and a fitting to the stacking fault energy, the calculated number being 32.0 mJ/m^2 for gold [12]. A MD simulation with 4200 MD- and 2440 quench-time-steps, in total 33.2 ps, resulted in a partial separation of 60 Å which is somewhat larger than the Cotterill-Doyama elasticity estimate 40 Å. The distances from the fixed boundaries were only 17 Å. Therefore it was possible that the partials could experience the attractive forces towards their (original) images beyond the fixed surface regions [15].

In order to eliminate the effects of the surfaces we performed a second set of simulations with the ATVF-model by starting with the initial geometry shown in Fig. 2. The x-dimension of the moveable region was increased from 90 Å to 180 Å whereas the y-dimension was decreased from 65 Å to 40 Å (4694 atoms out of which 408 fixed).

Figure 3 shows the development of the partial separation in two such simulations. In simulation I the separation increases gradually up to 110 Å

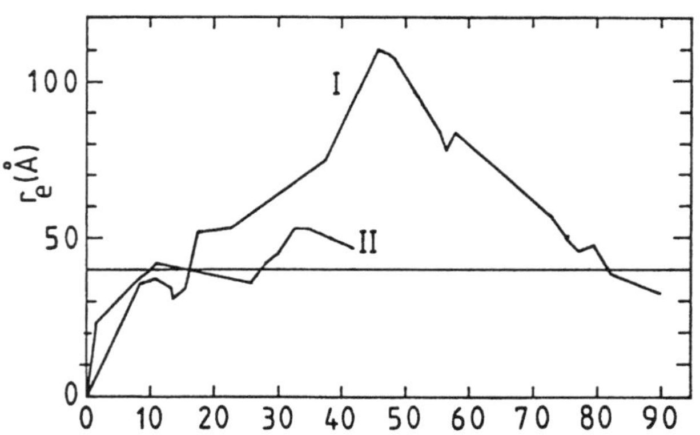

Figure 3. The development of the partial separation. t denotes the time and r_e the partial separation. Simulation I consists of the constant energy part until t=45.6 ps and the constant temperature part with T=0.1 K for 45.6 ps<t. Simulation II is performed at constant temperature T=0.1 K.

during the constant energy simulation and decreases during the constant temperature part (T=0.1 K) until 33 Å. In simulation II performed at constant temperature T=0.1 K the separation increases and remains oscillating around the Cotterill-Doyama elasticity estimate 40 Å. Both MD-separations are in reasonable agreement with this estimate although the MD-separation clearly depends on the way it is obtained and on the simulation time.

6. Structure of Shockley partials

The final structure of the left Shockley partial obtained from simulation I at t=89.4 ps is shown in Fig. 4. The hcp stacking fault ribbon can be seen on the right hand side of Fig. 4(b). The fcc stacking changes continuously to the hcp stacking fault in the region of about 20 Å around the Shockley partial. Notice the vertical shifts of the atoms with increasing x in Fig.4(b) (atoms of planes 2 and 3 downwards, atoms of plane 1 upwards).

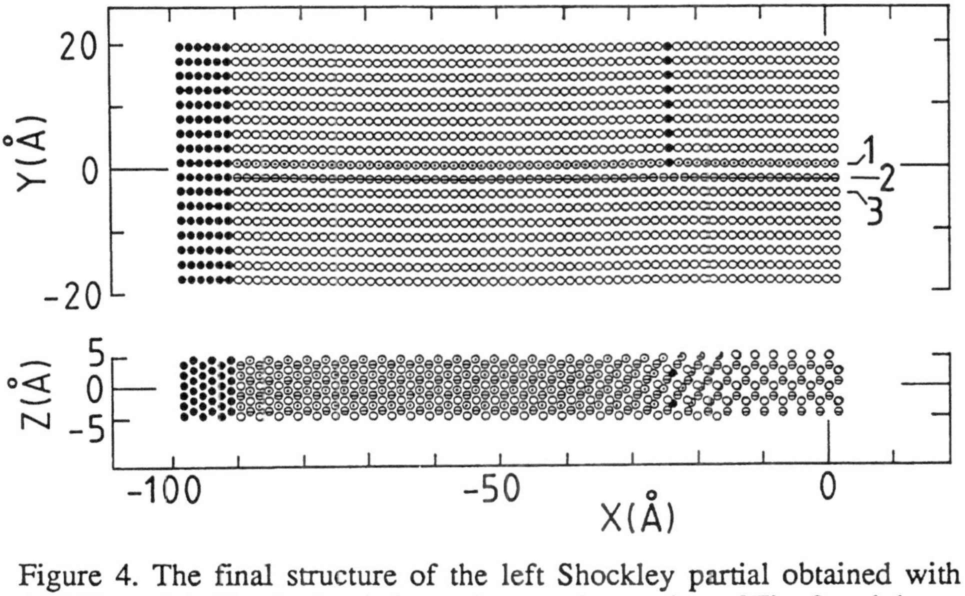

Figure 4. The final structure of the left Shockley partial obtained with the ATVF-model. For further information see the caption of Fig. 2 and the text.

7. Discussion

We have briefly sketched the initial steps in developing tools for realistic simulation of dislocations in metals. It is generally true that with increasing computer capacity, molecular dynamics simulation is beginning to shed new light on the atomistic nature of complex material properties. It is also becoming possible to test observed structural quantities, such as the partial separation in the present problem, against the assigned force laws for the many-atom interactions. While it is necessary to go beyond nearest neighbor interactions, the goal in the approximate techniques is to keep the description reasonably simple. This would then open up the possibility to perform large-scale simulations for such important time-dependent problems as dislocation glide, scattering by impurities etc.

Acknowledgements

The authors would like to thank H. Häkkinen, M. Manninen, G.J. Ackland and G. Tichy for illuminating discussions and F. Ercolessi for providing us with the computer programs of the glue model.

References

1 Institute of Mechanics
2 Laboratory of Physics
3 R.M.J. Cotterill and M. Doyama, Phys. Rev. **145**, 465-78 (1966)
4 M.S. Daw, M.I. Baskes, C.L. Bisson and W.G. Wolfer, in Modeling Environmental Effects on Crack Growth Processes, edited by R.H. Jones and W.W. Gerberich (The Metallurgical Society, Warrendale, PA, 1986)
5 M.S. Daw and M.I. Baskes, Phys. Rev. Lett. **50**, 1285-88 (1983)
6 M.S. Daw and M.I. Baskes, Phys. Rev. B**29**, 6443-53 (1984)
7 S.M. Foiles, Phys.Rev. B**32**, 3409-15 (1985)
8 S.M. Foiles, M.I. Baskes and M.S. Daw, Phys. Rev. B**33**, 7983-91 (1986)
9 F. Ercolessi, E. Tosatti and M. Parrinello, Phys. Rev. Lett. **57**, 719-22 (1986)
10 F. Ercolessi, M. Parrinello and E. Tosatti, Phil. Mag. A**58**, 213-26 (1988)
11 M.W. Finnis and J.E. Sinclair, Phil. Mag. A**50**, 45-55 (1984)
12 G.J. Ackland, G. Tichy, V. Vitek and M.W. Finnis, Phil. Mag. A**56**, 735-56 (1987)
13 M. Manninen, Phys. Rev. B**34**, 8486-95 (1986)
14 K.W. Jacobsen, J.K. Nørskov and M.J. Puska, Phys. Rev. B**35**, 7423-42 (1987)
15 J. Friedel, Dislocations, Pergamon Press 1964
16 W.C. Swope, H.C. Andersen, H.C. Berens and K.R. Wilson, J. Chem. Phys. **76**, 637-49 (1982)
17 M.P. Allen and D.J. Tildesley, Computer Simulation of Liquids, Clarendon Press, Oxford, 1987
18 H. Häkkinen, S. Mäkinen and M. Manninen, Europhys. Lett. **9**, 809-814 (1989)

One-Electron Energy Corrections in the Effective-Medium Theory: Application to Copper Clusters

S. Myllyrinne[1], *M. Manninen*[2], *and J.K. Nørskov*[3]

[1]Laboratory of Physics, Helsinki University of Technology, SF-02150 Espoo, Finland
[2]Department of Physics, University of Jyväskylä, SF-40100 Jyväskylä, Finland
[3]Laboratory of Applied Physics, Technical University of Denmark, DK-2800 Lyngby, Denmark

Abstract. The effective-medium theory has been extended to include the one-electron energy correction for the valence electrons in an approximate way. The formalism is applied to compute the properties of small copper clusters containing upto 22 atoms. The results show that in the small clusters the one electron energy correction is essential in determining cluster geometries and magic numbers.

1. Introduction

The effective medium theory (EMT) of Jacobsen, Nørskov and Puska[1] has shown to give a quantitatively good description of many bulk and surface properties of metals[2 – 4]. The theory has been extended to include the interactions beyond the nearest neighbours[5, 6]. In its simplest form the final formulas of EMT are very similar to those of related models[7 – 9]: The total energy depends on the local electron density at each atomic site and of a pair interaction. The single electron eigenvalues of the system do not have any explicit effect on the total energy or other properties of the system. There are, however, phenomena which depend intimately on the one-electron energy levels. For example, the Peierls-transition, the Kohn-anomalies and the Jahn-Teller-effect can be most easily understood in terms of the single-energy energy levels, and can*not* be results of the effective medium theory or any related theory if the details of the one-electron energy spectrum are neglected.

Most profoundly the effects arising from the one-electron energy eigenvalues are seen in small systems with discrete energy levels. In metal clusters the stability and the geometry of the smallest clusters are determined almost solely by the filling of the "energy shells" of the valence electrons[10]. This leads to magic numbers which give the sizes of the most stable clusters. For all alkali and noble metal clusters these magic numbers are 2, 8, 20, 34, 40, etc. corresponding to the filling of energy levels in a spherical potential.

In the present paper we have studied the applicability of the effective medium theory to study the properties of small copper clusters. We have extended the theory to include corrections from the one-electron energy levels and

Editors: R.M. Nieminen · M.J. Puska · M.J. Manninen

found out that for the small clusters the variation in the one-electron correction as a function of the cluster size is large enough to give the correct magic numbers. Also the cluster geometries for the smallest clusters become very similar to the results of pseudopotential[11] or Hartree-Fock[12] calculations. Moreover, due to the computational simplicity of the EMT we are able to study the stability of the clusters at finite temperatures. The results show that the enhanced stability of the magic cluster exists up to temperatures close to the melting temperature of bulk copper.

The method of including the one-electron energy correction is explained in Section 2 and the results for copper clusters are given in Section 3. Section 4 concludes the paper.

2. One-electron energy correction

The effective medium theory has been well documented by Jacobsen *et al*[1, 2]. The one-electron energy correction can be written as

$$\begin{aligned} E_{o-e} = & \sum_i \int d\epsilon\epsilon \int_{a_i} d\mathbf{r}(\rho(\mathbf{r},\epsilon) - \rho_0(\mathbf{r},\epsilon)) \\ & - \sum_i \int d\epsilon\epsilon \int_{a_i} d\mathbf{r}(\rho^h(\mathbf{r},\epsilon) - \rho_0^h(\mathbf{r},\epsilon)), \end{aligned} \tag{1}$$

where $\rho(\mathbf{r},\epsilon)$ is the local density of states in the atomic sphere a_i when the atom i is in the sphere and $\rho_0(\mathbf{r},\epsilon)$ is the density of states when the atom i is missing from the sphere. ρ^h and ρ_0^h are the corresponding densities of states in the homogeneous reference system. We first notice that the first term in Eq.(1) can be written as

$$\int d\epsilon\epsilon \sum_i \int_{a_i} d\mathbf{r}\rho(\mathbf{r},\epsilon) = \int d\epsilon\epsilon \int_{all\ space} d\mathbf{r}\rho(\mathbf{r},\epsilon) = \sum_i \epsilon_i, \tag{2}$$

where ϵ_i is an energy eigenvalue for the whole system. For the bulk metal the one-electron energy correction is approximated to vanish[1]. By denoting the density of states in the bulk metal by $\rho^b(\mathbf{r},\epsilon)$ we can approximate

$$\int d\epsilon\epsilon \sum_i \int_{a_i} d\mathbf{r}\rho^h(\mathbf{r},\epsilon) \approx \int d\epsilon\epsilon \sum_i \int_{a_i} d\mathbf{r}\rho^b(\mathbf{r},\epsilon) \approx \frac{3}{5}N\epsilon_F^h, \tag{3}$$

which is the kinetic energy for the homogeneous electron gas. The second approximation comes from the fact that in the bulk metal the integration over the atomic sphere represents the integration of the neutral Wigner-Seitz-cell: i.e. the average of the global density of states which in simple metals is very similar to that of a homogeneous electron gas. The two remaining terms are smaller because they present integrations over "empty" spheres (vacancies) where the electron density is small. These two terms will neglected (their contribution to the one-electron energy correction would be an order of magnitude smaller than that of the two included terms).

In calculating the sum over the one-electron eigenvalues in Eq. (2) we have to take care that the average potential outside the sphere in question is the same as in the reference system (which for the jellium equals to zero). This is estimated by equalizing the band centers in the two systems. For the free-electron metals (one electron per atom) we assume the band center to be at the Fermi energy (this is roughly the result of a tight binding model). With the above approximations we get the following one-electron energy correction:

$$E_{o-e} = \sum_i \epsilon_i - N\left(\frac{\epsilon_{min} + \epsilon_{max}}{2} - \frac{2}{5}\epsilon_F^h\right), \tag{4}$$

where ϵ_{min} and ϵ_{max} are the lowest and highest energy levels of the cluster.

For computing the valence electron energy levels for the clusters we use a simple tight binding model. The Hamiltonian is written as

$$\begin{aligned} H_{ii} &= 0 \\ H_{ij} &= -\beta_0 e^{-\eta_a(r_{ij}-r_{nn})}\left(\frac{1}{1+e^{-\eta_b(r_{ij}-r_0)}}\right), \end{aligned} \tag{5}$$

where r_{ij} is the distance between atoms i and j and r_{nn} is the distance between the nearest neighbours in the perfect fcc-lattice. β_0 has the value which gives for the perfect lattice the correct band width. In practice β_0 is estimated to be $\beta_0 = \frac{1}{8}\epsilon_F^h$. (The band width is $16\beta_0$ and the Fermi energy is close to the band center). The distance dependence is assumed to be exponential with the same decay coefficient η_a as the atomic wave function (for copper we use the value $\eta_a = 1.25$, consistent with the value used for the electron density decay). The fermi function with the parameters η_b and r_0 is introduced only to prevent the energy from diverging if two atoms come close to each other (we use values $\eta_b = 5$ and $r_0 = 0.8r_{nn}$). The nearest neighbour distance r_{nn} and the Fermi energy ϵ_F^h are calculated from the average density of the cluster, i.e. we determine the average atomic sphere radius which is used to give the density of the reference system. This guarantees that for the bulk fcc structure the one-electron energy correction is always zero, independent of the volume.

3. Results

The total energies and geometries of copper clusters from 1 to 22 atoms were calculated using Monte Carlo method and simulated annealing. Starting from an arbitrary geometry and from a high temperature the cluster was cooled down to zero temperature. Of the order of 20000 Monte Carlo steps were used in each run. Several runs were made for each cluster size to make sure that the ground state was observed. Figure 1 shows the total energy per atom as

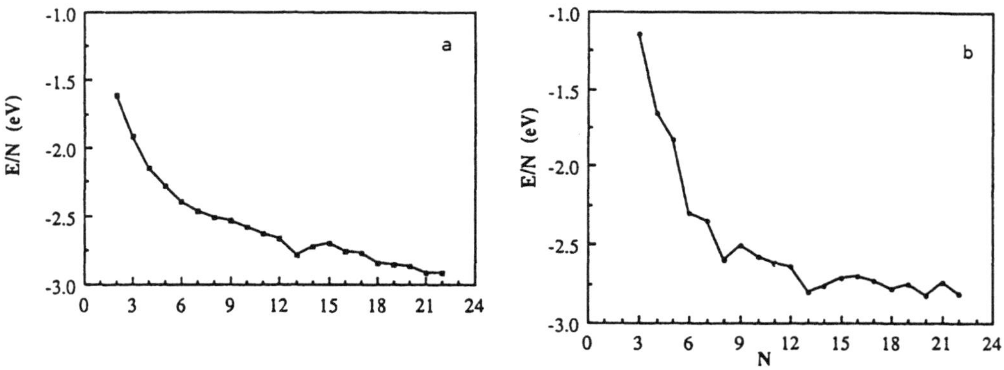

Figure 1. Total energy of the cluster per atom as a function of the cluster size calculated (a) without and (b) with the one-electron energy correction.

a function of the cluster size. Without the one-electron corrections there is a clear minimum at the icosahedral 13 atom cluster. The one electron corrections add new minima at cluster sizes 8, 18 and 20 which correspond to full valence electron shells.

The geometries of the smallest clusters are shown in Figure 2 and compared to the results obtained without the one-electron correction. The 4, 5 and 6 atom clusters are nearly planar in agreement with the results of *ab initio* calculations for alkali metals[11, 12]. These are very different from the results without the one-electron correction which are all very symmetric and compact structures. The geometries are intimately related to the symmetries of the one-electron eigenfunctions (as explained e.g. Manninen[11]).

The one electron energy eigenvalues show clearly the tendency of grouping the energy levels into energy shells which corresponds to the eigenvalues for independent electrons in a spherical potential well. This is most clearly seen in the magic clusters where there is a large energy gap between the last occupied and the first unoccupied energy level. The even-odd systematics in the total energy curve is also a result of the one-electron energy correction: For odd clusters there is an unpaired electron in the last occupied level.

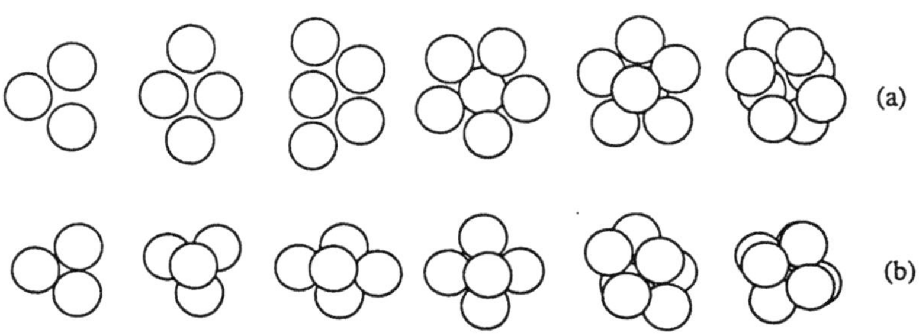

Figure 2. Ground state geometries of the smallest clusters calculated (a) with and (b) without the one-electron energy correction.

We have also computed the average internal energy of the clusters at elevated temperatures. The results show that the clusters 8 and 20 remain "magic" even at 1300 K which is already very close to the melting point of copper. This explains why the magic numbers are seen so easily in experiments, where the internal energy of the clusters could correspond to an elevated temperature.

4. Conclusions

In calculating properties of small clusters the inclusion of the one-electron energy correction is essential. For nearly free electron metals it is enough to include only the valence electrons. The tight-binding approximation of the one-electron energy correction is easy to implement at least in Monte Carlo computations. The results for small copper clusters are in excellent qualitative agreement with the experimental results. The stability of the clusters is determined almost solely by the one-electron energy correction. At elevated temperatures, even close to the melting point, the magic numbers are still clearly visible in the total energy curve.

Acknowledgements

We are grateful to Per Stoltze and Hannu Häkkinen for useful discussions.

References

1. K. W. Jacobsen, J. K. Nørskov and M. J. Puska, Phys. Rev. B **35**, 7423 (1987).
2. K. W. Jacobsen, Comments Cond. Mat. Phys. **14**, 129 (1988).
3. P. Stoltze, J. K. Nørskov and U. Landman, Phys. Rev. Lett. **61**, 440 (1988).
4. P. Stoltze, K. W. Jacobsen and J. K. Nørskov, Phys. Rev. B **36**, 5035 (1988).
5. H. Häkkinen and M. Manninen, J. Phys.: Condens. Matt. (in press).
6. H. Häkkinen, in this proceedings.
7. M. S. Daw and I. M. Baskes, Phys. Rev. Lett. **50**, 1285 (1983) and M. S. Daw, in this proceedings.
8. M. W. Finnish and J. E. Sinclair, Phil. Mag. **50**, 45 (1984).
9. F. Ercolessi, E. Tosatti and M. Parrinello, Phys. Rev. Lett. **57**, 719 (1987).
10. W. A. de Heer, W. D. Knight, M. Y. Chou and M. L. Cohen, Solid State Physics **40**, 93 (1987).
11. M. Manninen, Phys. Rev. B **34**, 6886 (1986).
12. B. K. Rao, P. Jena, Phys. Rev. B **32**, 2058 (1985).

Dynamic Methods for Electronic Problems

R. Virkkunen and R.M. Nieminen

Laboratory of Physics, Helsinki University of Technology,
SF-02150 Espoo, Finland

Abstract. Molecular dynamical simulations and density functional theory have greatly contributed to the present understanding of condensed matter physics. These traditionally fairly separate methods can be combined to form a powerful dynamic technique for simulations of physical systems. The unified approach to density functional theory and molecular dynamics is discussed. The advantages of introducing the density-functional wave functions, the Hartree potential and the nuclear coordinates as dynamical variables to be solved from the equations of motion are pointed out. The usefulness of the simulated annealing optimization is argued. The correspondence between the Poisson equation and minimization of the energy functional of the Hartree potential is derived. Finally, the annealing method for minimizing the energy functional is illuminated by examples of calculating the Hartree potential and the electronic wave functions as a solution to a dynamic optimization problem.

1. Introduction

Molecular dynamical techniques are able to simulate finite temperature properties and dynamical processes of physical systems [1]. However, the reliability of molecular dynamical simulations depends on to what extent the force laws between interacting particles can be specified. Empirical few-body potentials have so far been used as model potentials of real materials. While these simple force laws can be feasible for rare gases and polar molecules, they are not universally applicable in metallic or covalent systems, where the bonding configurations are strongly influenced by the details of the electronic structure. In addition, molecular dynamical calculations do not convey any information of the electronic properties.

A proper description of the electron-mediated forces is available in the density-functional theory [2]. Unfortunately, the computational burden of the density-functional wave functions has prevented its application to the computation of interatomic forces in molecular dynamical simulations and to the study of very large and disordered systems. Therefore, the combination of density functional theory and molecular dynamics is only possible if a powerful computational method is available.

2. Method

2.1 Combined molecular dynamics and density functional theory

Car and Parrinello first suggested a unified approach to molecular dynamics and density functional theory [3]. They consider the electronic density-functional Hamiltonian and the

Springer Proceedings in Physics, Vol. 48 Many-Atom Interactions in Solids
Editors: R.M. Nieminen · M.J. Puska · M.J. Manninen

ionic kinetic energy as an objective function to be dynamically minimized with respect to both ionic and electronic degrees of freedom. The optimization was carried out with an efficient simulated annealing technique. In the Car-Parrinello method a fictitious kinetic energy is added to the electronic fields and equations of motion are written for the electronic fields and the ionic coordinates. The set of coupled equations is then simultaneously solved and the time evolution of system is studied as in standard molecular dynamical simulations. The kinetic energy of system is associated with the temperature. When the simulated annealing minimization is used, the temperature can be slowly reduced in order to find the T=0 equilibrium state.

The dynamical method can also be applied to the Hartree potential of the system [4]. Instead of solving the Poisson equation directly, the Hartree potential can also be considered as a dynamical variable to be optimized simultaneously with the electronic wave functions and the ionic coordinates. Then the equations of motion are derived from the total Lagrangean of the system

$$
\begin{aligned}
L = \frac{1}{2}\mu \int \sum_i \dot{\psi}_i^*(r,t)\, \dot{\psi}_i(r,t)\, dr + \frac{1}{2}\mu_H \int \dot{V}_H^2(r,t)\, dr + \frac{1}{2}\sum_I M_I \dot{R}_I^2 \\
- E(\psi_i, V_H, R_I) + \sum_{ij} \Lambda_{ij} \left\{ \int \psi_i(r,t)\, \psi_j^*(r,t)\, dr - \delta_{ij} \right\},
\end{aligned}
\tag{1}
$$

where $\psi_i(r,t)$'s are the classical fields, $V_H(r,t)$ is the Hartree potential and R_I 's are the ionic positions. μ and μ_H are mass parameters and the M_I 's are the ionic masses. Λ_{ij} are Lagrange multipliers imposed by the orthogonality constraint of the electron states.The energy functional $E(\psi_i, V_H, R_I)$ is defined as

$$
\begin{aligned}
E = -\frac{1}{2}\sum_i \int \psi_i^*(r)\, \nabla^2 \psi_i(r)\, dr + E_{XC} + \int V_{el}(r)\, \rho(r)\, dr \\
+ \frac{1}{8\pi} \int V_H(r)\, \nabla^2 V_H(r)\, dr + \int V_H(r)\, \rho(r)\, dr,
\end{aligned}
\tag{2}
$$

where E_{XC} is the exchange and correlation energy functional, $V_{el}(\mathbf{r})$ is the external potential and $\rho(\mathbf{r})$ is the electronic density. Now equation (1) includes real dynamics of the ions as well as the fictitious kinetic energy terms of the electronic fields and the Hartree potential. The Lagrangean yields the equations of motion

$$
\mu \ddot{\psi}_i(r,t) = -\left\{ -\frac{1}{2}\nabla^2 + V_H(r) + V_e(r) + V_{XC}(r) \right\} \psi_i(r,t) + \sum_j \Lambda_{ij} \psi_j(r,t), \tag{3}
$$

$$
\mu_H \ddot{V}_H(r,t) = -\rho(r) - \frac{1}{4\pi} \nabla^2 V_H(r,t), \tag{4}
$$

$$
M_I \ddot{R}_I = -\nabla_R E, \tag{5}
$$

which generate the trajectories of the system. Physical quantities, like the kinetic energy, are obtained as temporal averages over the trajectories. The classical kinetic energy reveals how far the system lies from the T=0 equilibrium state. When the second time derivatives of the electronic fields and of the Hartree potential vanish, the right-hand side of equation (5) describes the density-functional forces acting on the ions and the Lagrangean (1) describes a real physical system. At equilibrium all the forces acting on the system vanish i.e. the left-hand sides of equations (3)-(5) are zero and the energy functional of the system (2) is at the minimum.

There are several advantages of introducing an auxiliary field that represents the Hartree potential to the Lagrangean [4]. First of all, one avoids integrating the Hartree potential from the charge density i.e. solving the Poisson equation directly. The amount of operations is reduced, as it becomes possible to work entirely in real space and the fields $\psi_i(\mathbf{r},t)$ and $V_H(\mathbf{r},t)$ can be represented in terms of their amplitudes. As no basis set expansion is required, it is easier to introduce a non-uniform mesh, which provides the neighbourhood of the ion cores with a larger amount of lattice points. Finally, almost all the operations can be made vectorial and parallel i.e. the method is particulary suitable for the present computational facilities.

2.2 Simulated annealing optimization

The simulated annealing algorithm for complex optimization problems is used in order to find the global minimum of the energy functional of the system [5,6]. The method is based on the analogy between optimization of a cost function and experimental annealing, in which a material is heated to a high temperature above its melting point, and then cooled slowly. The more slowly the material is cooled, the more likely it forms a perfect crystal that is the state of minimum energy for the system. In the simulated annealing method one starts annealing the system from a high temperature initial state by gradually damping the dynamics associated to the variables. If the standard Metropolis Monte Carlo algorithm were used, each new state of the system would be chosen randomly from the neighbourhood of the present state. In the molecular dynamical approach the suggested configurations are determined by the equations of motion. In many atomic systems this is more efficient than the Monte Carlo method, because all the particles are updated at the same time step to the direction of the atomic forces [7]. When a random Langevin force is added to the equations of motion, the system is able to move uphill in energy in the same way as in Monte Carlo simulations. The acceptance of the state depends on the temperature and on the energy difference of the successive configurations. As long as there is some kinetic energy left the system is allowed to move to a higher energy state, but the probability of such moves decreases when the temperature is reduced. Also, if the energy of the suggested state is much higher than the one of the present state, the transition probability is vanishingly small. Thus, if the cooling schedule is carefully planned, the algorithm avoids trapping in local minima and produces low-lying energy states of the system.

The dynamical simulated annealing method used by Car and Parrinello not only finds the ground state of the atomic system but allows the study of finite temperature properties [3]. Because the velocities associated with the electronic and ionic parameters can be manipulated separately, the kinetic energy of the electronic fields can be made very small, which means that the electrons are close to their ground state and the ions move under the action of the correct interionic forces. Then the equations of motion (3)-(5) represent the real dynamics of the physical system.

The dynamical annealing approach is a very general and flexible method. The wide variety of problems for which simulated annealing has been used includes such as image processing, circuit design and Boltzmann machines [6]. Although the method is basically developed to large optimization problems, the idea can be clarified by simple dynamical calculations.

3. Examples

To illustrate the dynamical minimization method we present two examples. Instead of solving the Poisson and the Schrödinger equations directly, we consider them as optimization problems, where the minimum is found by annealing techniques.

3.1 Poisson equation

The Hartree potential due to a charge density $\rho(\mathbf{r})$ is determined by the Poisson equation

$$\nabla^2 V_H(r) = -4\pi\rho(r). \tag{6}$$

If the charge density is known, the Hartree potential can be integrated directly from equation (6). However, in conjunction with the Car-Parrinello method it is more efficient to convert the Poisson equation into an optimization problem and iterate the Hartree potential simultaneously with the electronic and ionic degrees of freedom. Mathematically this means that the static boundary-value problem (6) is replaced by a time-dependent initial value problem (4). To justify the replacement, we show the correspondence between the solutions of the optimization problem and the Poisson equation and create appropriate dynamics to the Hartree potential.

Poisson equation as an optimization problem. In order to calculate the Hartree potential as an optimization problem, we search for the minimum of the energy functional

$$U = \frac{1}{2}\int \rho(r)\, V_H(r)\, dr, \tag{7}$$

which is the potential energy associated to the Hartree potential. Using the Poisson equation (6) the potential energy can also be written as

$$U = \int \rho(r)\, V_H(r)\, dr + \frac{1}{8\pi}\int \nabla^2 V_H(r)\, V_H(r)\, dr. \tag{8}$$

At the minimum the energy functional U does not change when the Hartree potential is slightly varied. If we disturb the Hartree potential by a $\delta V_H(\mathbf{r})$, we can calculate the energy change $\delta U = U(V_H(\mathbf{r}) + \delta V_H(\mathbf{r})) - U(V_H(\mathbf{r}))$ from equation (8).When the change in the potential is very small, the second and higher order terms of the change $\delta V_H(\mathbf{r})$ are ignored and the variation in the potential energy is approximately

$$\delta U = \int \rho(r)\, \delta V_H(r)\, dr + \frac{1}{8\pi}\int \left\{ \nabla^2 V_H(r)\, \delta V_H(r) + \nabla^2 \delta V_H(r)\, V_H(r) \right\} dr. \tag{9}$$

If we then integrate partially inside the brackets and remember that the Hartree potential vanishes at infinity, we get

$$\delta U = \int \rho(r)\, \delta V_H(r)\, dr + \frac{1}{8\pi}\int \left\{ -2\nabla^2 V_H(r)\, \delta V_H(r) \right\} dr. \tag{10}$$

After we have integrated partially once more, we take the variation of the Hartree potential as a common factor and get the final formula for the energy change which is to be zero at the minimum:

$$\int \delta V_H(\mathbf{r}) \left\{ \rho(\mathbf{r}) + \frac{1}{4\pi} \nabla^2 V_H(\mathbf{r}) \right\} d\mathbf{r} = 0. \tag{11}$$

Because the variation of the Hartree potential $\delta V_H(\mathbf{r})$ is arbitrary, we see that the potential $V_H(\mathbf{r})$ that fulfils condition (11) and thus minimizes the energy functional U is the solution of the Poisson equation (6).

Equation of motion for the Hartree potential. In order to solve the Poisson equation dynamically we assign a fictitious velocity to the Hartree potential. Then we can define the Lagrange function from potential energy (8) and the kinetic energy associated with $V_H(\mathbf{r})$ as

$$L = \frac{1}{2} \mu_H \int \dot{V}^2(\mathbf{r},t)\, d\mathbf{r} - \int \rho(\mathbf{r})\, V_H(\mathbf{r},t)\, d\mathbf{r} - \frac{1}{8\pi} \int \left\{ \nabla^2 V_H(\mathbf{r},t)\, V_H(\mathbf{r},t) \right\} d\mathbf{r}. \tag{12}$$

According to Euler's equation

$$\frac{d}{dt} \left(\frac{\partial L}{\partial \dot{V}_H(\mathbf{r},t)} \right) - \frac{\partial L}{\partial V_H(\mathbf{r},t)} = 0 \tag{13}$$

we generate the equation of motion to the Hartree potential

$$\mu_H \int \ddot{V}_H(\mathbf{r},t)\, d\mathbf{r} = - \int \rho(\mathbf{r})\, d\mathbf{r} - \frac{1}{4\pi} \int \nabla^2 V_H(\mathbf{r},t)\, d\mathbf{r}, \tag{14}$$

which corresponds to equation (4) when the integration signs are omitted.

The equation of motion is integrated numerically as in standard molecular dynamical simulations. The ∇^2 operator is discretised. An appropriate integration algorithm is derived from the Taylor expansion [8]. For the Hartree potential we choose the Verlet formula

$$V_H(\mathbf{r},t) = -V_H(\mathbf{r},t-2\Delta t) + 2V_H(\mathbf{r},t-\Delta t) + \ddot{V}_H(\mathbf{r},t-\Delta t)\Delta t^2, \tag{15}$$

$$\dot{V}_H(\mathbf{r},t-\Delta t) = \frac{1}{2\Delta t} \left\{ V_H(\mathbf{r},t) - V_H(\mathbf{r},t-2\Delta t) \right\}. \tag{16}$$

Thus at each time-step Δt the potential and its time derivative are calculated approximately from equations (15)-(16). A damping factor is added to the equation of motion in order to reduce the velocity of the potential. Here we use a simple constant cooling schedule, which should not be automatically used in more complex problems [6]. At the end the first and the second time derivatives of the potential are zero and equation (4) fulfils the Poisson equation.

Results. Figure 1 shows the calculated Hartree potential due to a uniform charge density inside a circle (radius r=1) placed at the origin. The two-dimensional density creates a potential surface that resembles a paraboloid in the middle and goes logarithmically down to zero at the boundaries. The calculated potential function differs about two percent from the analytical results. Replacing the Verlet algorithm with another integration algorithm or the changing from a three point formula of the gradient to a five point formula do not improve the results. A better accuracy is achieved with a shorter time step and a denser grid as long as

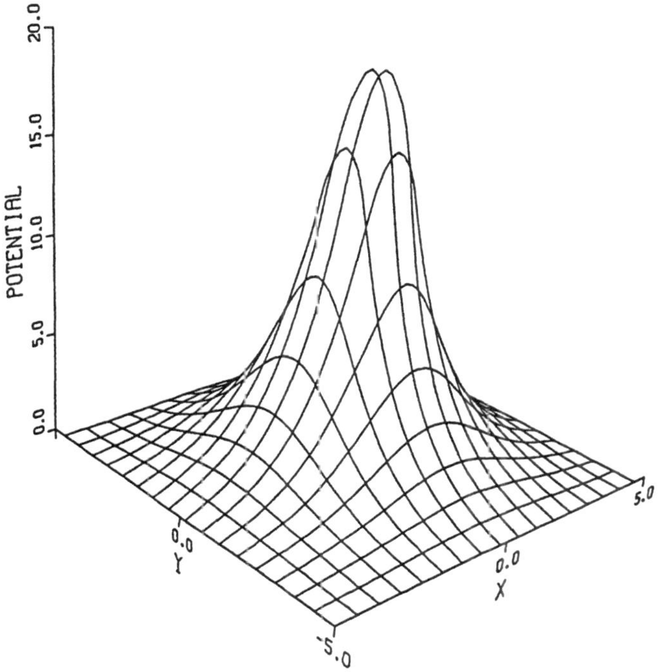

Fig. 1. Annealing techniques applied to the Poisson problem. The solution is the potential surface due to a uniformly charged circle placed at the origin.

this does not lead to undesirably long convergence times or to numerical instabilities. For a more efficient calculation the constant cooling rate could be replaced with a more powerful schedule.

3.2 Schrödinger equation

Equation of motion for the electronic wave function. The solution of the Schrödinger equation can be calculated dynamically from the equation of motion (3), which is derived with the same formalism as the one for the Hartree potential. Here we omit the Lagrangean multipliers associated to the orthogonality condition of the electronic states and use the Gram-Schmidt orthonormalization procedure. Thus the equation for the electronic wave function is

$$\mu \ddot{\psi}_i(\mathbf{r},t) = -\eta \dot{\psi}_i(\mathbf{r},t) - \left\{ -\frac{1}{2}\nabla^2 + V_H(\mathbf{r}) + V_{el}(\mathbf{r}) + V_{XC}(\mathbf{r}) - \varepsilon_i(t) \right\} \psi_i(\mathbf{r},t), \tag{17}$$

where η is the damping parameter and $\varepsilon_i(t)$ is the electronic eigenvalue. Again, the damping factor is simply a constant. For the wave-function and its time derivative we use the Beeman algorithm

$$\psi_i(\mathbf{r},t) = \psi_i(\mathbf{r},t-\Delta t) + \dot{\psi}_i(\mathbf{r},t-\Delta t)\,\Delta t + \frac{1}{6}\left\{ 4\ddot{\psi}_i(\mathbf{r},t-\Delta t) - \ddot{\psi}_i(\mathbf{r},t-2\Delta t) \right\} \Delta t^2, \tag{18}$$

$$\dot{\psi}_i(\mathbf{r},t) = \dot{\psi}_i(\mathbf{r},t-\Delta t) + \frac{1}{6}\left\{ 2\ddot{\psi}_i(\mathbf{r},t) + 5\ddot{\psi}_i(\mathbf{r},t-\Delta t) - \ddot{\psi}_i(\mathbf{r},t-2\Delta t) \right\} \Delta t\,. \tag{19}$$

Results. The solution of the Schrödinger equation is calculated in a time-independent potential valley $V(x)=-4.375\cosh^2 x$. Figure 2 shows the potential function and the three lowest electronic states. The convergence of the energy eigenvalues is shown in figure 3. The final eigenvalues $E_1=-3.12$ Ry, $E_2=-1.13$ Ry and $E_3=-0.11$ Ry are obtained with an accuracy of 0.01 Ry.

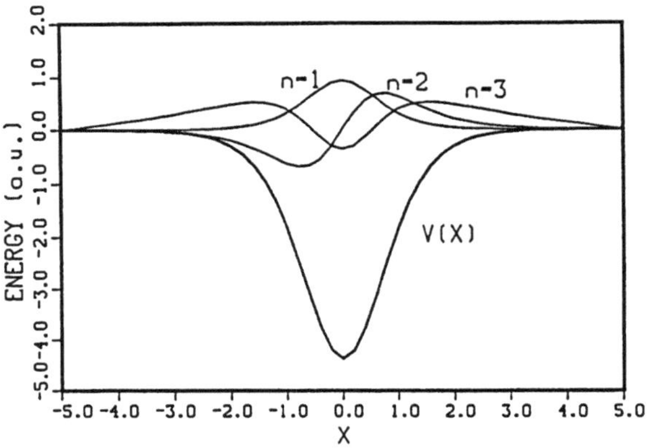

Fig. 2. Dynamical annealing of the Schrödinger equation. The three electronic eigenvalues are calculated in a simple potential $V(x) = -4.375\cosh^2 x$.

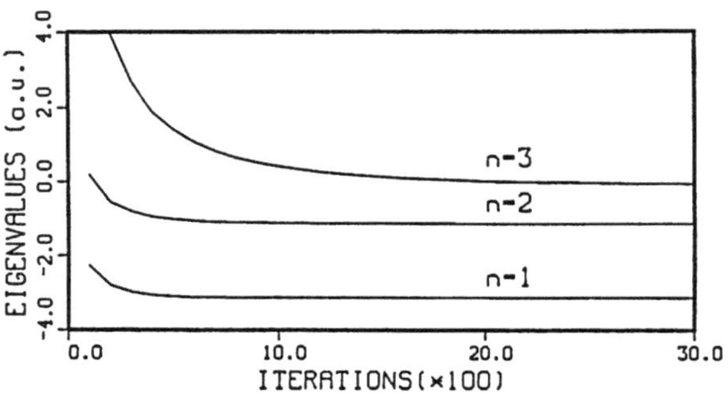

Fig. 3. The convergence of the solution of the Schrödinger equation when deterministic dynamics is used. The final eigenvalues are $E_1=-3.12$ Ry, $E_2=-1.13$ Ry and $E_3=-0.11$ Ry.

4. Conclusions

The dynamic annealing method is a powerful tool for electronic problems. In combined optimization of the density-functional electronic wave functions and the ionic positions annealing techniques provide a way of simulating the finite temperature properties of the ions with interactions derived from electronic structure. If also the Poisson problem is formulated as an optimization problem, it is possible to work entirely in real space. As no basis set representation is needed, the computational cost of the electronic part of the problem reduces remarkably. Thus these fully numerical calculations will be superior to conventional methods for very large electronic structure problems.

References

1. Nieminen, R.M., Phys. Scripta T19, 320 (1987).
2. *Theory of the Inhomogeneous Electron Gas*, edited by Lundquist, S. and March, N. H. (Plenum, New York, 1983).
3. Car, R. and Parrinello, M., Phys. Rev. Lett. 55, 2471 (1985).
4. Car, R. and Parrinello, M., Solid State Com. 62, 403 (1987).
5. Kirkpatric, S., Gelatt, C. D., Jr. and Vecchi, M. P., Science 220, 671 (1983).
6. *Simulated Annealing: Theory and Applications*, van Laarhoven, P.J.M. and Aarts,E.H.L. (D. Reidel Publishing Company, 1987).
7. Biswas, R. and Hamann, D. R., Phys. Rev. B 34, 895 (1986).
8. *Computer Simulation Methods in Theoretical Physics*, Heermann, D.W. (Springer-Verlag Berlin Heidelberg 1986).

Part III

Semiconductors and Covalent Materials

Simulated Annealing Studies of Sulphur and Selenium Clusters

R.O. Jones and D. Hohl

Institut für Festkörperforschung der Kernforschungsanlage Jülich,
D-5170 Jülich, Fed. Rep. of Germany

Abstract. The structure of a molecule or solid can be determined by calculating the total energy of the system of ions and electrons for all geometries. In this lecture we review the problems inherent in such calculations, and show that the combination of density functional calculations with molecular dynamics techniques is an appropriate response. The method is applied to structural determinations in clusters containing group VI A elements. The ground state structures of S and Se clusters are described very well. In addition, the method gives very good results in cases where there are large structural changes involving large barriers (S_7O) and small energy differences with energy barriers on a thermal scale (isomers of Se_xS_y). The last of these results can be understood in terms of a very simple model.

1. Introduction

Molecular and solid state structures are easy to determine in principle. One chooses a structure and determines the total energy E of the system of electrons and ions. If one repeats this calculation for all possible geometries, then the ground state structure is that with the lowest energy. In the first part of this lecture, we show that this method is impracticable in all but the smallest clusters. Not only is the calculation of the total energy difficult, but the number of local minima in the energy surface increases dramatically with increasing cluster size.

The combination of density functional (DF) methods, with a local density approximation, and molecular dynamics (MD) with "simulated annealing" addresses both of these problems. After outlining the important features of the approach, we show in simple examples that reliable geometries are found for small clusters of sulphur and selenium. In the case of heterocyclic clusters of the form Se_xS_y, the method gives a remarkably good description of the energy differences between different isomers. Furthermore, molecular dynamics allows us to follow changes in the molecular structures, as we show for S_7O and Se_xS_y.

Springer Proceedings in Physics, Vol. 48 Many-Atom Interactions in Solids
Editors: R.M. Nieminen · M.J. Puska · M.J. Manninen

2. Determination of Structural Properties

2.1 Calculation of the total energy, E

From the exact wave function, Ψ, of an interacting system of electrons and ions, one can determine many quantities of interest, including the total energy

$$E = \langle\Psi|\widehat{\mathcal{H}}|\Psi\rangle / \langle\Psi|\Psi\rangle, \tag{1}$$

where $\widehat{\mathcal{H}}$ is the Hamiltonian of the system. Calculations of this type can be carried out for very few systems, the best-known example being the hydrogen atom. Even for the helium atom it is not possible to determine the wave function and energy in closed form, and approximations for Ψ are unavoidable.

Approximate solutions for Ψ are usually based on the variational principle of Rayleigh and Ritz, i.e. if $|\Phi\rangle$ is an approximate wave function, then

$$\langle\Phi|\widehat{\mathcal{H}}|\Phi\rangle / \langle\Phi|\Phi\rangle \geq E_{GS}. \tag{2}$$

The simplest form of the many-particle wave function is that of Hartree (1928), who represented Φ as a product of single-electron functions:

$$|\Phi\rangle = \phi_1(\mathbf{r}_1)\phi_2(\mathbf{r}_2)...\phi_n(\mathbf{r}_n), \tag{3}$$

where the functions ϕ_i each satisfy a Schrödinger equation whose potential term is given by the mean field of the other electrons.

$$\left(-\frac{1}{2}\nabla^2 + V_{ext} + \varphi\right)\phi_i(\mathbf{r}_i) = \varepsilon_i\phi_i(\mathbf{r}_i), \tag{4}$$

where V_{ext} is the external field of the nuclei, and the Coulomb potential φ is determined by solving Poisson's equation [we use atomic units, $e = m = \hbar = 1$]. The incorporation of Fermi statistics by replacing the product by a single (Slater) determinant leads to an additional potential term (the "exchange" potential), without changing the single-particle picture. This Hartree-Fock approximation has been the basic method of atomic and molecular physics for the past 60 years.

Experience has shown, however, that a single determinant is not generally sufficient to obtain a satisfactory total energy. While a representation of the wave function as a linear combination of determinants ("configuration interaction", CI) should lead in principle to an exact wave function and energy, it also leads to an explosive increase in the numerical effort required as the number of electrons increases. In practice it is still difficult to obtain accurate energies for systems with more than ~ 20 electrons.

2.2 Determination of the lowest-lying minima

These arguments apply to a *single* geometry, i.e. for a particular set of "internal coordinates" of a system. If we have N atoms in a cluster, the total number of internal coordinates is $3N - 6$, since we obtain the same energy if we rotate or translate the cluster. In a 10-atom cluster, for example, there are 24 independent interatomic degrees of freedom and the magnitude of the problem is apparent: (1) Finding the energy minimum by calculating the energy for all possible geometries is obviously impracticable, since 10 calculations per coordinate would lead to 10^{24} calculations in a cluster where *one* seems to be too difficult. (2) If one takes any particular geometry and locates the *next* minimum, there is still little prospect of success. As we now show, the number of local minima in the energy surface can be very large.

Table 1: Number of minima in cluster interacting with Lennard-Jones potential (Hoare and McInnes, Ref. [1]).

N	LJ
6	2
7	4
8	8
9	18
10	57
11	145
12	366
13	988

Hoare and McInnes [1] have studied clusters where the atoms interact with each other with a simple, pairwise Lennard-Jones potential:

$$V_{Lennard-Jones} = 4\epsilon_0\left[\left(\frac{\sigma}{r}\right)^{12} - \left(\frac{\sigma}{r}\right)^{6}\right] \tag{5}$$

For small clusters it is possible to locate all the minima, and the results are shown for clusters of different sizes in Table 1. The number of minima increases rapidly with increasing N, with signs that the increase is exponential. For clusters of identical atoms interacting with a pairwise potential, Wille and Vennik [2] have shown, in fact, that there is no known algorithm that grows with time as power of N and with which one can determine the ground state energy and structure. Such problems are classified as "$NP-hard$" [3] and are considered "intractable". It is a sobering moment when mathematicians say that the solution of an important problem is unattainable. Moreover, parametrized force laws are inherently semi-empirical, and it is difficult to know how the choice of parameters affects the final results. It is well known that *no* pairwise force can stabilize the diamond structure, and there are many other examples where three- and four-body forces are essential.

The above discussion has shown that there are two distinct problems to be solved, and the problems are *multiplicative*. Fortunately, we have methods that address both aspects. First, the density functional formalism provides a parameter-free, numerically efficient scheme for calculating the *total energy* of the system of ions and electrons. We avoid the problem of calculating the exact wave function of the interacting system of electrons and ions. Second, we introduce the strategy of simulated annealing as a procedure for determining low-lying energy minima, i.e. for finding solutions to the minimization problem that are *almost* optimal.

2.3 Energy calculations using density functional theory

The two basic theorems of the density functional formalism were derived by Hohenberg and Kohn [4]. They showed that:

(1) Ground state (GS) properties of a system of electrons and ions in an external field, V_{ext}, can be expressed as functionals of the electron density, $n(\mathbf{r})$, i.e. they are determined by a knowledge of the density *alone*. The total energy, E, is an example.

(2) $E[n]$ satisfies the variational principle $E[n] \geq E_{GS}$, and the density for which the equality holds is the ground state density, n_{GS}.

A simple and general proof of these assertions has been provided by Levy [5].

The *applicability* of this scheme results from the observation by Kohn and Sham [6] that the minimization of $E[n]$ is simplified if we write:

$$E[n] = T_0[n] + \int d\mathbf{r}\; n(\mathbf{r})\Big(V_{ext}(\mathbf{r}) + \frac{1}{2}\varphi(\mathbf{r})\Big) + E_{xc}[n], \tag{6}$$

where T_0 is the kinetic energy that a system with density n would have in the absence of electron-electron interactions, $\varphi(\mathbf{r})$ is the Coulomb potential, and E_{xc} defines the exchange-correlation energy. The variational principle now yields

$$\frac{\delta E[n]}{\delta n(\mathbf{r})} = \frac{\delta T_0}{\delta n(\mathbf{r})} + V_{ext}(\mathbf{r}) + \varphi(\mathbf{r}) + \frac{\delta E_{xc}[n]}{\delta n(\mathbf{r})} = \mu, \tag{7}$$

where μ is the Lagrange multiplier associated with the requirement of constant particle number. For a given set of atomic coordinates $\{\mathbf{R}_I\}$, the minimization of $E[n]$ in Eq. (6) is usually performed by solving the Kohn-Sham equation

$$\Big(-\frac{1}{2}\nabla^2 + V_{ext}(\mathbf{r}) + \varphi(\mathbf{r}) + V_{xc}(\mathbf{r})\Big)\; \psi_i(\mathbf{r}) = \varepsilon_i\; \psi_i(\mathbf{r}), \tag{8}$$

where $V_{xc} = \delta E_{xc}/\delta n(\mathbf{r})$. The density can be constructed from the eigenfunctions of this equation, $n(\mathbf{r}) = \sum_i |\psi_i(\mathbf{r})|^2$, with i running over all occupied states. The procedure must be continued until self-consistency is reached.

A comparison of equations (4) and (8) shows that the problem has been reduced *exactly* to the solution of a single-particle equation of Hartree form. The solution of the equations leads to the energy and density of the lowest state, and all quantities derivable from them. It shares the main advantage of the HF method, namely the single-particle interpretation of the results. However, in contrast to the Hartree-Fock potential, the effective potential has a *local* dependence on the density.

All terms in the energy expression (6) are straightforward to evaluate, apart from the exchange-correlation energy, E_{xc}. This term is relatively small, at least for atoms and molecules, and there is an exact expression for it [7]:

$$E_{xc} = \frac{1}{2} \int d\mathbf{r}\; n(\mathbf{r}) \int d\mathbf{r}' \; \frac{1}{|\mathbf{r} - \mathbf{r}'|} n_{xc}(\mathbf{r}, \mathbf{r}' - \mathbf{r}), \tag{9}$$

where

$$n_{xc}(\mathbf{r}, \mathbf{r}' - \mathbf{r}) \equiv n(\mathbf{r}') \int_0^1 d\lambda \left(g(\mathbf{r}, \mathbf{r}', \lambda) - 1 \right) . \tag{10}$$

E_{xc} can then be understood as the interaction between $n(\mathbf{r})$ and the exchange-correlation hole, $n_{xc}(\mathbf{r})$. The function $g(\mathbf{r}, \mathbf{r}', \lambda)$ is the pair-correlation function of a system with density $n(\mathbf{r})$ and electron-electron interaction $\lambda\varphi$. In general it is only possible to evaluate such correlation functions via the exact wave function, so that it is necessary to develop and test approximations for E_{xc}.

The most widely used approximation is the local spin density (LSD) approximation

$$E_{xc}^{LSD} = \int d\mathbf{r}\; n(\mathbf{r})\; \varepsilon_{xc}[n_\uparrow(\mathbf{r}), n_\downarrow(\mathbf{r})]. \tag{11}$$

Here $\varepsilon_{xc}[n_\uparrow, n_\downarrow]$ is the exchange and correlation energy per particle of a homogeneous, spin-polarized electron gas with spin-up and spin-down densities $n_\uparrow$ and $n_\downarrow$, respectively. We use the parametrization of Vosko *et al.* [8]. This approximation is free of any adjustable parameters, but its application to atoms, molecules and solids cannot be justified by small departures from homogeneity. For a detailed discussion, see the review article by Jones and Gunnarsson [9].

There have been many DF applications to small molecules using this approximation [9], including group VI A molecules [10]. Ozone (O_3) [11] and thiozone (S_3) [12], for example, are described very poorly by a single determinant wave function, and Hartree-Fock calculations give a ground state with the wrong symmetry. By contrast, the LD-approximation gives a good description of both the ground state geometry and the energy difference between the ground state and the lowest state with bond angle $\alpha = 60°$. In S_3, there are two states with quite different geometries that are almost degenerate, and this prediction has been confirmed by CI-calculations.

2.4 The molecular dynamics/ density functional approach

In systems where the ground state is unknown or there are many local minima, it is necessary to develop alternative methods for finding solutions that are near to optimal. Kirkpatrick *et al.* [13] noted the connection between statistical physics and the minimization of a function of many variables, and suggested "simulated annealing" based on a Monte Carlo sampling as a generally applicable way of finding such solutions. It has been shown recently [14] that simulated annealing can lead to *nearly* optimal solutions of special NP-hard problems in polynomial average time.

Molecular dynamics (MD) provides an alternative to the Monte Carlo approach, and Car and Parrinello [15] showed that it could be combined with the density functional (DF) scheme for calculating total energies. The result is a parameter-free method for calculating electronic properties that makes no assumptions about ground state geometries. Finite temperature MD techniques allow an efficient sampling of the potential energy surface, and the DF method [9], with the local spin density (LSD) approximation for the exchange-correlation energy, does not require the parametrization of the interatomic forces common in MD schemes [16]. As shown below, MD also allows us to study chemical dynamics.

The scheme and some of its applications have been described in the contribution of R. Car, and details of the calculations are given in Ref. [17]. There are *two* minimization problems to be solved: The total energy must be minimized for each geometry, and we must find the geometry with the lowest energy. This is done conveniently by viewing E as a function of two interdependent sets of degrees of freedom, $\{\psi_i\}$ and $\{\mathbf{R}_I\}$, and using standard MD techniques [15]. The energy E is determined from the density via the orbitals ψ_i, $n(\mathbf{r},t) = \sum_i |\psi_i(\mathbf{r},t)|^2$ and the density functional expression (6, 11) for the total energy E. The artificial second order Newton's dynamics for the electronic degrees of freedom ensure that the electrons remain close to their ground state. If we define a "temperature" T by the mean classical kinetic energy of the atoms, the atomic configurations observed during the annealing process then represent physical molecules at elevated temperatures.

3. Application to clusters of group VI A elements

The group VI A elements provide some of the best characterized of small atomic clusters. These elements are unique in that many allotropes comprise regular arrays of well-separated ring molecules, and X-ray structure analyses have been performed for $S_n, n = 6-8, 10-13, 18, 20$ [18,19] and $Se_n, n = 6, 8$ [18,20]. S_9 has been found as microcrystals, and S_{18} exists in two distinct forms. Several mixed crystals of the form Se_nS_m [21] and a range of sulphur oxides (S_nO, $n =$ 5–10; S_7O_2, $S_{12}O_2$) and ions are also known. The preparation of these clusters is reviewed by Steudel [19]. S_7, for example, can be prepared by the reaction of titanocene pentasulphide with dichlorodisulphane in organic solvents at 0° C, yielding

$$(C_5H_5)_2TiS_5 + S_2Cl_2 \rightleftharpoons S_7 + (C_5H_5)_2TiCl_2 \ . \tag{12}$$

The structures can usually be determined accurately, although the packing of the molecular units can result in structures slightly different from those of the molecules in free space.

Sulphur and selenium have much in common and it is not surprising that mixed isomers can coexist. In addition to the intrinsic interest in Se_xS_y molecules, an understanding of their structures should provide insight into the complex structures of related liquid and amorphous materials, including Se itself. Mixed molecules have been studied in the vapour phase [22, 23], as liquids and as solid solutions [21, 24, 25]. The different species crystallize together, leading to structures with sulphur and selenium atoms distributed over the atomic sites [26]. The possible complexity is evident from the example of eight-membered rings, where 30 different crown-shaped isomers can occur.

In this section, we describe MD/DF calculations for some of these systems. We use a large unit cell [constant volume 1000 Å^3; fcc lattice constant 15.9 Å] with periodic boundary conditions, and this guarantees a weak interaction between the clusters. The O, S and Se atoms are described by non-local pseudopotentials [27], using pseudopotential components for $v_\ell(r), \ell = 0, 1, 2$. The energy cut-off for the plane wave expansion of the eigenfunctions, ψ_i, was taken to be 5.3 or 7 Hartree a.u., corresponding to $\sim$ 4000–6000 plane waves, except for S_7O, where a cut-off of 22.5 a.u. (36000 plane waves) was used. We use a single point $k = 0$ in the f.c.c. Brillouin zone. The "mass" of the electronic degrees of freedom was taken to be 300 a.u., and the time step for the simulations was $\Delta t = 1.7 \times 10^{-16}$ s (7.3×10^{-17} s for S_7O). Further details of the calculations are found in Ref. [17].

3.1 Ring molecules Se_n, S_n.

As a first example of this method we show in Fig. 1 the change in structure of a nearly linear chain of five Se atoms [28]. The time between successive plots corresponds to 500 time steps of 3×10^{-16} s each. There are many structures with local minima in the energy surface, and the ability to vary the average kinetic energy of the ions enables one to find the structure with lowest energy in a relatively straightforward fashion.

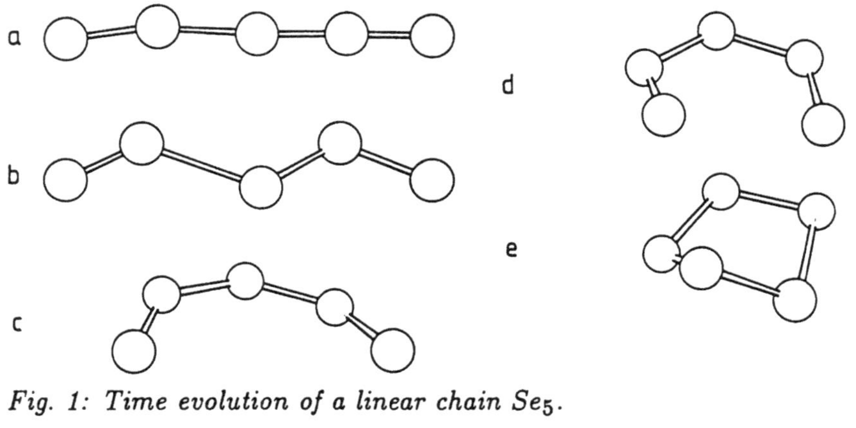

Fig. 1: Time evolution of a linear chain Se_5.

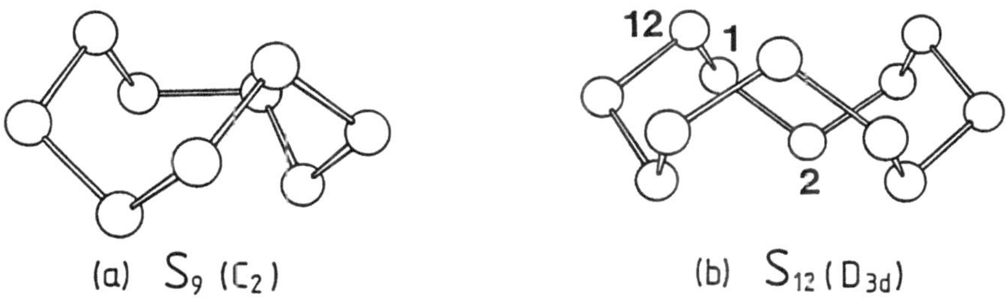

Fig. 2: Structures of (a) S_9 and (b) S_{12}.

This example shows that it is possible to determine low-lying energy minima even if one starts from a geometry that is quite different. This is also true in the case of sulphur clusters, S_2 to S_{13} [17]. In S_{3-6} we started from almost linear chains and in S_{7-13} from nearly planar rings. By repeatedly heating and cooling the system, and ultimately cooling the molecule to $T = 0$, the final structures agreed well with experiment in all cases where X-ray data were available. As an example we show in Fig. 2 the structure of S_{12}. The D_{3d} symmetry is reproduced very closely, and the structural parameters (bond length $d = 3.97$ a.u., bond angle $\alpha = 106°$, dihedral angle $\gamma = 88°$) are in good agreement with measured values (3.88 a.u., 106.2°, 87.2°, respectively).

Perhaps even more interesting is the application of the method to cases where the ground state structures are not yet known. In the case of S_9, for example, the crystallites that can be prepared are unfortunately too small for X-ray diffraction analysis. We find a ground state isomer with C_2 symmetry [Fig. 2] as well as several local minima with C_s symmetry with energies > 0.2 eV higher. The calculated ground state structure is consistent with measured Raman spectra, which indicate almost equal bond lengths and bond angles, as well as dihedral angles in the range $70 - 130°$.

3.2 Structural change in S_7O.

A particularly interesting example of a structural optimization is provided by the S_7O molecule [29]. In the course of the simulation the molecule changed from a stable local minimum with a ring structure similar to that in S_8 to a structure that is topologically unrelated, namely an oxygen atom outside the S_7 structure. The latter is the experimental ground state, and it is remarkable that this complicated structural change took place in only ~ 8000 time steps ($< 10^{-12}$ s), although the energy is only ~ 0.2 eV lower than the energy of the ring structure and the energy barrier between ~ 5 eV. It appears that the molecular dynamics technique is a particularly efficient generator of geometrical configurations in the simulated annealing procedure. The final structure is in very good agreement with experiment [29]. [Fig. 3].

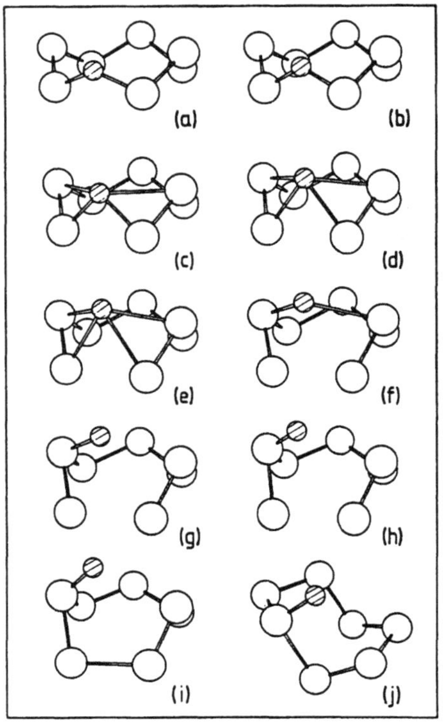

Fig. 3: Evolution of S_7O from ring (a) to ground state structure (j).

3.3 Isomers of Se_2S_6, Se_6S_2.

Early kinetic work [22, 23] on S_nSe_{8-n} ring molecules excluded the possibility of $Se - Se$ bonds, and this was attributed [22] to the weakness of the $Se - Se$ bond compared with both $S - S$ and $Se - S$ bonds. Raman spectroscopy has shown, however, that $Se - Se$ stretching vibrations can readily be identified in such structures [30], even for mixtures with very low Se concentrations. On the other hand, the $S - S$ stretching band was absent if the nominal sulphur content dropped below that in Se_6S_2, suggesting that the formation of $S - S$ homonuclear bonds is not favoured in mixed crystals with low sulphur content. The apparent asymmetry between $S-$rich and $Se-$rich molecules was also traced to the weakness of $Se - Se$ bonds [30].

The last decade has seen considerable use of Raman and NMR spectroscopies to study these systems. The structures [31] of the eight-membered systems $1,2,3-Se_3S_5$ [32] and $1,2,5,6 - Se_4S_4$ [33] have been determined qualitatively, and the first rings with less than eight atoms [$1,2 - Se_2S_5$ and SeS_5] prepared and characterized [34]. Ring molecules with a majority of Se atoms (Se_5S, Se_5S_2, Se_6S_2) [35] have been studied, and the most stable isomer of the last of these contained adjacent Se atoms. ^{77}Se NMR spectroscopy has been used [36] to identify isomers of Se_nS_{8-n} obtained from molten mixtures of the elements. The main components were SeS_7 and $1,2-Se_2S_6$. For a given chemical composition, the most abundant isomers were those with all Se atoms adjacent to each other.

Predictions of the relative stability of different isomers have been difficult to make, since the interconversion of various Se_xS_y molecules and ring-chain equilibration processes involve the rupture and formation of $S-S$, $Se-Se$ and $Se-S$ bonds. Reactions involving the groups

$$-S-S- \quad + \quad -Se-Se- \quad \longrightarrow \quad 2\ -Se-S- \tag{13}$$

have been studied recently using Hartree-Fock techniques with a minimal Gaussian basis [37]. The energy changes of various rearrangement reactions related to (13) were found to be very small (between 0 and -2 kJ mol^{-1}). The stated uncertainties in these calculations (10–20 kJ mol^{-1}) are substantially greater than the energy differences themselves, and the exothermic reaction is not consistent with calorimetric measurements in the liquid [for an equimolar solution, the reaction enthalpy for (13) was close to the enthalpy of mixing $\Delta H_m = +1.3$ kJ mol^{-1}] [38]. For diatomic molecules in the gas phase [39], the enthalpy change is $+7.9 \pm 6.7$ kJ mol^{-1} and the sum of the dissociation energies on the left side of (13) is greater than the sum on the right.

We have performed extensive calculations on seven- and eight-membered rings of the type Se_xS_y [40]. In each case we started with the ground state geometry of S_8 (D_{4d}), and determined the ground state geometry by alternating steepest descents techniques ($\Delta t = 1.0 \times 10^{-15}$ s, $\mu_i = 50000$ a.u.) with MD at $T = 200$ K ($\Delta t = 8.5 \times\ 10^{-17}$ s, $\mu_i = 300$ a.u.) In the simulated annealing runs, the average instantaneous temperature $\langle T \rangle$ – the mean kinetic energy of the ions – is kept constant with 10° tolerance by rescaling the ionic velocities uniformly with a factor $\sqrt{\langle T \rangle / T}$ [41]. Each structure determination required a total of about 4000 time steps, and the system departs very little from the BO surface. The electronic system is kept within $\ll 0.01$ eV of its ground state, so that the molecular configurations observed represent physical entities at elevated temperatures. Within the framework of the LSD approximation and the basis described above, this approach allows us to determine the relative energies of the minima to better than 1×10^{-3} eV. Each time step required approximately 1 second of CPU time on a Cray X-MP/416 computer.

The cyclic structures possible in Se_2S_6 and Se_6S_2 are shown in Fig. 4 and the structural parameters for Se_2S_6 are given in Table 2. The structures show several remarkable features. In all eight structures, the three bond lengths ($S-S$, $S-Se$, $Se-Se$) are not only the same within 0.01 Å (2.08 Å, 2.23 Å, 2.35 Å), but the $S-S$ and $Se-Se$ bonds are the same as found in our previous calculations for the ring structures of S_8 and Se_8. All three are simply the sums of the corresponding atomic radii. The idea of transferability between different structures is enhanced by the observation that the *average* of the structural parameters is the same for each of the four structures [$\langle d_{ij} \rangle = 2.15$ Å, $\langle \alpha_i \rangle = 108.0°$, $\langle \gamma \rangle = 98.7°$ in Se_2S_6; $\langle d_{ij} \rangle = 2.29$ Å, $\langle \alpha_i \rangle = 107.4°$, $\langle \gamma \rangle = 99.5°$ in Se_6S_2]. If these parameters

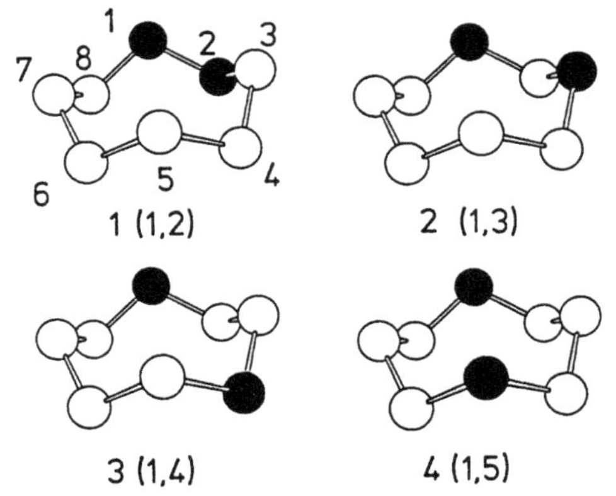

Fig. 4: Structures of Se_2S_6 (and Se_6S_2).

Table 2. Molecular parameters d, α and γ for all isomers of Se_2S_6 [see Fig. 1]. Bond lengths d_{ij} (between atoms i and j) in Å, bond angles α_i (at atom i) and torsion angles γ_{ij} (at bond ij) in degrees.

	(1,2) [C_2]	(1,3) [C_s]	(1,4) [C_2]	(1,5)[C_{2v}]
d_{12}	2.34	2.22	2.22	2.23
d_{23}	2.23	2.22	2.07	2.08
d_{34}	2.07	2.22	2.23	2.08
d_{45}	2.08	2.08	2.23	2.23
d_{56}	2.09	2.08	2.08	2.23
d_{67}	2.08	2.08	2.07	2.07
d_{78}	2.07	2.08	2.08	2.08
d_{81}	2.23	2.22	2.23	2.23
α_1	105.9	106.1	106.7	106.7
α_2	105.8	108.1	108.2	108.9
α_3	108.1	106.2	108.2	108.6
α_4	108.8	108.4	106.7	108.4
α_5	109.1	108.7	108.5	106.5
α_6	109.2	109.2	108.7	109.2
α_7	108.8	108.7	108.8	108.7
α_8	108.2	108.5	108.6	108.5
γ_{12}	96.8	98.0	98.3	99.1
γ_{23}	97.2	96.6	97.8	96.6
γ_{34}	97.6	96.1	96.3	98.2
γ_{45}	100.6	100.4	99.1	100.5
γ_{56}	102.9	102.1	101.1	99.0
γ_{67}	100.5	100.4	98.9	96.5
γ_{78}	97.5	99.0	99.1	97.7
γ_{81}	97.3	97.6	99.1	100.4

are extrapolated to Se_0S_8 and Se_8S_0, we obtain structures close to the known structures of S_8 and Se_8.

The energy calculations show that the (1,2) structures are the ground states in both Se_2S_6 and Se_6S_2. The energies of the other structures are *degenerate* to within 3 meV and lie 0.08 eV above the ground state in each case. We note here that the (1,2)-isomer is the most prominent detected in the NMR spectra of Se_2S_6 – there are smaller amounts of the other three – and the only one found in NMR spectra of Se_6S_2 [36]. The calculations presented in the previous section show some remarkable regularities, particularly in predicting ground states with adjacent minority atoms in each case. We now show that the bonding trends can be understood in terms of a very simple model.

The eight-membered rings are characterized by small deviations from the crown-shaped $[D_{4d}]$ structures familiar from S_8 and Se_8. We have also seen that there are three bond lengths in the clusters, d_{S-S}, d_{S-Se} and d_{Se-Se}. The well-known relationship between bond length and bond strength [42] suggests that we associate with each bond type in these cyclic molecules corresponding contributions to the binding energy, E_{S-S}, E_{S-Se}, and E_{Se-Se}. In the case of the (1,2) structure in Se_2S_6, this leads to

$$E_{12} = E_{Se-Se} + 2E_{S-Se} + 5E_{S-S} \; . \tag{14}$$

For *all three* remaining structures, we find

$$E_{13,14,15} = 4E_{S-Se} + 4E_{S-S} \; . \tag{15}$$

From this simple argument we may then expect the energies of three structures to be equal, and the fourth (E_{12}) to differ by the amount

$$\Delta E = E_{Se-Se} + E_{S-S} - 2E_{S-Se} \; . \tag{16}$$

If we apply the same argument to the Se_6S_2 structures, Eq. (14-15) still apply, with S and Se interchanged. Again we find the (1,2) structure separated from the other three, and the energy difference is precisely that given in Eq. (16). This very simple argument explains why the energy orderings in Se_2S_6 and Se_6S_2 are the same, and is consistent with the most stable (1,2) isomers being separated from three structures with almost identical energies. The separation in the present calculations is approximately 0.08 eV (7.4 kJ mol^{-1}), which is consistent with the endothermicity of the reaction (2) and in reasonable agreement with the measured heats of reaction in the gas and liquid phases.

The results of the simple model agree with trends found in recent measurements, in particular with the relative abundance of structures with adjacent minority atoms. The symmetry between the energy ordering of the isomers of Se-rich and $S-$rich molecules is in direct contrast to earlier discussions of these systems [22,30]. This symmetry is perturbed, of course, by differences between the elements such as the size, but the essential features are not changed. The trend to "segregation" of minority components could also occur in the disordered liquid and amorphous states.

3.4 Dynamics of Se_2S_5 – "pseudorotation".

Mixtures of different molecules in solution and in the liquid phase (whether isomers or conformers) are often viewed as comprising molecules that are essentially static: Apart from vibrations around the equilibrium geometry, the individual molecules are not seen as changing entities. However, solutions and liquids are usually kept at higher temperatures (often $T = 300$ K) and other dynamical phenomena could be evident. In Se_2S_5, the ^{77}Se NMR signal in CS_2 solution has a single peak at room temperature, so that the Se atoms are magnetically equivalent on the NMR time scale [43]. This is evidence for "pseudorotation", by which all ring atoms become equivalent within a certain time interval due to the simultaneous rotation about several of the ring bonds. This implies small energy barriers, consistent with the torsional barrier of $S-S$ bonds in sulphur rings, which has been estimated from the heat of formation of S_7 to be ≤ 24 kJ mol^{-1} [44]. To study the dynamical properties of Se_2S_5, we have coupled it to a heat bath of $T = 300$ K, simulating the interaction with the environment by a rescaling of the atomic velocities.

We began with the structure with the *highest* energy of those with neighbouring $Se-Se$ atoms, and continued the molecular dynamics at a temperature of 300 K. The geometry changes occurring in the first 3.0×10^{-13} s (3500 time steps, $\Delta t = 8.5 \times 10^{-17}$ s) are shown in Fig. 5. The energy rises initially and

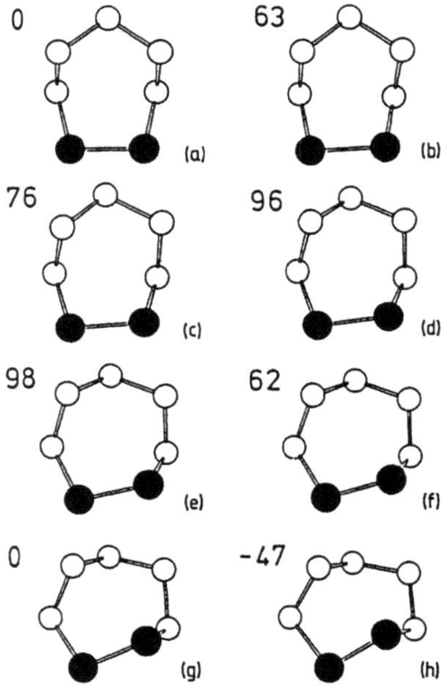

Fig. 5: Molecular dynamics of Se_2S_5 at $T = 300$ K, showing the change from the (4,5)-structure (a) to the (3,4)-ground state structure (h). The interval between each plot is $500\Delta t$, with $\Delta t = 8.5 \times 10^{-17}$ s. The energies relative to the starting energy are shown in meV.

passes through two pairs of local maxima and minima to a saddle point in the energy surface 0.12 eV above the starting energy. The corresponding geometry is close to the one shown in Fig. 5(e). After crossing this barrier, the molecule has a structure with the same pattern of torsion angles as the ground state, and the energy falls rapidly to a local minimum with structure shown in Fig. 5(h). If we now locate the bottom of the energy minimum and the corresponding $T = 0$ energy, we find precisely the ground state energy and structure. The energy is 0.22 eV [20 kJ mol^{-1}] below the saddle point, similar to the value of the torsion barrier quoted above for S_7 [44].

A comparison of the structures 5(a) and 5(h) shows that the geometry changes are substantial. The $Se - Se$ bond length changes from 2.50 Å to 2.26 Å, and the dihedral angle from zero to 82.8°. In fact, the bond length *increases* to 2.52 Å before decreasing to its final value. The bond angle changes are much smaller. Fig. 5 shows that the change in structure involves a concerted change of the positions of all atoms, with the largest motion apparent one of the Se atoms. During this simulation, the maximum instantaneous temperature fluctuations $\langle T - \langle T \rangle \rangle$ for the individual atoms are as high as 750°, i.e. the kinetic energy is distributed over the atoms in a far from uniform fashion. At any instant in time, individual atoms can have a kinetic energy of ~ 0.1 eV, sufficient to surmount energy barriers such as the structural change in Fig. 5 .

In order to study the dynamics further, we have carried out an extended simulation of 60000 time steps from the structure of the most stable isomer. The system passed through many (~ 200) extrema in the energy surface, with a maximum energy ~ 0.12 eV above the ground state energy. This is insufficient for the system to cross a barrier to one of the other stable local minima, and the system remained in the basin of attraction of the ground state structure throughout. This was also true in a second simulation performed at 500 K (10000 time steps) starting from the structure and velocities found at the top of the energy barrier between the initial and final structures in Fig. 5. In Fig. 6 we show the variation in the instantaneous "temperature" (kinetic energy) of each of the seven atoms in Se_2S_5 during the first 25000 steps in the $T = 300$ K simulation. The kinetic energy available to the atoms is distributed over the molecule in a very non-uniform fashion. The energy supplied to an individual atom can be up to four times the average, thus helping the system to overcome potential energy barriers that would be insurmountable under equilibrium conditions.

These simulations suggest that in solutions or in the liquid phase statistical averages over structurally different molecules dominate the properties in question. The individual constituents can change their shapes constantly as dictated by the temperature of the bath and the shape of the energy surface. In the seven-membered rings basins of attraction (local minima) of different extent are separated by barriers that can be overcome at an externally imposed kinetic temperature of $T = 300$ K. While the time required for the conversion in Fig. 5 in Se_2S_5 (4×10^{-13} s) is much shorter than the NMR time scale, our maximum

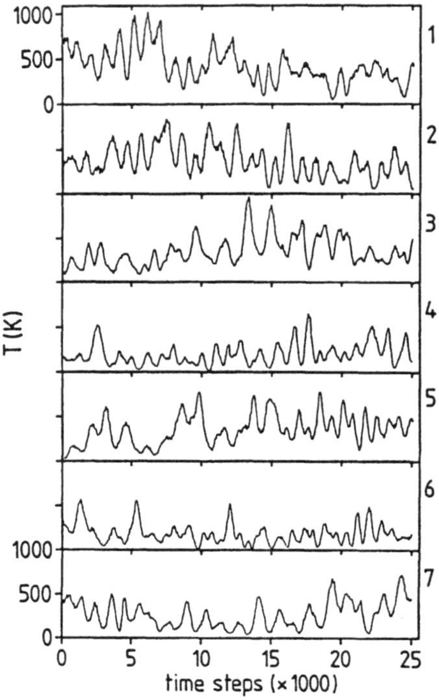

Fig. 6: Fluctuation in instantaneous temperature of different atoms in Se_2S_5 at $T = 300K$. Atoms 1 and 2 are Se atoms.

simulation period (5 ps) was too short to observe further structural changes. The basin of attraction for the ground state is not only deeper by ≥ 0.05 eV, but is apparently more extended.

4. Concluding Remarks

The combination of molecular dynamics and density functional methods has given excellent results for small sulphur and selenium clusters, and useful results in S_7O, where the geometry changes and the energy barrier between the ring and ground state structures are both very large. In the heterocyclic ring molecules Se_xS_y we have studied the opposite extreme – structures with very small energy differences and energy barriers that are small on a thermal scale. The results are very encouraging. While no method can guarantee finding the absolute energy minima in the low-symmetry clusters of this size, the results demonstrate further the value of this approach.

Four different factors are important in enabling the MD simulated annealing procedure to locate the optimal energy minima, in spite of the complexity of the energy surfaces and large energy barriers. (1) The continuous supply of kinetic energy via the external heat bath is essential to overcome potential energy barriers that are large on a thermal scale (> 5 eV at $\langle T \rangle = 2000$ K in the case of S_7O). This process is effective unless the energy surface is so steep that the kinetic energy loss cannot be compensated at the imposed $\langle T \rangle$. (2) The kinetic energy can be partitioned among the individual atoms of the molecule in a very non-uniform way, allowing one atom in a small cluster to acquire essentially all available kinetic energy. (3) The trajectories generated by the MD equations of motion appear to be very successful in locating reaction pathways to low-lying minima. (4) Our experience with group VI A clusters has shown that the lowest-lying minima are significantly more extended than higher ones. However, it is conceivable that the reverse could be true in other systems, and an extension to group V A elements is in progress.

We thank the German Supercomputer Centre (Höchstleistungsrechenzentrum, HLRZ) for a generous grant of computer time on the Cray X-MP 416 in the Kernforschungsanlage Jülich.

References

[1] M.R. Hoare, J.A. McInnes: Adv. Phys. **32**, 791 (1983).

[2] L.T. Wille, J. Vennik: J. Phys. A **18**, L419, L1113 (1985).

[3] M.R. Garey, D.S. Johnson: *Computers and Intractability: A Guide to the Theory of NP-Completeness* (Freeman, San Francisco, 1979).

[4] P. Hohenberg, W. Kohn: Phys. Rev. **136**, B864 (1964).

[5] M. Levy: Proc. Nat. Acad. Sci. USA **76**, 6062 (1979).

[6] W. Kohn, L.J. Sham: Phys. Rev. **140**, A1133 (1965).

[7] O. Gunnarsson, B.I. Lundqvist: Phys. Rev. B **13**, 4274 (1976). See also J. Harris, R.O. Jones: J. Phys. F **4**, 1170 (1974).

[8] S. Vosko, L. Wilk, M. Nusair: Can. J. Phys. **58**, 1200 (1980).

[9] For a recent survey of this formalism and some of its applications, see R.O. Jones, O. Gunnarsson: Rev. Mod. Phys. **61**, 689 (1989).

[10] For a discussion of O_3, SO_2, SOS and S_3, see R.O. Jones: Adv. Chem. Phys. **67**, 413 (1987), and references therein.

[11] R.O. Jones: J. Chem. Phys. **82**, 325 (1985).

[12] R.O. Jones: J. Chem. Phys. **84**, 318 (1986).

[13] S. Kirkpatrick, C.D. Gelatt Jr., M.P. Vecchi: Science **220**, 671 (1983).

[14] G.H. Sasaki, B. Hajek: J. Assoc. Comput. Machin. **35**, 387 (1988).

[15] R. Car, M. Parrinello: Phys. Rev. Lett. **55**, 2471 (1985).

[16] F. Stillinger, T.A. Weber, R.A. LaViolette: J. Chem. Phys. **85**, 6460 (1986).

[17] D. Hohl, R.O. Jones, R. Car, M. Parrinello: J. Chem. Phys. **89**, 6823 (1988).

[18] J. Donohue: *The Structures of the Elements* (Wiley, New York, 1974), Chap. 9.

[19] R. Steudel: In *Studies in Inorganic Chemistry, Vol. 5*, ed by A. Müller, B. Krebs (Elsevier, Amsterdam, 1984), p. 3.

[20] R. Steudel, E.-M. Strauss: Adv. Inorg. Chem. Radiochem. **28**, 135 (1984).

[21] R. Steudel, E.M. Strauss: In *The Chemistry of Inorganic Homo- and Heterocycles, Vol. 2*, (Academic, London, 1987), p. 769.

[22] R. Cooper, J.V. Culka: J. Inorg. Nucl. Chem. **29**, 1217 (1967).

[23] M. Schmidt, E. Wilhelm: Z. Naturforsch. **25 B**, 1348 (1970).

[24] R. Steudel, R. Laitinen: Topics in Current Chemistry, **102**, 177 (1982).

[25] *Selenium: Gmelin Handbuch der Anorganischen Chemie*, 8. Aufl., Ergänzungsband B2, ed by H. Bitterer (Springer, Berlin, 1984).

[26] R. Laitinen, L. Niinistö, R. Steudel: Acta Chem. Scand. A **33**, 737 (1979).

[27] G.B. Bachelet, D.R. Hamann, M. Schlüter: Phys. Rev. B **26**, 4199 (1982).

[28] D. Hohl, R.O. Jones, R. Car, M. Parrinello: Chem. Phys. Lett. **139**, 540 (1987).

[29] D. Hohl, R.O. Jones, R. Car, M. Parrinello: J. Am. Chem. Soc. **111**, 825 (1989).

[30] H.H. Eysel, S. Sunder: Inorg. Chem. **79**, 2626 (1979).

[31] The numbers before a chemical formula indicate the positions of the minority atoms in a sequentially numbered ring.

[32] R. Laitinen, N. Rautenberg, J. Steidel, R. Steudel: Z. Anorg. Allg. Chem. **486**, 116 (1982).

[33] D.M. Giolando, M. Papavassiliou, J. Pickardt, T.B. Rauchfuss, R. Steudel: Inorg. Chem. **27**, 2596 (1988).

[34] R. Steudel, E.M. Strauss: Angew. Chem. **96**, 356 (1984).

[35] R. Steudel, M. Papavassiliou, E.M. Strauss, R. Laitinen: Angew. Chem. **98**, 81 (1986).

[36] R.S. Laitinen, T.P. Pakkanen: Inorg. Chem. **26**, 2598 (1987).

[37] R. Laitinen, T. Pakkanen: J. Mol. Struct. (Theochem) **91**, 337 (1983); **124**, 293 (1985).

[38] T. Maekawa, T. Yokokawa, K. Niwa: Bull. Chem. Soc. Japan **46**, 761 (1973).

[39] J. Drowart, S. Smoes: J. Chem. Soc., Faraday Trans. II **73**, 1755 (1977).

[40] R.O. Jones, D. Hohl: J. Am. Chem. Soc. (submitted).

[41] L.V. Woodcock: Chem. Phys. Lett. **10**, 257 (1971).

[42] For a discussion of the relationship between core size, valence eigenfunctions and bond strengths, see J. Harris, R.O. Jones: Phys. Rev. A **19**, 1813 (1979).

[43] R. Steudel, M. Papavassiliou, D. Jensen, K. Seppelt, K: Z. Naturforsch. **43 B**, 245 (1988).

[44] R. Steudel: Angew. Chem. **87**, 683 (1975); Z. Naturforsch. **38 B**, 543 (1983).

Effective-Medium Theory: Considering Corrections and Delimitations

A. Zwartkruis and J.E. van Himbergen

Institute for Theoretical Physics, University of Utrecht, P.O. Box 80.006, 3508 TA, Utrecht, The Netherlands

Abstract. We calculate corrections to the simplest form of EMT both for hydrogen in semiconductors and oxygen in $YBa_2Cu_3O_7$. From these examples we infer limitations on the applicability of the theory.

1. Introduction

Nowadays computational schemes exist that handle the electronic properties of low-symmetry solid state problems in a self-consistent way. These approaches provide accurate results. However, their nature is not modular: a final result cannot be seen as a composition of terms that are separately well-defined. In case the low symmetry has to be combined with relaxation of the structure, first-principles calculations can be prohibitively time-consuming. Therefore it is useful to precede or supplement them by approximate schemes that exploit underlying physical principles. Approximate methods allow calculations for a wide range of atomic configurations.

Obviously for approximations there is a need to check assumptions and to assess accuracy from testcases. This paper considers corrections to the Effective-Medium Theory (EMT), especially when it is applied to hydrogen in semiconductors or oxygen impurities in a High-T_c superconductor. Originally EMT [1,2,3] was designed for application to simple metals. As it was quite successful also for non-simple metals as e.g. the transition metals [4,5], it is interesting to explore to what extent it is applicable to more extreme cases.

2. Theory

We want to calculate heats of absorption $\tilde{\Delta} E$ and equilibrium positions for extra atoms in inhomogeneous host systems by linking the problem to the embedding in a homogeneous jellium. If the impurity potential is well localised, the *local* host electron density forms the dominant quantity for determining the energy of embedding. Then there exists a region *a* near the nucleus of the impurity where both the potential and the electron density are very similar for an impurity in a real host system or one in a homogeneous jellium. However, there is also a region *b* where both potential and electron density of the combined host-impurity system resemble more the ones in the pure host.

Springer Proceedings in Physics, Vol. 48 Many-Atom Interactions in Solids
Editors: R.M. Nieminen · M.J. Puska · M.J. Manninen

To enable perturbation theory in small differences only, Nørskov [2] has pointed out that one can employ an expansion of the energy functional that is different for regions a and b. In region b we expand in small adatom induced differences $\tilde{\Delta}$ (real host) and Δ (jellium). In region a we only consider differences that are caused by the difference in host system: δ^0 (differences between real host and jellium without impurity) and δ (such differences with impurity). So in total we consider four different systems, each divided into two different spatial regions. Potential and density for the real host with and without impurity are denoted $(\tilde{v},\tilde{n})$ and $(\tilde{v}^0,\tilde{n}^0)$ respectively. Their counterparts in jellium by (v,n) and (v^0,n^0).

For details of the derivation we refer to [1,6]. If we choose the jellium density by sampling the host density (usually taken as a superposition of atomic densities) over the adatom induced electrostatic potential

$$n^0 \quad = \quad \int \tilde{n}_a^C(\mathbf{r})\Delta\phi(\mathbf{r})d\mathbf{r} \;/\; \int \Delta\phi(\mathbf{r})d\mathbf{r}, \tag{1}$$

the outcome of EMT in its most simplified form is

$$\tilde{\Delta}E \quad = \quad \Delta E(n^0) - (\alpha_{at} + \alpha_v)n^0 \equiv \Delta E_{eff}^{hom}(n^0). \tag{2}$$

Here $\tilde{\Delta}E$ is the embedding energy in the real host, ΔE is the embedding energy in jellium as a function of n^0. It is known for many atoms from the work of Puska et al. [7]. For most applications ΔE is the dominant contribution to the true embedding energy. The quantities $\alpha_{at}n^0$ and $\alpha_v n^0$ describe in an approximate way electrostatic and exchange contributions.

All quantities in formula 2 are known or can be calculated. Note that in this case the absorption energy in the real system is dependent only on the sampled host density. Except for the integal in the sampling procedure, all quantities occurring are spherically symmetric. It appears as if the screening has removed all reference to the true geometry of the system. However, this is not the case in a more precise analysis of $\tilde{\Delta}E$ [6]. One then obtains

$$\begin{aligned} \tilde{\Delta}E \quad &= \quad \Delta E + \int\int \frac{\tilde{\Delta}\rho_a(\mathbf{r})\delta\rho_b(\mathbf{r}')}{\mid \mathbf{r}-\mathbf{r}' \mid} d\mathbf{r}\, d\mathbf{r}' + \int \tilde{\Delta}\rho_a \delta v_{a\rightarrow a}^{\text{ext}} d\mathbf{r} \\ &+ \quad \sum_{i,\text{occ}} \epsilon_i \Big|_{v_a,\tilde{v}_b^0} - \sum_{i,\text{occ}} \epsilon_i \Big|_{v_a,v_b^0} - \sum_{i,\text{occ}} \epsilon_i \Big|_{v_a^0,\tilde{v}_b^0} + \sum_{i,\text{occ}} \epsilon_i \Big|_{v_a^0,v_b^0} . \end{aligned} \tag{3}$$

Here $\tilde{\Delta}\rho_a$ is the real induced charge density. $\delta v_{a\rightarrow a}^{\text{ext}}$ signifies the Coulomb potential from positive charges in a in as far as they act on charges in a. Finally the ϵ_i denote one-electron energies. Expression 3 can be written as the contribution given by formula 2 together with some correction terms

$$\begin{aligned}\widetilde{\Delta}E &= \Delta E^{hom}_{eff}(n^0) + \{\alpha_v n^0 + \int \widetilde{\Delta} v^{xc}_a(\widetilde{n}^0_a - n^0_a)d\mathbf{r}\} \\ &\quad +\{\textit{core-corrections}\} + \int (\widetilde{\Delta}\rho_a - \Delta\rho_a)\delta v^{\mathrm{ext}}_{a\to a} d\mathbf{r} \\ &\quad + \int\int \frac{(\widetilde{\Delta}\rho_a(\mathbf{r}) - \Delta\rho_a(\mathbf{r}))\rho^0_b(\mathbf{r}')}{\mid \mathbf{r}-\mathbf{r}'\mid} d\mathbf{r}\, d\mathbf{r}'. \end{aligned} \tag{4}$$

In expression 4 the terms in curly braces have been considered before, see e.g. [4]. In the usual case of constructing the host electron density by superposing atomic densities, they have the form of pair potentials. The first term in curly braces replaces the approximately linear behaviour of an exchange contribution in a by its real value. The core-corrections arise because one cannot treat core-electrons as free electrons. One takes care of them by a reversed embedding procedure employing embedding potentials for rare gas atoms.

The fourth and fifth term on the left-hand side were recognised only recently [6]. By an appropriate choice of region a, the fourth term can usually be made to vanish. However the fifth term does not vanish, and it shows that even to first order, part of the embedding energy consists of dipole and higher multipole contributions. Unlike the other terms it cannot be used to actually improve the accuracy of EMT. Additional input from self-consistent calculations would be needed. However we can estimate the error of leaving out this term for some cases, and thereby infer the validity of using EMT for a class of applications. Also it is important to find out whether leaving out this term gives just a fixed offset, or an error that varies strongly with position.

3. Hydrogen and semiconductors

We investigate in some detail the magnitude of the corrections for hydrogen atoms in the semiconductor GaAs (see also [6]). It is well known [8] that, as regards the electronic levels or the screening, interstitial hydrogen in semiconductors is not a "hydrogenic" impurity. Its impurity potential is well localised, and it makes sense to investigate whether the corrections in expression 4 do not invalidate a description by EMT.

As stated before, the actual induced quantities $\widetilde{\Delta}$ can only be obtained from fully self-consistent calculations. In case the quantities $\widetilde{\Delta}$ are not available, the best approximation is to replace them by the spherically symmetric quantities Δ from jellium calculations. Using these, the last two corrections in expression 4 drop out. Figure 1 gives contour maps for the potential energy of H in GaAs. Figure 1a is obtained using the $\Delta E^{hom}_{eff}(n^0)$ term only. Then we see that in contrast to metals, semiconductors have an interstitial density that is low enough to allow the impurities to enter the bulk without having to overcome any potential barrier. Tunnels of lowest potential lead directly from the surface into the bulk. For these tunnels in the bulk, the addition of correction terms leads to

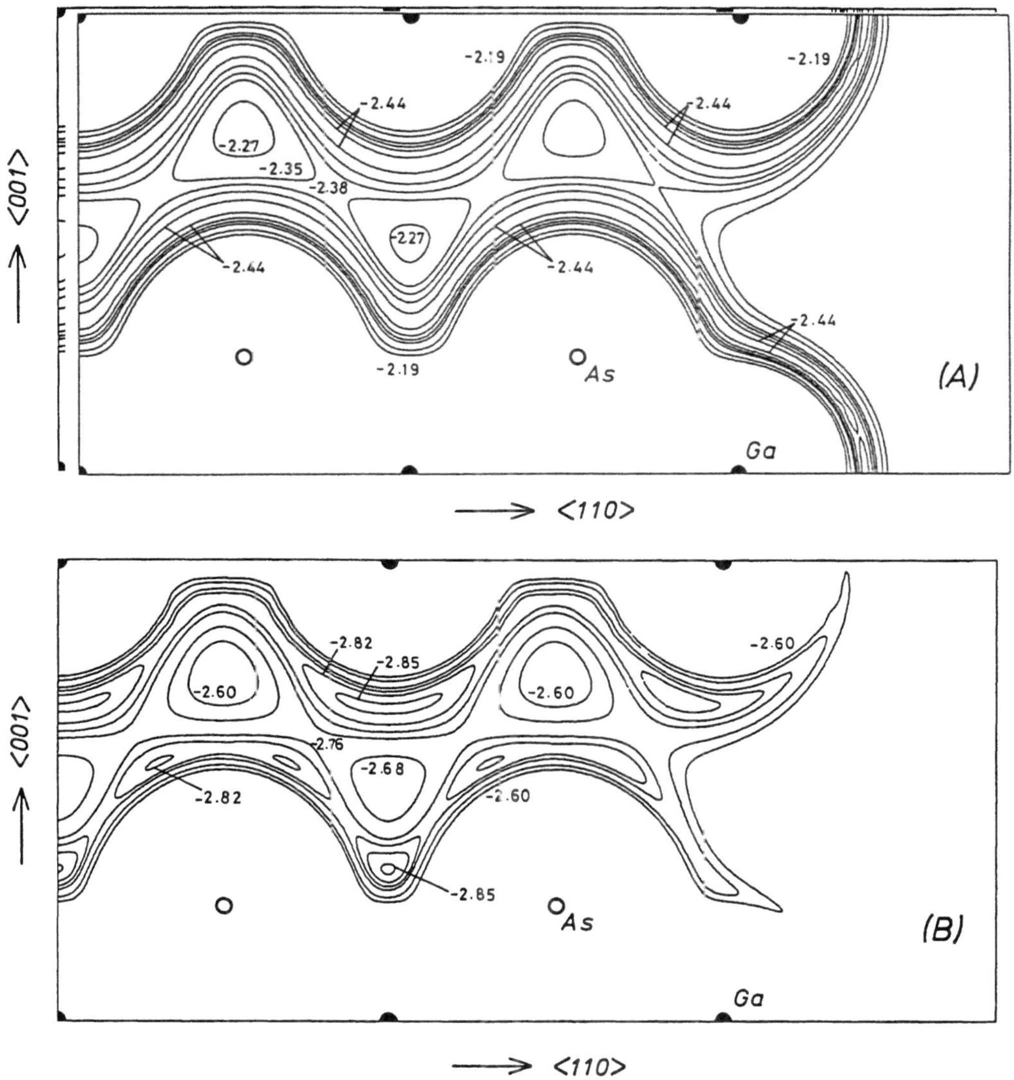

Figure 1: Contour maps of the potential energy of H in GaAs. Figure a is calculated from the $\Delta E_{eff}^{hom}(n^0)$ term only, figure b includes the corrections in braces. The $(1\bar{1}0)$ plane shown is terminated on the right by Ga atoms from a (110) surface. The six contours bridge a total of 0.25 eV.

slight undulation of the potential energy (see figure 1b), but the picture remains essentially the same. The surface becomes less favourable for absorption as there are less atoms that can contribute the small negative corrections.

3.1 Exchange and core corrections

Denoting corrections between braces by ΔE_v and ΔE_c respectively, figure 2 shows the form of these 'pair potential' corrections. It is not possible to use only the correction ΔE_v (solid curve). Being too large it definitely causes unphysical absorption positions (too close to the host atoms), signifying that core electrons cannot be described in the free electron way. Therefore core electrons are taken out, and treated by a reverse embedding procedure [4]. Accordingly, ΔE_v is calculated with the valence electrons only. After adding the new ΔE_v (dotted line) and ΔE_c (dashed-dotted line), the dashed curve in figure 2 is obtained. It equals the solid ΔE_v curve for host atom-impurity distances $R > 3.3\ a_0$. So for distances R that are approximately equal to or larger than $3.3\ a_0$ one can trust the combination of the corrections between braces, as for these distances they are very robust under variations of their precise construction.

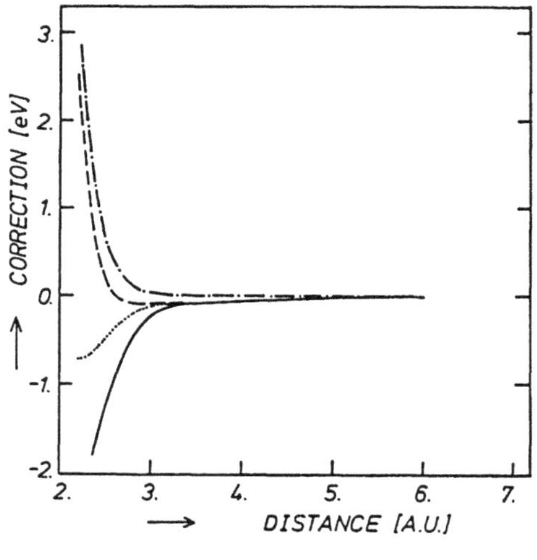

Figure 2: Corrections as function of distance to the nucleus of the Ga atom.

In the bulk the electron density is provided by many atoms, each of them lying rather far from the impurity. Following the above arguments, we can then safely apply the corrections $\Delta E_v + \Delta E_c$. However, near the open (110) surface the electron density is mainly supplied by just one atom and impurity-host distances can be as small as $2.85\ a_0$. Then the role of ΔE_c is not just to keep the impurity away from the range where ΔE_v is less reliable, but it actually adds contributions to the chemisorption energy itself. Since it is generated by a rather heuristic construction, the precise value of the correction for this region should then be regarded with some more caution.

3.2 Dipole and higher multipole contributions

To have an idea about the magnitude of the last term in formula 4, which represents a Coulomb interaction between $\widetilde{\Delta}\rho_a - \Delta\rho_a$ and the host density ρ_b^0, we take a simple dipolar form for the actual induced density $\widetilde{\Delta}\rho_a = \Delta\rho_a(1 + \epsilon\cos\vartheta)$. Here ϑ is the angle with the axis between the impurity and its nearest neighbour. ϵ is estimated from some typical self-consistent calculations of hydrogen atoms on semiconductor surfaces. Both $\widetilde{\Delta}\rho_a$ and $\Delta\rho_a$ integrate to zero, but this does not apply for ρ_b^0 : $\int_b \rho_b^0(\mathbf{r})dr \neq 0$. So the interaction is just singly screened, and it comes as no surprise that at the GaAs surface, on the position of equilibrium determined from the $\Delta E_{eff}^{hom}(n^0)$ term, this correction can be as large as 0.5 eV. For equilibrium positions in the GaAs bulk the corrections considered are less alarming. More host atoms than just one take part in the interaction with $\widetilde{\Delta}\rho_a - \Delta\rho_a$. This leads to a charge density ρ_b^0 in region b that is more symmetrical than in the case of a surface or any other situation with dangling bonds. This greater symmetry causes both a decrease of ϵ and brings about some systematic cancellations in the low multipole parts of the Coulomb interaction. A second important point is that the required electron density can be collected from more host atoms, whereby the separation of the hydrogen and its nearest neighbours will be larger than for the dangling bond situation. This also reduces the magnitude of the total interaction because the decrease in each of the separate interactions outweighs the effect of a greater number of neighbours. For GaAs the ratio between interaction on the surface and in the bulk is typically found to be 5 : 1.

3.3 Conclusions and comparison for hydrogen impurities

For semiconductors we conclude that due to small impurity-host atom distances, an EMT description of chemisorption to surfaces or dangling bonds is difficult. Both corrections $\Delta E_v + \Delta E_c$ and the dipole and higher multipole terms become large. However for the crystalline semiconductor bulk a proper treatment of neutral, interstitial hydrogen is not ruled out a priori. There also, for the low density regions that are favoured by the $\Delta E_{eff}^{hom}(n^0)$ term, the existence of a gap, causing localised spin-polarised dangling bonds and reconstruction phenomena, is of less importance.

For GaAs we have not found theoretical calculations for comparison. The experimental value 0.62 eV of the barrier for hydrogen diffusion [9] is several times the value that follows from our calculation (< 0.1 eV). The authors of [9] however keep open the possibility that the diffusion process is more intricate and involves e.g. the codiffusion of a vacancy. The heat of adsorption to the surface of GaAs is 2.56 eV [10]. Remembering the pitfalls in EMT for a semiconductor surface, the agreement with the value 2.60 eV that we read from the surface contour in figure 1 is probably fortuitous.

For Si we performed similar calculations [6]. For that case, the small barriers (0.05 eV) in the regions of low density are comparable to the ones found in CNDO calculations [11], and they are slightly smaller than the ones (0.2 eV) calculated with MNDO [12]. Van der Walle et al. [13] have performed self-consistent LDA calculations for H in Si. Besides a low-density path with small barriers, they find a high-density path still 0.2 eV lower in energy. In all theoretical treatments the calculated activation barrier for diffusion is much smaller than the experimental value of 0.48 eV [14].

As regards metals, the surfaces are more close-packed. Thus almost always the host density favoured by the $\Delta E_{eff}^{hom}(n^0)$ term is collected from several atoms. Each of these is at a distance that is large enough to trust the accompanying corrections $\Delta E_v + \Delta E_c$ and to leave out the dipole terms.

4. Oxygen and $YBa_2Cu_3O_7$

Next we consider impurities that have higher atomic numbers than H. As an explicit example we investigate diffusion of interstitial O in the High-T_c superconductor $YBa_2Cu_3O_7$. This specific example has been investigated before by Ronay and Nordlander [15], however leaving out corrections and using the $\Delta E_{eff}^{hom}(n^0)$ term only. We recalculate the potential energy along the path 1-2-3 in figure 3, employing both corrections and relaxation. The latter is needed because the path traverses both regions of high electron density and large core-corrections. Then relaxation of the host atoms brings the electron density closer to the value favoured by the $\Delta E_{eff}^{hom}(n^0)$ term (0.013 a_0^{-3}) and also moves the host cores away from the impurity. Besides the question of robustness of corrections and dependence on the poorly known elastic constants, there is now also the question of interplay between relaxation and corrections.

To have an approximate description of the relaxation about an interstitial O impurity, we use the continuum Green's function for a cubic lattice [16]. This Green's function is characterised by three elastic constants, taken to be $C_{11} = 2.38\ 10^{12}$, $C_{12} = 1.80\ 10^{12}$ and $C_{44} = 0.26\ 10^{12}$ dyn/cm^2 respectively. The force causing the relaxation of some host atom is calculated from the gradient of $\widetilde{\Delta} E$ with respect to the coordinate of that particular host atom. The procedure of recalculating forces for each new configuration of the host atoms is repeated until one obtains self-consistency between forces and positions.

Again we compose the host electron density from a superposition of atomic densities. Like for GaAs it is necessary to calculate and use a core-correction term. Dividing host electrons into a 'core' and a 'jellium' part, this can easily be done. However, we are interested in robustness of corrections and after comparing atomic energy levels with the Fermi energy 0.53 Ry of a jellium with density 0.013 a_0^{-3}, we might consider also two different subdivisions for two of the host atoms. For O that has valence levels $2S^2$ (-2.14 Ry) and $2P^4$

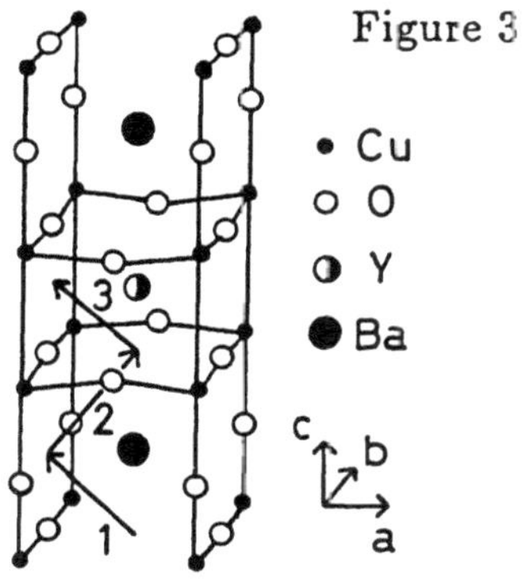

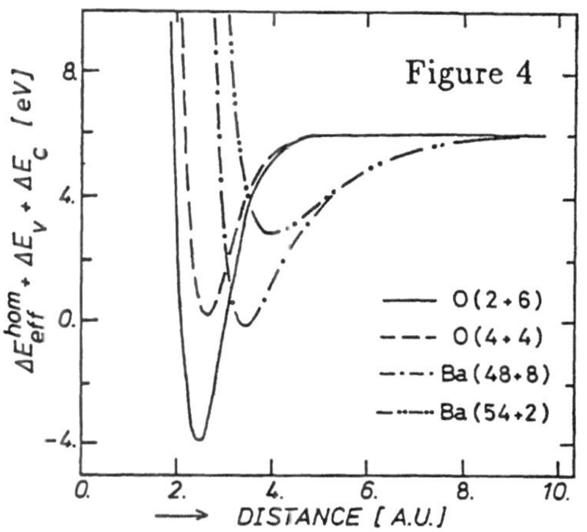

Figure 3: Paths 1–2–3 in $YBa_2Cu_3O_7$ along which the O embedding energy is calculated.

Figure 4: 'Pairpotentials' $\Delta E_{eff}^{hom} + \Delta E_v + \Delta E_c$ for host atoms O and Ba as function of distance to their nuclei.

(-1.04 Ry) we take subdivisions O(2+6) and O(4+4). For Ba with a core level $5P^6$ (-1.68 Ry) and a valence level $6S^2$ (-0.33 Ry) we investigate Ba(48+8) and Ba(54+2). For these cases it follows from our calculations that taking only electrons from the highest levels as 'jellium electrons', gives the smallest total corrections. This can for instance be seen from figure 4, where we show the combination of these corrections with a $\Delta E_{eff}^{hom}(n)$ term, the density n being the density of the single atom (so although we do not use it in an actual calculation, we show here some sort of 'pair potential' arising from a single host atom). It is seen that for small distances the different subdivisions result in differences of several electronvolts. It is important to find out how these differences appear in e.g. diffusion barriers.

In figure 3 we indicate the path 1-2-3 in the (010) plane along which we calculate the embedding energy of O. Not including corrections and using all electrons, we get a barrier 1.5 eV (not shown here) for path 3 agreeing with the one calculated by Ronay and Nordlander [15]. However for paths 1 and 2 we find barriers (0.7 eV) that are only half as large as theirs.

Including corrections for three different combinations of subdivisions into 'core' and 'jellium' electrons we obtain the diffusion barriers as in figure 5. They can be understood well if only one keeps in mind that the elastic forces are rather weak in comparison to the forces that come from ΔE_{eff}^{hom} (in case of high electron densities) or the correction terms. Then by large relaxations it is possible to find host configurations with almost optimal density on the position of the impurity, and only small *positive* corrections. If on the other hand for a

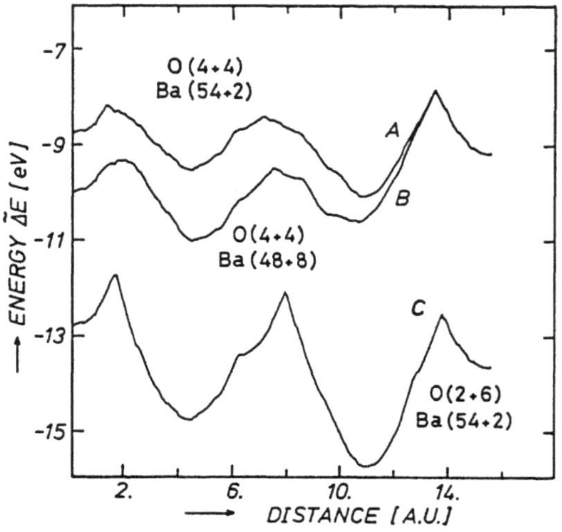

Figure 5: Relaxed embedding potential of O for different subdivisions of host electrons.

particular host atom the minimum of the 'pair potential' $\Delta E_{eff}^{hom} + \Delta E_v + \Delta E_c$ is *negative relative to the minimum value of* ΔE_{eff}^{hom} the configuration tries to profit as much as possible of this negative contribution. So positive corrections are masked in the final results but negative corrections come out clearly.

We now illustrate these principles by indicating the consequences of the graphs in figure 4 for those in 5. The only difference in input for curves A and B in figure 5 is the different subdivision of the Ba electrons. In unrelaxed state one Ba atom is close to the impurity for paths 1 and 2, and it is far from it for path 3. Consequently a modest part (1.0–1.5 eV) of the difference (3 eV) between the positive minima of the two pairpotentials for Ba, appears in the almost constant difference between the curves A and B. On the other hand, the difference between the curves A and C is rather large (> 4 eV). So we see that almost the full difference between the negative minima of the pairpotentials of O(4+4) and O(2+6) reappears in the final result for the potential energy. In fact, especially for path 3, there are also some small negative contributions from next nearest O host atoms. Finally we note that the magnitude of the diffusion barriers correlates strongly with the most negative minimum of $\Delta E_{eff}^{hom} + \Delta E_v + \Delta E_c$ belonging to host atoms that are close to the impurity in the unrelaxed state. For curves A and B this is the Cu atom (minimum of pairpotential: -1.8 eV), for curve C this is the O atom (see figure 4, minimum: -4 eV).

The conclusions remain the same in case we increase all elastic constants by 50 percent. All potential energy curves are shifted rigidly to somewhat higher values. We conclude that for oxygen impurities corrections are large. If the minimum of $\Delta E_{eff}^{hom} + \Delta E_v + \Delta E_c$ lies below the minimum of ΔE_{eff}^{hom}, corrections greatly influence final total energy results. One should be very cautious in combining an EMT description of impurities of high atomic number and relaxation.

Acknowledgements. We thank M.J. Puska for generously supplying the computer codes for the jellium embedding problem. This work is financially supported by the "Stichting voor Fundamenteel Onderzoek der Materie (FOM)".

References

[1] J.K. Nørskov and N.D. Lang: Phys. Rev. B**21**, 2131 (1980)

[2] J.K. Nørskov: Phys. Rev. B**26**, 2875 (1982)

[3] M.J. Stott and E. Zaremba: Phys. Rev. B**22**, 1564 (1980)

[4] P. Nordlander, S. Holloway and J.K. Nørskov: Surface Sci. **136**, 59 (1984)

[5] B. Chakraborty, S. Holloway and J.K. Nørskov: Surface Sci. **152/152**, 660 (1985)

[6] A. Zwartkruis and J.E. Van Himbergen: Surface Sci. **214**, 1 (1989)

[7] M.J. Puska, R.M. Nieminen and M. Manninen: Phys. Rev. B**24**, 3037 (1981)

[8] S.T. Pantelides: Hyperfine Interactions **6**, 145 (1979)

[9] J.M. Zavada, H.A. Jenkinson, R.G. Sarkis and R.G. Wilson: J. Appl. Phys. **58**, 3731 (1985)

[10] W. Mokwa, D. Kohl and G. Heiland: Phys. Rev. B**29**, 6709 (1984)

[11] A. Mainwood and A.M. Stoneham: Physica B **116**, 101 (1983)

[12] S. Tiansheng, S.N. Sahu, J.W. Corbett and L.C. Snyder: Scientia Sinica (Peking) **27**, 98 (1984)

[13] C.G. van de Walle, Y. Bar-Yam and S.T. Pantelides: Phys. Rev. Letters **60**, 2761 (1988)

[14] A. van Wieringen and N. Warmoltz: Physica **22**, 849 (1956)

[15] M. Ronay and P. Nordlander: Physica C **153-155**, 834 (1988)

[16] V.K. Tewary: Adv. Phys. **22**, 757 (1973)

Improved Tight Binding Parameters for Silicon

L. Goodwin and A.J. Skinner

Department of Mathematics, Imperial College of Science, Technology and Medicine, 180 Queen's Gate, London SW72BZ, UK

We present a method [1] for improving the overlap integrals and the pair wise potential in a widely used semi empirical tight binding model. The precise forms of the overlap integrals and the pair wise potential are adjusted to fit the calculated equilibrium volumes and equilibrium energies of some bulk phases to the results of ab-initio calculations. The transferability of the fitted parameters is tested by calculating the energetically most favourable geometries of silicon microclusters containing up to ten atoms. The energetically most favourable geometries are seen to resemble close packed lattice fragments in agreement with the results of ab-initio calculations, with bond lengths typically calculated to within five per cent of the more accurate data.

1 Introduction

Semi empirical tight binding (SETB) is a powerful and computationally cheap method of studying cohesion on the atomic level which has been widely applied to the study of transition metals, sp bonded elements and compounds. Large numbers of atoms can be studied allowing the study of extremely complex materials science problems. For an overview of SETB models, and a summary of other methods available, the reader is referred to the review article by Finnis et al. [2], and for examples of recent calculations to the papers by Majewski & Vogl [3] and Paxton & Sutton [4].

1.1 Semi Empirical Tight Binding

We divide the binding energy of a collection of atoms into a quantum mechanical bond energy, which incorporates the covalent bond energy and the cost of promoting electrons to higher energy orbitals, a repulsive energy, which incorporates the non-orthogonality, electrostatic and exchange contributions, and a constant energy shift C which is largely composed of the spin polarisation energy of a free atom [2]. The energy expression is as follows

$$E = \sum_{i\alpha} \int_{-\infty}^{\varepsilon_F} (\varepsilon - \varepsilon_{i\alpha})\, n_{i\alpha}(\varepsilon)\, d\varepsilon + \sum_{i\alpha} \varepsilon_{i\alpha} \Delta n_{i\alpha} + \sum_{\substack{i,j \\ i<j}} \phi(r_{ij}) + C N_{at} \quad (1)$$

where i denotes an atomic site, α an atomic orbital, $n_{i\alpha}(\varepsilon)$ the local density of states, $\varepsilon_{i\alpha}$ a Hamiltonian diagonal element, $\Delta n_{i\alpha}$ the change in the occupation relative to the free atom, $\phi(r_{ij})$ a repulsive pair wise potential between atoms i and j separated by a distance r_{ij}, and N_{at} the number of atoms in the system. The model Hamiltonian is constructed in a minimal orbital

Springer Proceedings in Physics, Vol. 48 Many-Atom Interactions in Solids
Editors: R.M. Nieminen · M.J. Puska · M.J. Manninen

basis consisting of the four valence s and p orbitals. Three centre integrals and crystal field terms are neglected in the usual two centre approximation. The two centre overlap integrals are constructed following the method of Slater and Koster [5] from the pure ssσ, spσ, psσ, ppσ and ppπ overlap integrals. In the charge neutral approximation the on-site integrals $\varepsilon_{i\alpha}$ are set equal to constants and are then perturbed about these values until each atom has exactly four valence electrons.

1.2 Harrison's SETB model

In Harrison's nearest neighbour parameterisation of SETB [6] the overlap integrals vary as the inverse square of inter-atomic separation by analogy with a nearly free electron-like system [7] with the equilibrium values fitted to accurate band structure calculations. The fitted values at equilibrium are as follows

$$\begin{aligned} h_{ss\sigma}(r_o) &= -1.82 \text{ eV} \\ h_{sp\sigma}(r_o) &= 1.96 \text{ eV} \\ h_{ps\sigma}(r_o) &= -h_{sp\sigma}(r_o) \\ h_{pp\sigma}(r_o) &= 3.06 \text{ eV} \\ h_{pp\pi}(r_o) &= -0.87 \text{ eV} \end{aligned}$$

The sp energy splitting has the value 7.03 eV. The repulsive pair wise potential varies with the inverse fourth power of inter-atomic separation with the equilibrium value fitted to the experimental lattice constant [6]. Energy volume curves calculated for four representative lattices using these parameters are shown in figure 1 with ab-initio results [8] included for comparison. The local densities of states were obtained with the recursion method [9] with the continued fraction taken to five exact levels (equivalent to ten exact moments) and terminated following the method of Beer and Pettifor [10]. Second neighbour interactions were included in the β-tin calculations since the first and second neighbour distances were very close. The model correctly reproduces the relative ordering of the equilibrium lattice energies,

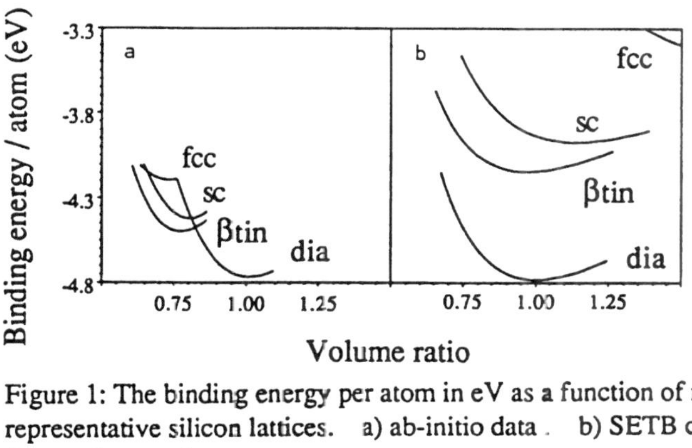

Figure 1: The binding energy per atom in eV as a function of relative volume for four representative silicon lattices. a) ab-initio data. b) SETB calculations using parameters from reference 6.

but overestimates their separations. The diamond fcc energy separation is approximately twice the ab-initio value. The model does not reproduce the relative ordering of the equilibrium volumes. The fcc equilibrium volume is calculated to be approximately twice the diamond equilibrium volume. The ab-initio calculations give a value of two thirds the diamond equilibrium volume. The bulk modulus is calculated to be 0.66 Mbar compared with the experimental value of 0.99 Mbar. It is this poor reproduction of the equilibrium energy separations, of the equilibrium volumes and of the bulk modulus which we remedy in the following sections.

2 Improved tight binding parameters

2.1 Improving the equilibrium energy separations

Harrison's pairwise potential is as follows

$$\phi(r_{ij}) = \phi(r_o)\left(r_o / r_{ij}\right)^m \tag{2}$$

with m chosen on the basis of extended Hückel theory to be equal to four [6]. Recent calculations by Paxton and Sutton [4] show that increasing the value of m leads to improved equilibium energy separations as well as a slight improvement in the equilibrium volumes. We shall treat m as a free parameter to be fitted. The sp energy splitting, $\Delta\varepsilon_{sp} = \varepsilon_{is}-\varepsilon_{ip}$, is a function of the atomic environment. In Harrison's parameterisation of SETB it is approximated by a constant obtained from the Hartree-Fock atomic term values [6] with the equilibrium overlap integrals subsequently fitted to the equilibrium diamond lattice. We shall adjust the sp energy splitting to obtain the best *global* fit to to the ab-initio data for bulk phases. $\Delta\varepsilon_{sp}$ and m were simultaneously adjusted to give the best fit to the diamond fcc equilibrium energy difference and the simple cubic diamond energy difference and the constant C adjusted to give the correct diamond equilibrium energy. The fitted values are m = 4.54, $\Delta\varepsilon_{sp}$ = 8.295 eV and C = 0.508 eV with $\phi(r_o)$ = 3.4581 giving the correct equilibrium volume. The fitted value of $\Delta\varepsilon_{sp}$ is eighteen per cent larger than Harrison's value.

2.2 Improving the equilibrium volumes

In the nearest neighbour SETB scheme the binding energy of an ideal lattice is a functional of the nearest neighbour distance r. Defining a new energy functional by replacing r with some function of r

$$E^{\text{rescaled}}(r) = E\left[\, f(r) \,\right] \tag{3}$$

preserves the calculated equilibrium energies while modifying the equilibrium bond lengths and hence volumes. Denoting an unscaled equilibrium bond length as r_{eq}

$$r_{eq}^{\text{rescaled}} = f^{-1}\left(r_{eq}\right) \tag{4}$$

where $f^{-1}(r)$ denotes the inverse function. For a fit to ab-initio data we require that

$$r_{eq}^{\text{ab-initio}} = f^{-1}\left(r_{eq}\right) \tag{5}$$

The condition that the diamond equilibrium volume be preserved is

$$f(r_o) = r_o \tag{6}$$

where r_o denotes the diamond equilibrium bond length. Equations (5) and (6) define the value that the function should take for four values of the operand (we are considering only four representative lattices). A practical constraint on the function is that it have a simple algebraic form so that an analytic expression for the atomic forces can be found. The best candidate we have found for the function is as follows

$$f(r) = r \exp\left[\left(r^{n_c} - r_o^{n_c}\right) / r_c^{n_c}\right] \tag{7}$$

Equation (7) defines a smoothed step function with the position and the sharpness of the step set by two parameters r_c and n_c. Substituting this function in to the SETB parameters we get

$$h_\alpha(r_{ij}) = h_\alpha(r_o)\left(r_o / r_{ij}\right)^n \exp\left\{-n\left(r_{ij}^{n_c} - r_o^{n_c}\right)/ r_c^{n_c}\right\} \tag{8a}$$

$$\phi(r_{ij}) = \phi(r_o)\left(r_o / r_{ij}\right)^m \exp\left\{-m\left(r_{ij}^{n_c} - r_o^{n_c}\right)/ r_c^{n_c}\right\} \tag{8b}$$

where n is equal to two. The bulk modulus rescales as follows

$$B_o^{rescaled} = B_o\left(1 + n_c\left(r_o / r_c\right)^{n_c}\right)^2 \tag{9}$$

We adjusted r_c and n_c, under the constraint that the diamond bulk modulus be given exactly, to obtain the optimum fit to the fcc bulk modulus and equilibrium volume. The fitted values are $r_c = 3.67$ Å and $n_c = 6.48$.

We show in figure 2 the rescaled overlap integral and the original overlap integral as a function of interatomic separation. On the figure are marked the fcc nearest neighbour and the diamond next nearest neighbour separations. The fcc next nearest neighbour separation,

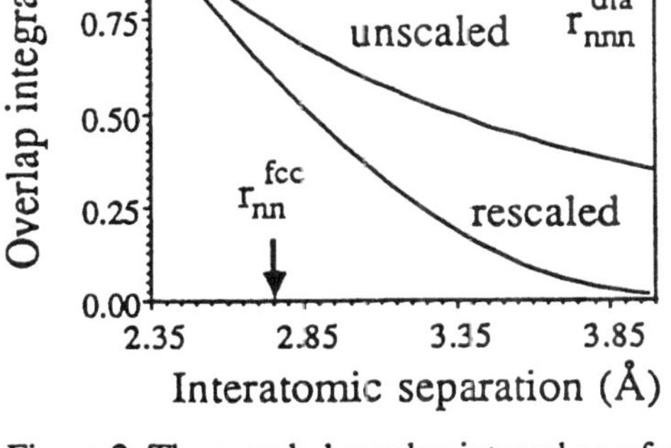

Figure 2: The rescaled overlap integral as a function of interatomic separation and the original inverse square form. The fcc nearest neighbour and diamond next nearest neighbour separations are marked. The integrals are normalised to unity at the diamond equilibrium bond length of 2.35 Å.

which is not marked on the figure, lies close to the diamond next nearest neighbour separation. The figure illustrates the rapid decay of the rescaled overlap integral, so much so that we can drop the nearest neighbour constraint on its use. The pair wise potential tends to zero even more rapidly. In all future calculations all non-zero interactions are included.

2.3 Results

Energy volume curves calculated with the rescaled parameters are shown in figure 3b. The TB calculations were performed in an identical fashion to those described earlier in this paper. Ab-inito results to which the rescaled parameters were fitted are shown in figure 3a. The improvement in the modelling of bulk phases is seen, from comparison between figure 1 and figure 3 (note the different scales), to be dramatic. The fitting has reproduced the qualitative features of the ab-initio results [8], including the high pressure diamond to β-tin transition, the relative ordering of the equilibrium energies, and the equilibrium lattice constants which are fitted to within 5% of the ab-initio values.

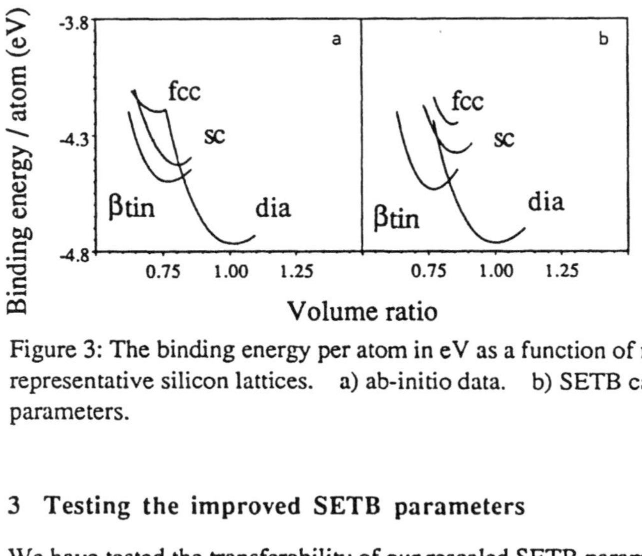

Figure 3: The binding energy per atom in eV as a function of relative volume for four representative silicon lattices. a) ab-initio data. b) SETB calculations using the new rescaled parameters.

3 Testing the improved SETB parameters

We have tested the transferability of our rescaled SETB parameters by performing extensive molecular dynamics quenches on silicon microclusters containing up to ten atoms. The eigenstates, and hence the density matrix, were found by explicit diagonalisation of the full Hamiltonian matrix, with atomic charge neutrality enforced throughout the molecular dynamics. To find a charge neutral solution we first used linear perturbation theory to get the required site energy perturbations [2]. In this scheme the number of computer operations scales as the number of atoms to the fourth power. Later calculations used a more efficient scheme in which the site energy perturbations were taken as proportional to the corresponding excess atomic charge. The constant of proportionality was found by trial and error to be roughly one electron volt per electron, the optimum value depending on the system under study. In this simpler scheme the number of computer operations scales linearly with the number of atoms. Using the output site energies from the previous timestep as the input for the current timestep with the new charge neutrality algorithm convergence to within ± 0.01

electrons per site was achieved within one or two iterations. The molecular dynamics were performed using the Verlet algorithm [11] as follows

$$\mathbf{r}_i(t+h) = \mathbf{r}_i(t) + \mu\big(\mathbf{r}_i(t) - \mathbf{r}_i(t-h)\big) + h^2\mathbf{f}_i(t) \tag{10}$$

where the position of an atom i at a time t is denoted by $\mathbf{r}_i(t)$, a small increment of time by h, and the force on atom i at time t by $\mathbf{f}_i(t)$. The force on each atom was calculated from the analytic derivative of the energy expression [2]. Initial heating and thermal equilibration were carried out by setting μ greater than or equal to unity respectively. To quench the system the v-component (v equal to x, y or z) of an atoms velocity was set to zero whenever the corresponding force component was in the opposite direction, i.e. acting to slow the atom down [12]. We found this quench to be considerably more efficient than the renormalisation quench (μ less than 1). Each starting configuration was first heated, until on the verge of evaporating, by renormalizing the velocities at each time step, followed by an equilibration period, and then quenched [12] through up to 1000 time steps until each Cartesian component of the atomic forces was less than 0.001 eV per Å. Rather fortuitously we found for the dimer a bond length of 2.32 Å in perfect agreement with ab-initio calculations [13,14]. We found an isosceles triangle for the trimer, a paralellopiped for the tetramer, a trigonal biprism for the pentamer and an edge capped trigonal biprism for the hexamer. The hexamer can be thought of as an octahedron which has undergone a Jahn-Teller distortion, the energy difference between these two structures being less than 0.02 eV per atom. Tomanek and Schlüter [14] found only the undistorted structure. Interestingly the equilateral triangle (three atoms) and the tetrahedron (four atoms) are unstable as they also undergo Jahn-Teller distortions although in these cases the distortions do not lead to the lowest energy structures. The remaining clusters containing from seven to ten atoms adopted geometries corresponding to face capped octahedrons. These geometries agree with those found by Raghavachari and Logovinsky [13] and Tomanek and Schlüter [14] with bond lengths calculated to within ten per cent of the more accurate values. The geometries of four of these clusters are shown schematically in figure 4 with some of the bond lengths marked on the figure. The bond lengths calculated with our TB method are given first and in brackets the corresponding ab-initio values [13].

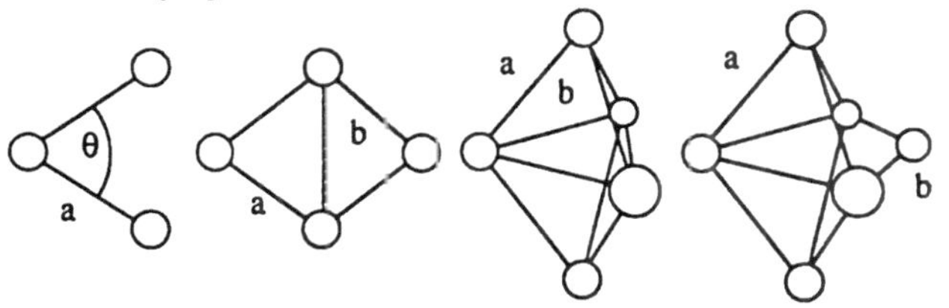

a = 2.25 (2.17)	a = 2.36 (2.30)	a = 2.39 (2.34)	a = 2.44 (2.36)
θ = 77.9 (77.8)	b = 2.56 (2.40)	b = 3.38 (3.26)	b = 2.47 (2.44)

Figure 4: The lowest energy geometries of four silicon microclusters found using SETB and molecular dynamics quenching. Bond lengths calculated with an ab-initio method [13] are in brackets.

4 Summary

In summary we have addressed within the charge neutral approximation a widely used nearest neighbour parameterisation of SETB for silicon. We have shown that the calculated values of the equilibrium energies of bulk phases can be improved by adjusting the pair potential exponent and the sp energy splitting. We have shown, via a rescaling argument, that the corresponding calculated equilibrium volumes can be improved by introducing a smooth truncation of the overlap integrals and the pair wise potential. The resulting rescaled SETB parameters were sufficiently short ranged that the nearest neighbour constraint on the model could be relaxed to include all non zero interactions permitting their use in molecular dynamics simulations. We have presented an efficient algorithm for enforcing the charge neutrality constraint. We have demonstrated the transferability of the rescaled parameters by performing molecular dynamics quenches on silicon microclusters containing up to ten atoms. The predicted lowest energy geometries were in excellent agreement with the results of ab-initio calculations. We are at present engaged in extending this work to other systems and to other elements.

5 Acknowledgements

It is with pleasure that we thank David Pettifor, Mike Finnis, Adrian Sutton, José Cressoni and Roger Davies for interesting conversations and helpful suggestions. LG would like to acknowledge the financial support of the U.K. Science and Engineering Research Council. AJS would like to thank the U.S. Department of Energy, Energy Conversion and Utilization Technologies (ECUT) Materials Program for support under the subcontract No. 19X-55992V through Martin Marietta Energy Systems Inc..

6 References

[1] L.Goodwin, A.J.Skinner and D.G.Pettifor,Europhys.Lett.,9,7,701 (1989).
[2] For a review see M.W.Finnis, A.T.Paxton, D.G.Pettifor, A.P.Sutton and Y.Ohta, Phil.Mag.A,58,1,143 (1988).
[3] J.A.Majewski and P.Vogl,Phys.Rev.Lett.,57,11,1366 (1986).
[4] A.T.Paxton and A.P.Sutton,accepted for publication in Acta Met. (1989).
[5] J.C.Slater and G.F.Koster, Phys.Rev.B,94,6,1498 (1954).
[6] W.A.Harrison,Phys.Rev.B,27,6,3592 (1983).
[7] S.Froyen and W.A.Harrison, Phys.Rev.B,20,6,2420 (1979).
[8] M.T.Yin and M.L.Cohen,Phys.Rev.B,26,5668 (1982);Phys.Rev.Lett.,45,1004 (1980)
[9] V.Heine, R.Haydock, D.W.Bullett and M.J.Kelly,Solid State Physics 35 (1980).
[10] N.Beer and D.G.Pettifor, Electronic Structure of Complex Systems, eds P.Phariseau and W.Temmerman, proc NATO Adv.Study Inst., New York: Plenum (1984).
[11] L.Verlet,Phys.Rev.,159,98 (1967).
[12] J.B.Gibson, A.N.Goland, M.Milgram and G.H.Vineyard, Phys.Rev.,120,4,1229 (1960).
[13] K.Raghavachari and V.Logovinsky,Phys.Rev.Lett.,55,26,2853 (1985).
[14] D.Tomanek and M. Schlüter ,Phys.Rev.Lett.,56,1055 (1986).

A Pair Potential Model of Covalent Bonding in Silicon

G. Ackland

Dept. of Materials Science and Engineering, University of Pennsylvania, Philadelphia, PA 19104, USA

It is shown that both structural polytypes and surface reconstructions in silicon can be understood via a simple, intuitive, analytic model. The model incorporates ionic core repulsions written as a pairwise repulsion between ions, and a cohesive energy written as a sum over valence electron energies. The valence electrons are represented by localised, two-centre orbitals, which are in turn represented by a simple pair potential. The resulting model has a clearer interpretation and fewer parameters than previously published empirical potentials. It can be regarded as a classical equivalent of tight-binding, in so much as the electrons are treated as distinct entities with a restricted set of possible orbitals. We find that representing cohesion in silicon in terms of classical electrons in an intuitive way, and not by a continuous function of the atomic positions as has been done previously, the structural properties are well reproduced and are almost independent of the parameterisation.

1. Introduction

There has been an enormous amount of recent work directed toward developing empirical interatomic potentials for silicon [1-9]. All of this work is based on the premise that the energy can be written as a sum of one-atom energies associated with each site, where the energy at each site is written in terms of a continuous function of the relative positions of all the other atoms in a surrounding region. This general approach is borrowed from the pair-potential / embedded atom ideas which have been successful in metallic systems [10-12], but it does not account for the saturation of the bonding which is reached when the valence band is filled.

In the present model we represent cohesion as being due to the electrons in the solid being in lower energy states than in the free atoms. We assume that silicon is covalently bonded, so the valence electrons are taken to be in orbitals around two nuclei. Each orbital contains two electrons of opposite spin. Each atom contributes four electrons to bonding states, and we assume that the exclusion principle prevents more than eight electrons (four antispin pairs) sharing the same atom as one of the foci of their orbitals. The important difference between this model and previous ones lies in using the basic premise that electronic states are highly localised, and that the number of such states associated with a given atom can become saturated.

We shall show that this electron-based picture of covalent bonding leads to an energy in the form of a simple pair-potential whose cohesive term is restricted to four neighbours. Furthermore, with sensible parameterisation, this model successfully reproduces most of the experimentally known data for bulk and surface structures.

2. The Model

We take a simple parameterisable model for a tetravalent semiconductor, in which each valence electron pair is located in a two-centre orbital (covalent bond). We assume that the energy of these electrons relative to their atomic levels depends only on the separation of the two ions. On each ion no more than four two-centred states can be occupied by different electron pairs (since others would have to be antibonding states). The remaining interactions between the nuclei and the core electrons are represented by a simple pairwise repulsion. The model thus consists of a sum over all N atoms of a pair potential and a sum over all $4N$ valence electrons of their respective orbital energy:

$$E = \frac{1}{2}\sum_{i=1}^{N}\sum_{j=1}^{N} V(r_{ij}) - \sum_{k=1}^{4N} \phi(r_k) \tag{1}$$

where the k sum runs over all valence electrons, and r_k is the separation between the two ions on which the kth electron's orbital is centred.

If we make a Born-Oppenheimer type approximation that the electronic configuration equilibriates much faster than the ionic configuration, and assume that the electrons are independent, then we can associate the binding energy of one electron in each bond with each of the atoms that define the position of that bond. Since the bond energy depends only on the bond length, it can be written as a pair potential between the two atoms, provided that no atom has more than four such bonds (if an atom has less than four bonds, then the extra electron associated with it is regarded as a dangling bond, and gives no contribution to the cohesion). This restriction to four bonds arises naturally from the exclusion principle for localised 3s3p orbitals.

In one case we relax the independent electron approximation: If two under-coordinated atoms are bonded neighbours of one another, we allow the possibility that the extra electrons can form a double bond between the atoms (always provided that neither atom is then associated with more than four bonds). This occurs when forming the double bond allows the system to reduce its energy by more than it could by forming single bonds to more distant atoms, or remaining as a 'dangling bond'. For the double bond the mutual electrostatic repulsion of the electrons will be especially great, and we attempt to account for this. It is well known that the valence energy of a delocalised band scales as the square root of the coordination. By analogy we assume that the doubly bonded silicon has an energy of $\sqrt{2}$ times the single bond. In the present model this leads to a reasonable dimer energy and is also important in considering certain defect and surface configurations.

Thus, while we have included the electronic nature of covalent bonding in our model, by partitioning in energy in a particular way we have managed to obtain a model in which the energy can be broken down into a sum of individual atom energies consisting of a sum of interatomic pair potentials. This makes it suitable for use in computer simulation.

It is interesting to note that the diamond structure cannot be stabilised by a simple pair potential without complex, long-range oscillations. Previous empirical approaches have tried many variants on a simple pair potential. The most common are the use of three-body terms [1-3] or embedding-type functions [6-8]. We have shown here that the interpretation via electrons in a filled band gives a different recipe. At low co-ordination there are similarities with the embedding idea, and it is interesting that one empirical 'embedding'-type function (of an effective coordination) saturates at a co-ordination approximately equal to four [7].

3. Parameterisation

In choosing a parameterisable model we try to minimise the number of parameters. This will enable us to see the results due to theoretical input rather than judicious parameterisation. While the functional forms used may not be optimised, this is compensated for by the ease with which we can test the dependence of our model on the fitting procedure.

We represent the pairwise repulsion by a simple exponential, and use the radial part of the hydrogenic p-wavefunction to represent the strength of the covalent bond. These choices are arbitrary, but alternative parameterisation schemes using cubic splines or power series functions give essentially similar results:

$$E = \frac{1}{2}\sum_{i=1}^{N}\sum_{j=1}^{N} Ae^{-\alpha r_{ij}} - \frac{1}{2}\sum_{i=1}^{N}\sum_{k=1}^{4} Br_{ik}e^{-\beta r_{ik}} \quad . \tag{2}$$

This model has four fitting parameters. We can fit these to various experimental data. We choose to fit the cohesive energy, lattice parameter and bulk modulus of diamond silicon. The fourth parameter is initially left unfitted so that we can examine a family of potentials to determine what the effects of the physics of the model are, and which effects are sensitively dependent on the fitting. Eventually, we will fit this parameter to give quantitative agreement with density functional calculation of structural polytypes (table 1).

Table 1. A parameterisation using exponential functions (cf. equation 2).

Parameter	Value (eV and Å)
A	208442.8
B	16.63588
α	5.673585
β	1.144811

For convenience in evaluating the sum over electrons, we assign one electron from each bond to each of the two ions with which the bond is associated. A, B, α and β are fitting parameters. The choice of bonded neighbours k will be determined by minimising the total energy. This is the equivalent of the Born-Oppenheimer approximation.

4. Calculation of Properties

Crystal Structures. The energies and lattice parameters of various structures have previously been examined using density functional total energy calculations [13] (DF). This technique is expected to be highly accurate, and for comparison with our model we shall assume that these predictions are 'correct'. We find that the present scheme reproduces the trends in lattice parameter and relative energy (see figure 1), and that these trends are insensitive to the parameterisation scheme. However, the quantitative energy differences between the structures depend sensitively on the parameterisation, and in particular on the absolute value of the two terms. The absolute value of the bonding term for a full band is related to the position of the Fermi level relative to the band centre. Our tacit assumption is

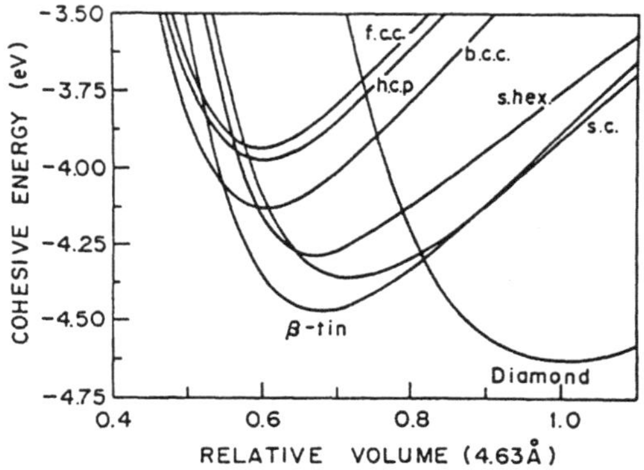

Figure 1: Graph of energy against volume for various crystal structures

that the band has zero width (all orbitals in the perfect lattice have equal energy). Thus the energy per bond is equal to the energy difference between the band and the Fermi level.

The three fitted properties (cohesive energy, lattice constant and bulk modulus of diamond) are unaffected by adding a constant to both attractive and repulsive parts, and hence are not good for determining the absolute values of each term. For further studies we select the potential which gives the best qualitative agreement with the DF results, effectively fitting the fourth parameter to the Fermi energy - band centre energy difference (table 1).

The agreement between this potential and the DF work is closer than other empirical potentials [1-8], including those which use many more parameters to fit explicitly to this data. The high pressure stability of the β-tin structure is well reproduced, the transformation occurring via the softening of the tetragonal phonon mode. The higher pressure simple hexagonal [14,15] phase is not obtained with this model, a failing which appears to be common to all other empirical potentials. The overall agreement, and in particular the independence of choice of fitting scheme, suggest that the physical basis of the model is correct.

Point Defects. We also evaluate point-defect energies by molecular statics relaxations of blocks of 1000±1 atoms. The actual configuration of point defects in silicon is not well-known experimentally, but from diffusion data the sum of the formation and diffusion energies is known to be about 4eV. Moreover, from the existence of amorphous silicon we can deduce that defect formation energies are likely to be lower than in metals which can not be made amorphous. We attempt to compare our results with DF calculations [16-18]. Most DF calculations consider highly symmetric configurations. The reason for this is partly because such configurations can be easily treated using supercells, and partly because the intuition about point defects gained from work in metals suggests that the most likely configurations are the highly symmetric ones (although even in metals there have been exceptions [19]).

If we make the normal assumptions regarding defect configurations: highly symmetric structures with radial relaxation, and corresponding long- or dangling-bonds, then our model is in agreement with DF calculations. If some asymmetric rearrangement of the bonding is allowed, and the atomic positions are relaxed within the new bonding arrangement, the interstitial and vacancy energies all become small. These reconstructions are shown in figures 2-4 and the defect energies are summarised in table 2.

The re-bonding in the vacancy produces four five-membered rings (the perfect diamond structure has only six-membered rings). The 'interstitial' is found to be spread over a large

Table 2 . Point Defect Energies (eV)

Defect	Unrelaxed	Relaxed	LDF [16-18]
Vacancy	4.63	4.62	4.7, 5.5±0.5
Reconstructed Vacancy	3.05	0.27	
Hexagonal Interstitial	8.19	5.52	7.0±1.0
Tetrahedral Interstitial	5.81	5.52	6.0±0.5
Reconstruction A (tetra)	2.19	0.07	
Reconstruction B (hex)	2.10	0.03	

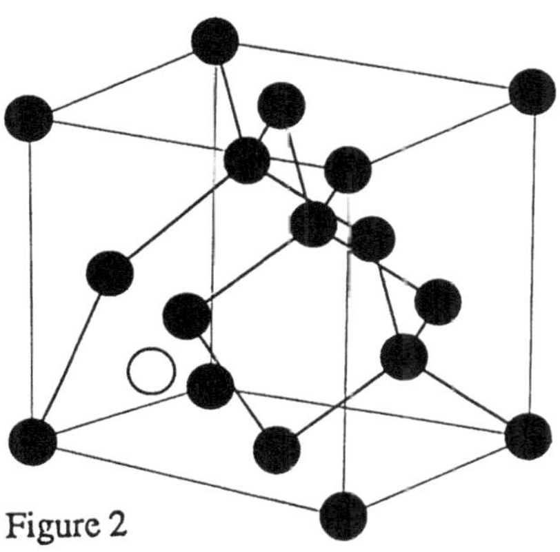

Figure 2

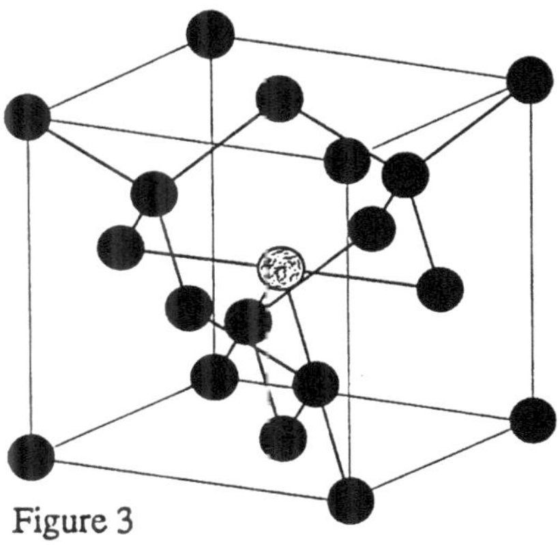

Figure 3

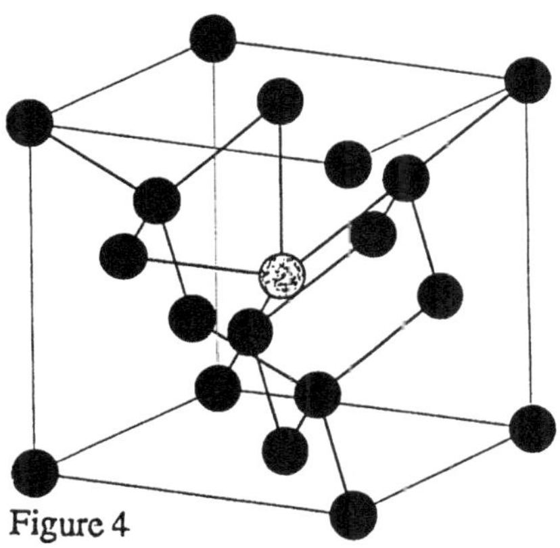

Figure 4

Figure 2:
Bonding configuration of diamond vacancy (unoccupied site shown by open circle)

Figure 3:
Bonding configuration of diamond interstitial type A

Figure 4:
Bonding configuration of diamond interstitial type B

volume, with two possible rebonding schemes giving very similar energies. It manifests itself by four, five and seven-membered rings. The interstitial reconstruction A starts in the tetrahedral configuration while B starts as hexagonal. The re-bonding in A produces one four-membered and two five-membered rings. In B the re-bonded relaxed structure has two four-membered rings. We believe that point defects in silicon should be thought of in terms of rings, since it is impossible to define any one atom in these reconstructions as 'the' interstitial.

We believe that while the DF calculations have found the correct energy for the systems which they investigate, they have chosen the wrong starting conditions because their choice was motivated by intuition more appropriate to metals than to semiconductors. We do not claim that our (very low) calculated energies are quantitatively correct, nor do we believe that our scheme is sufficiently accurate to differentiate between the various possible asymmetric interstitial reconstructions. The point which we emphasise is that the saturation of bonding which occurs in semiconductors (but not in metals) gives rise to complex asymmetric point defect configurations. At the present time this hypothesis is untested by experiment, but we believe that the very fact that the configuration has not been characterised provides a negative inference that the structure is rather complex.

Once again we find that while the quantitative values of the defect energies varies (although they never become negative for monotonic functions V and ϕ), the stability of the asymmetric configurations is insensitive to the details of the parameterisation. Defect asymmetry follows as a consequence of the formulation of the cohesion as a sum over one-electron energies.

Surfaces. The experimental situation is much better understood for surface structures [20-22]. In this case to simulate reconstruction we initially allow only symmetric relaxations from the truncation of the bulk, and then look for other minima by re-ordering the bonding. It is also important to allow for double bonding between adjacent undercoordinated atoms.

For the (001) surface we find that the threefold coordinated dimer structures (2x1 reconstruction) are much lower in energy than the fourfold (two near neighbour, two second neighbour) 1x1 surface. The surface dimer bond length is shorter than the bond length in the bulk (2.285Å), this is in good agreement with DF calculations. The model is unable to differentiate between the two possible arrangements of the dimers. The dimerised surface is so stable that, even if we assume that the dimers are singly bonded, the 2x1 reconstruction with dangling bonds is still more stable than the 1x1.

For the (111) surface we find that the π-bonded chain [22] is stable. In this case if no multiple bonding is allowed, the 1x1 unreconstructed structure is favoured. Thus we arrive at the unspectacular conclusion that the stability of the π-bonded chain is dependent on the π-bonding.

The numbers quoted in table 3 are for the parameterisation given in table 1. As with the point defects, the absolute values of the surface energies are rather sensitive to details of

Table 3. Surface Energies.

Surface Plane	Structure	Relaxed Energy mJm^{-2}
001	2x1 dimers	1023
001	2x1 dimers-singly bonded	1272
001	Truncation of bulk	1660
001	2x1 vertical ripples	unstable
111	π-bonded chain	765
111	chain without double-bonds	1454
111	Truncation of bulk	1452
111	2x1 dimers	1453

the parameterisation, but once again the preferred surface structures were the same for all the parameterisation schemes which were tried.

Clusters. A final, extreme test of the model is in cluster calculations. Previous empirical schemes have been unsuccessful in predicting the structure of small clusters. In the present scheme we can also anticipate problems: since there are no bond charge forces the energy of three- and four-membered rings is probably too low.

The cohesive energy of a doubly bonded Si=Si dimer is 1.7eV, in and the atomic separation is 2.287Å. The energy is in good agreement with DF calculation, and the contraction of the bond relative to the bulk lattice parameter is in agreement with experimental data [23], but at variance with the predictions of 3-body potentials. This gives us some encouragement that our rather arbitrary treatment of double bonds is reasonable.

We examined small clusters of up to seven atoms, and found in each case that the most closely packed arrangement was more stable than structures made up of 'pieces of diamond lattice'. The most stable structures were: equilateral triangle (3-atom); tetrahedron (4); triangular bipyramid (5); octahedron (6); pentagonal bipyramid (7). This is roughly correct, and at variance with models relying on three-body forces to stabilise the diamond lattice. However, the exact structures were not reproduced. As deduced from DF calculations, for clusters of three or more atoms, these are distortions from the high symmetry structures found above.

5. Discussion

We have derived a parameterisable model for covalently-bonded materials and applied it to the case of tetravalent silicon. Although the model includes electronic effects in a very simplified manner, its algebraic form can be reduced to that of a pair potential with a restricted number of cohesive neighbours. This form has been parameterised using sensible functions to represent the various terms, and thus is suitable for use in computer simulations.

Independent of the parameterisation scheme chosen, the model reproduces qualitative trends in the cohesive energy of various structures and the reconstructions of various surfaces. By empirical fitting to the experimental cohesive energy, atomic volume and bulk modulus of silicon and to the band centre, quantitative agreement is obtained. The model has also been applied to point defects and small clusters.

The derivation of the model also enables us to envisage in advance where it is likely to fail. This gives us some idea of which empirical data are suitable for use in the parameterisation. This is a crucial failing of purely empirical fitting schemes. The lack of an electron-electron bond charge force seems to be the most obvious missing physical input. This will be of importance in cases where either the bond angle is very small or the major effect is to change the bond angles and not the bond length. In the first category are some of the cluster structures, especially those containing equilateral triangles, and the interstitial structures involving four-membered rings. In the second category are the shear moduli, in particular the tetragonal shear modulus (since a tetragonal strain has only a second-order effect on the nearest neighbour bond length). The tetragonal shear modulus is indeed very low.

We believe that the present model will be of use in studies of plausible defect structures in the diamond lattice. Its success in all test cases and its ease of application suggest that it is suitable for setting up starting configurations for more precise electronic structure calculations (in simple cases such as point defects), and for examining more complex defects (surfaces, grain boundaries, the amorphous state).

References

1. P.N. Keating: Phys.Rev. **145**, 637(1966)
2. E. Pearson, T. Takai, T. Halicioglu, W.A. Tiller: J.Cryst. Growth, **70** 33 (1984)
3. F.A. Stillinger, T.A. Weber: Phys.Rev.B **31**, 5262 (1985)
4. D.W. Brenner, B.J. Garrison: Phys.Rev.B **34** 1304 (1986)
5. R. Biswas, D.R. Hamman: Phys. Rev. Lett. **55**, 2001 (1985)
6. M.I. Baskes: Phys. Rev. Lett. **59**, 2666 (1987)
7. J.Tersoff: Phys. Rev. Lett. **56**, 632 (1986); Phys. Rev. B **37**, 6991 (1988); Phys. Rev. B. **38**, 9902 (1988)
8. B.W. Dodson: Phys. Rev. B **35**, 2795 (1987)
9. K.E. Khor, S. Das Sarma: Phys. Rev. B **38**, 3318 (1988)
10. M.W. Finnis, J.E. Sinclair: Phil. Mag.A **50** 45 (1984)
11. M.S. Daw, M.I. Baskes: Phys.Rev.Lett. **50** 1285 (1983); Phys.Rev.B, **29** 6443 (1984)
12. G.J. Ackland, M.W. Finnis, V. Vitek: J.Phys.F **18** L153 (1988)
13. M.T. Yin, M.L. Cohen: Phys. Rev. Lett. **45**, 1004 (1980); Phys. Rev.B **26**, 5668 (1983); Phys. Rev. B. **29**, 6996 (1984)
14. H. Olijnyk, S.K. Sikka, W.B. Holzaptel: Phys. Lett. **103A**, 137 (1984)
15. J.Z. Hu I.L. Spain: Solid State Comm. **51**, 263 (1984)
16. R. Car, P.J. Kelly, A. Oshiyama, S.T. Pantelides: Phys. Rev. Lett. **52**, 1814 (1984) ; Phys. Rev. Lett. **54**, 360 (1985)
17. G.A. Baraff, M. Schluter: Phys. Rev. B. **30**, 3460 (1984)
18. Y. Bar-Yam, J.D. Joannopoulos: Phys. Rev. Lett. **30**, 1844 (1984)
19. R. Thetford: these proceedings (1989)
20. A. Zangwill: In *Physics at Surfaces*,Cambridge University Press (1988)
21. J.E. Northrup: M.L. Cohen, Phys. Rev. Lett. **49**, 1349 (1982)
22. K.C. Pandey: Phys.Rev.Lett. **47**, 1913 (1981); Phys.Rev.Lett. **49**, 223 (1982)
23. K.P. Huber, G. Herzberg: In *Molecular Spectra and Molecular Structure: Constants of Diatomic Molecules*. (Van Nostrand, New York 1979)

Derivation of Angular Forces for Semiconductors and Transition Metals

A.E. Carlsson

Department of Physics, Washington University, St. Louis, MO 63130, USA

Abstract. Using a tight-binding model Hamiltonian, the basic ingredients responsible for the angular dependence of interatomic forces are elucidated by a moment analysis of the electronic density of states. While analyses based on the width of the electronic density of states yield many-body interaction schemes without explicit angular dependences, the analysis of the crudest features of the **shape** of the density of states immediately give an angular dependence. The effects of the shape of the electronic density of states are included by a treatment utilizing the second-moment **matrix** on a particular site, as well as the scalar fourth moment. It is shown that the second-moment matrix can include both the covalent and metallic limits of bonding in semiconductors, and thus can obtain defect formation energies which are more accurate than those obtained by existing many-atom interaction schemes. The inclusion of this basic physics in an empirically fitted scheme for Si is discussed, and schematic results for a broader range of materials are presented.

1. Introduction

The faith of the materials physics community in the accuracy of computer simulations of bulk and defect properties of metals has grown considerably as a result of the development of "many-atom" interaction schemes[1–3] for obtaining interatomic forces. These interactions are typically based on a simplified treatment of the quantum-mechanical electronic degrees of freedom, using either the second moment of the electronic density of states[1] or the local "background" charge density[2,3] at the nucleus of a particular atom. They include, effectively, very high-order interactions, and are thus able to achieve a much better global description of broken-bond energies and defect relaxations than can be obtained by pair potentials alone. The computational speed is essentially that of pair potential calculations. However, simulations for semiconductors have typically not been grounded as solidly in quantum-mechanical analysis as the "many-atom" schemes for metals. Instead, assumed functional forms[4,5] based on physical intuition have been employed. In this paper we describe the lowest-order quantum-mechanical analysis[6,7] which yields angular forces suitable for semiconductors. As an example of this analysis, we treat the vacancy-formation energy in a tetrahedral semiconductor. We then discuss an empirically fitted scheme for Si based on this underlying physics. Finally, we discuss the chemical trends in the angular dependence of interatomic forces that are expected on the basis of the quantum-mechanical analysis.

2. Mathematical Form of Angular Forces

Our derivation of the angular forces is based on a tight-binding Hamiltonian of the following form:

$$H = \sum_{i,j} \sum_{\alpha\beta} h_{ij}^{\alpha\beta} |i,\alpha\rangle\langle j,\beta| \quad , \tag{1}$$

where (i,j) denote atomic positions and (α,β) are sp^3 orbital indices. The $h_{ij}^{\alpha,\beta}$ are assumed to have a two-center form with angular dependences given by the Slater-Koster relations. On-site terms are ignored to emphasize the covalent bonding effects. The interatomic couplings are responsible for the broadening of the atomic levels into bands of extended states in the solid environment. If the valence orbitals are partly filled, then the bandwidth provides the lowest-order description of bonding properties such as the cohesive energy[8–10] and surface energy.[11] Treatments based on an electronic density of states (DOS) of fixed shape, so that the width is the only variable, have provided basic understanding of the basic chemical trends in cohesive energies.[8] "Many-atom" interaction schemes[1] based on the bandwidth have proved successful in treating defect properties in metals, obtaining, for example reasonably accurate vacancy formation energies and surface relaxations. However, they have no explicit angular dependence, and cannot treat open structures such as the diamond structure.

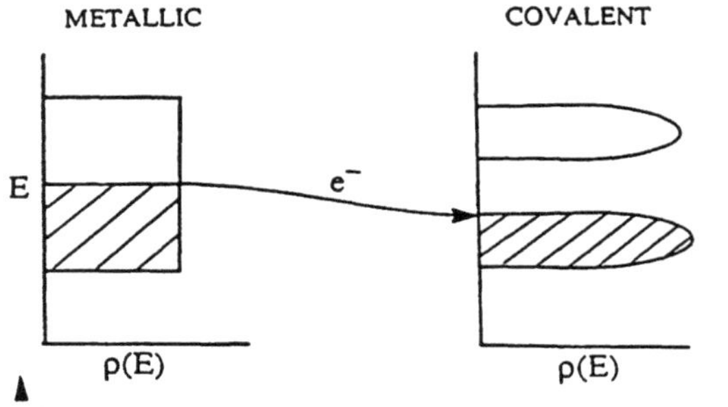

Fig. 1. Features in shape of electronic density of states (ρ(E)) responsible for angular forces.

Fig. 2. Configurations of three-body clusters of Group V atoms, with associated electronic densities of states ρ(E).

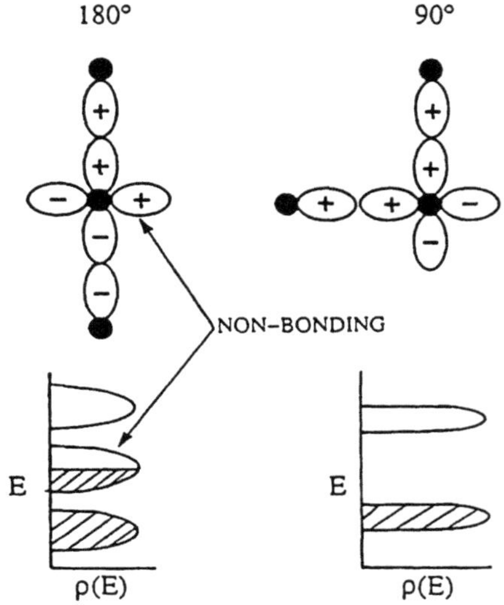

Beyond the width of the DOS, a very crucial feature which affects the bonding energy is whether it has a bimodal shape, characterized by bonding and antibonding peaks, or instead has a single peak or flat distribution of states. This is illustrated in Fig. 1, where we have used the terms "covalent"and "metallic" to describe the two types of shapes. In the case of a half-filled valence- and conduction-band complex, which is our main interest, the "covalent" shape is preferred. This is because, when a gap is created in the DOS, electron states move down from the center of the band into the bonding peak, lowering their energies. The relation of this effect to the angular dependence of the interatomic forces is most easily understood in the case of group V materials, which have a half-filled p-electron complex; for the qualitative analysis here, we will ignore the effects of the deep s–electron states. Figure 2 shows two possible arrangements of three group V atoms, along with sketches of the electronic density of states projected on the central atom. In the case of the 90° bond angle, both of the p–orbitals on the central atom form bonding and antibonding orbitals with "mates" on the neighboring atom (the third p–orbital, pointing out of the page, would also be saturated in a typical three-fold coordinated structure for a group V material). Thus the density of states is dominated by bonding and antibonding peaks. However, if the bond angle is 180°, then the horizontal orbital has no "mate" on a neighboring atom, and forms a non-bonding state. Thus the total DOS has both bonding and antibonding contributions from the vertical p–orbital, and a nonbonding contribution from the horizontal orbital. According to the physics discussed above, we would then expect the 90° bond angle to be favored. This is, in fact, consistent with the observed crystal structures of P, As, Sb, and Bi, which display three-fold coordination and bond angles of roughly 95°.

This effect can be approximately described by two types of terms[7] in a total-energy expression.

2.1. Second-moment matrix

We generalize existing second-moment treatments by considering the matrix $\hat{\mu}_2(i)$ defined by

$$\mu_2^{\alpha\beta}(i) = \langle i,\alpha|H^2|i,\beta\rangle = \sum_{j,\gamma} h_{ij}^{\alpha\gamma}\, h_{ji}^{\gamma\beta} \quad . \tag{2}$$

The scalar second moment is given by Tr $\hat{\mu}_2(i)$. In an sp^3 tight-binding model, $\hat{\mu}_2(i)$ measures the asymmetry of the local environment. As shown in Fig. 3, an atom at a surface, which lacks inversion symmetry, has non-zero values of $\mu_2^{s,p}$. If inversion symmetry is present, but the symmetry is tetragonal rather than cubic, then the p–p part of $\hat{\mu}_2(i)$ is not a multiple of the identity. If the environment is cubic, then $\hat{\mu}_2(i)$ is diagonal, with three p–eigenvalues and one s–eigenvalue.

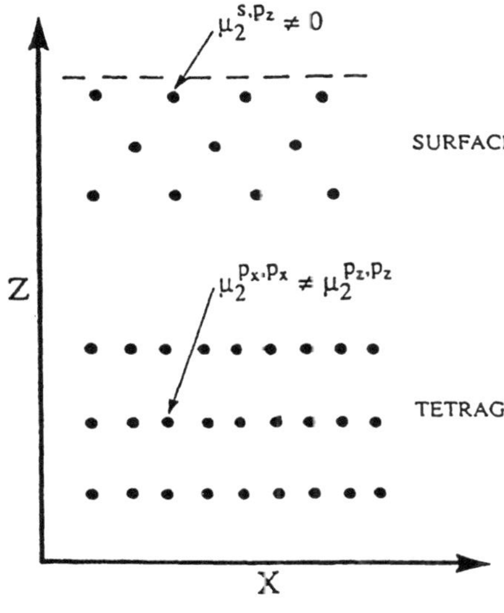

Fig. 3. Behavior of second-moment matrix $\hat{\mu}_2$ in various asymmetric geometries.

The energy associated with $\hat{\mu}_2(i)$ is obtained in two steps. First a model DOS is generated by the matrix recursion method,[13] truncated at the first level. This gives a DOS of the form

$$\rho_i(E) = -\frac{1}{\pi} \,\mathrm{Im\,Tr}[E - \hat{\mu}_2(i)/E]^{-1} \quad . \tag{3}$$

The DOS is integrated, in turn, to obtain the $\hat{\mu}_2$ contribution $E_{el}^{(2)}(i) = \int_{-\infty}^{0} E\rho_i(E)dE$ to the bonding energy. (For the symmetric, half-filled band here, $\varepsilon_F = 0$) From Eq. (3), we obtain that $E_{el}^{(2)}(i) = -\mathrm{Tr}[\hat{\mu}_2(i)]^{1/2}$.

To understand the content of our form for $E_{el}^{(2)}(i)$, the unrelaxed vacancy formation energy E_{vac}^{f} in the diamond structure is a convenient example. This is given by

$$E_{vac}^{f} = \sum_{i\neq 0}[E_{el}^{(2)}(i) - E_{el}^{(2)}(\mathrm{bulk})] \quad , \tag{4}$$

where "bulk" denotes the value in the ideal crystal and the vacancy is at the site $i = 0$. We summarize the calculation of E^f_{vac} below, showing a scalar second-moment calculation for comparison. This calculation is based on the bond-orbital model for the electronic structure, keeping only interactions between a given sp^3 hybrid and its "mate" on a nearest neighbor. Interactions between different sp^3 hybrids on the same site are ignored. The coupling between two hybrid mates is denoted by h; quantities labeled "vac" are characteristic of nearest neighbors to the vacancy. The exact value of E^f_{vac} is readily seen to be $4|h|$.

Matrix $\hat{\mu}_2$	Scalar μ_2
$\hat{\mu}_2(\text{vac}) = \begin{bmatrix} 0 & 0 & 0 & 0 \\ 0 & h^2 & 0 & 0 \\ 0 & 0 & h^2 & 0 \\ 0 & 0 & 0 & h^2 \end{bmatrix}$	$\mu_2(\text{vac}) = 3h^2$ (vs. $\mu_2(\text{bulk}) = 4h^2$)
$E^{(2)}_{el}(\text{vac}) = -3|h|$ $E^{(2)}_{el}(\text{bulk}) = -4|h|$	$E^{(2)}_{el}(\text{vac}) = -4\sqrt{\frac{3}{4}}\,|h| \approx -3.5|h|$ $E^{(2)}_{el}(\text{bulk}) = -4|h|$
$E^f_{vac} = 4|h| = |E^{(2)}_{el}(\text{bulk})|$	$E^f_{vac} \approx 4[-3.5 - (-4)]|h| = 2|h| = \frac{1}{2}|E^{(2)}_{el}(\text{bulk})|$

The main conclusion to be drawn from this calculation is that the incorporation of the **matrix** second moment correctly allows bond strengths to act independently at low coordination numbers; in contrast, in the scalar second moment treatments, the bonds never act independently. At high coordination numbers, such as in closely packed crystals both methods give similar results, with strong interaction between the bonds.

2.2. Fourth-moment Contributions

Unfortunately, the treatment outlined above is unable to distinguish reliably between different crystal structures having cubic symmetry; in this case, then the matrix character is secondary. The fourth moment essentially measures the fluctuations in E^2. If the density of states is simply a pair of δ–functions at opposite energies, then these fluctuations vanish; as the density of states moves away from the δ-function limit, then the fluctuations grow. Thus the "covalent" shape discussed above (*cf.* Fig. 1) corresponds to small values of μ_4, and the "metallic" shape to large values. Since the "covalent" shape is preferred for half-filled bands, a repulsive contribution to the energy is associated with large values of μ_4.

To establish the form of this contribution, we use the rigorous expression[11]

$$\mu_4(i) = \sum_{jkl} \sum_{\alpha\beta\gamma\delta} h^{\alpha\beta}_{ij}\, h^{\beta\gamma}_{jk}\, h^{\gamma\delta}_{jl}\, h^{\delta\alpha}_{li} \tag{5}$$

which gives $\mu_4(i)$ as a sum of two-, three-, and four-body paths. Thus an energy term proportional to μ_4 will contain up to four-body contributions. However, part of the contributions to $\mu_4(i)$ are already included in $E^{(2)}_{el}(i)$. The latter, in effect, includes the effects of all paths which hop back to the site i in the middle step of the path. To exclude these paths from double counting, one uses an expression of the form

$$E^{(4)}_{el}(i) = C\{\mu_4(i) - \text{Tr}[\hat{\mu}_2(i)]^2\}/\mu_2(i)^{3/2} \quad ,$$

where C is a positive constant and the normalizing factor of $\mu_2(i)^{3/2}$ is obtained from a quantum-mechanical analysis.[14]

2.3. Effective Potentials

It is important to establish the connection between the types of terms we have described here, and the types of interatomic potentials which have been used previously. The key new concept that enters here is that of the "reference environment". This is a uniform system, such as a particular bulk crystal structure, in which the values of $\hat{\mu}_2$ and μ_4 are fairly similar to those in the system of interest. In analogy with scalar treatments[1,15–17] one expands[7] the energy of the system under study in powers of the deviations of the moments from their reference values:

$$E_{el}(i) = E_{el}^{(2)}(i) + E_{el}^{(4)}(i) = E_{el}(\text{ref}) + \left.\frac{\partial E_{el}}{\partial \hat{\mu}_2}\right|_{\text{ref}} [\hat{\mu}_2(i) - \hat{\mu}_2(\text{ref})]$$

$$+ \left.\frac{\partial E_{el}}{\partial \mu_4}\right|_{\text{ref}} [\mu_4(i) - \mu_4(\text{ref})] + (\text{second–order}) \quad . \tag{6}$$

This results in effective pair, triplet, and four-body potentials for the model used here, providing some theoretical backing for the notion of effective angular potentials in covalent systems. However, it is important to note that the potentials are environmentally dependent, essentially because the derivatives in Eq. (6) are environmentally dependent. Thus they can not be used, by themselves, to simulate the properties of systems with strong inhomogeneities. It is also interesting to observe that the three-body potential has a separable form[6] similar to that used in some previously used potentials.[18,19]

3. Empirical Fit

The physical mechanisms discussed here have recently been incorporated in an empirically fitted method[7] for Si. In addition to the terms discussed above, a term proportional to $|\hat{\mu}_2^{s,p}(i)|^2$ was included to account for the electrostatic dipole energy and a repulsive pair potential was assumed. The method contained eight parameters describing the radial dependence and strength of the couplings, and the strength of the μ_4 and dipole contributions. The values of the parameters were obtained by a least-squares fit to a database including energies, lattice constants, and bulk moduli for the diamond, simple cubic, graphitic, and fcc structures; a phonon frequency and an elastic constant in the diamond structure; and the vacancy-formation energy in the diamond structure. It was found that a scalar second-moment treatment could describe the energies, lattice constants, and bulk moduli reasonably well, but that to obtain a sufficiently large vacancy-formation energy it was necessary to include the matrix second moment; to obtain accurate structural energies, it was necessary to include the four-body terms as well.

The resulting functional gave an effective three-body potential which resembles the "Stillinger-Weber" (SW) potential[18] quite closely. Its minimum occurs at 118° (vs. 109° for the SW potential), and it is roughly 40% weaker overall than the SW potential. Since various interatomic potential descriptions of Si can yield potentials which differ by an order of magnitude,[2] this can be considered to be very close agreement indeed. A few test results were also reported. The bcc bond length was found to be identical to the value obtained by density-functional (DF) theory, while the energy favoring the diamond structure relative to bcc was overestimated by roughly 20%. For tetrahedral- and hexagonal-site interstitials, the relaxations are comparable to but larger than the DF values, while the formation energies are roughly 40% smaller than the DF values. However, the calculated value of the elastic constant C_{44} is low by more than 50%. This may indicate that this elastic constant contains long-ranged electrostatic contributions which are not included in analysis of the present type.

4. Conclusions

We have seen that a low-order moment analysis can provide the beginnings of a basis for calculating explicitly angular forces in covalent materials. Although truncation at the fourth-moment level, as described here, cannot provide quantitative accuracy, it is capable of reproducing basic chemical trends in crystal structures. Therefore we conclude this paper by presenting a sketch of the form of the angular forces for a range of materials in the periodic table, in Fig. 4. At the top of the figure is H. Since the dominant valence orbital is the 1s orbital, one expects very little explictly angular character in the interatomic forces; however, three-body forces are present because of bond-competition effects obtained already at the scalar second-moment level.[1] Moving to P and Si, the angular variation of the valence p orbitals leads to corresponding variations in the three-body potential. The minimum for Si is shifted to larger angles because of the effects of s–p hybridization. Finally, in the case of Mo and Ni, the dominant contribution to structural energies comes from the four-body contributions[20] to μ_4. These are extremely rapidly oscillating because the atomic d-orbitals oscillate rapidly. The change in sign between Mo and Ni corresponds to a change in the sign of the energy difference between the fcc and bcc structures; the fcc structure contains many nearest-neighbor 90° angles, which are at a maximum of the Mo potential, but at a minimum of the Ni potential. It is to be hoped that terms of these types can be included, even if only in an empirical fashion, in simulations for many types of materials.

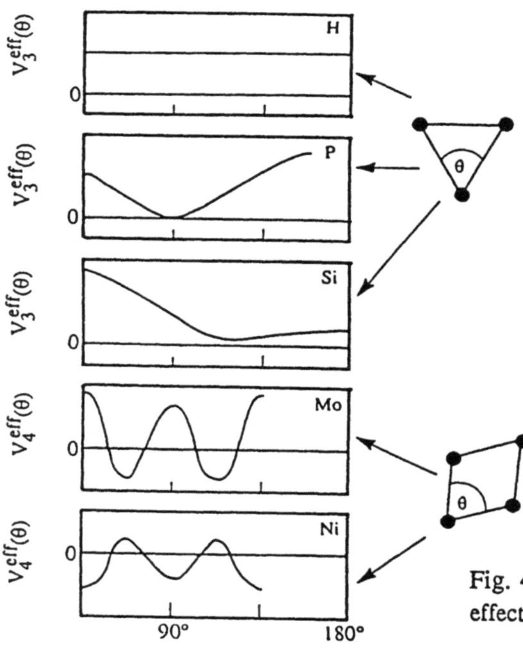

Fig. 4. Chemical trends in three- and four-body effective potentials.

Acknowledgements.

This work was supported by the Department of Energy under Grant No. DE-FG02-84ER45130.

References

1. M. W. Finnis and J. M. Sinclair, Phil. Mag. **A50**, 45 (1984); Phil. Mag. **A53**, 161 (1986).
2. M. S. Daw and M. I. Baskes, Phys. Rev. Lett. **50**, 12385 (1983); Phys. Rev. B **29**, 6443 (1984).
3. F. Ercolessi and E. Tosatti, Phys. Rev. Lett. **57**, 719 (1986); F. Ercolessi, M. Parrinello, and E. Tosatti, Phil. Mag. **A58**, 213 (1988).
4. For a discussion, see T. Haliocioglu, H. O. Pamuk, and S. Erkoc, Phys. Stat. Sol. **149**, 81 (1988).
5. See also A. E. Carlsson, in *Solid State Physics: Advances in Research and Applications, Vol. 43*, edited by H. Ehrenreich and D. Turnbull (in press).
6. A. E. Carlsson, in *Atomistic Modeling of Materials*, edited by V. Vitek and D. J. Srolovitz (Plenum, in press).
7. A. E. Carlsson, P. A. Fedders, and C. W. Myles, Phys. Rev. B (submitted).
8. J. Friedel, in *The Physics of Metals*, edited by J. Ziman (Cambridge University Press, 1969), Chapter 8.
9. C. D. Gelatt, Jr., H. Ehrenreich, and R. E. Watson, Phys. Rev. B **15**, 1613 (1977).
10. D. G. Pettifor, J. Phys. **F8**, 219 (1978).
11. F. Cyrot-Lackmann, J. Phys. Chem. Solids **29**, 1235 (1968).
12. J. Inoue and Y. Ohta, J. Phys. **C20**, 1947 (1987).
13. R. Jones and M. Lewis, Phil. Mag. **B49**, 95 (1984).
14. R. H. Brown and A. E. Carlsson, Solid State Commun. **61**, 743 (1987).
15. A. E. Carlsson and N. W. Ashcroft, Phys. Rev. B **27**, 2101 (1983).
16. S. M. Foiles, Phys. Rev. B **32**, 3409 (1985).
17. A. E. Carlsson, Phys. Rev. B **32**, 4866 (1985).
18. F. H. Stillinger and T. A. Weber, Phys. Rev. B **31**, 5262 (1985).
19. R. Biswas and D. R. Hamann, Phys. Rev. Lett. **55**, 2001 (1985).
20. The sketches and discussion for Mo and Ni are based on results from R. B. Phillips and A. E. Carlsson, in *Atomistic Modeling of Materials: Beyond Pair Potentials*, edited by V. Vitek and D. J. Srolovitz (Plenum, in press); see also R. B. Phillips, Ph.D. thesis (Washington University, 1989) (unpublished).

Molecular Dynamics and Tight-Binding

K. Laasonen and R.M. Nieminen

Laboratory of Physics, Helsinki University of Technology,
SF-02150 Espoo, Finland

Abstract. We present an extension of classical molecular dynamics (MD) to include the forces calculated from electronic degrees of freedom using the tight-binding (TB) approximation. The combined MD–TB problem is solved using the simulated annealing techniques. As an example we study the structures and energetics of small silicon clusters, containing up to 10 Si atoms.

1. Introduction

Classical molecular dynamics (MD) has become increasingly popular for solving various physical problems [1]. It is most useful for studying materials properties at non-zero temperatures, in cases where quantum–mechanical effects in the ionic degreees of freedom are small. The central problem in MD is usually to find a classical interatomic potential which produces all or at least most of the observable physical quantities, by mimicking the quantum mechanics of the electrons [2-4]. In the case of silicon these classical potentials do not give a fully satisfactory description of the properties of Si [5]. Also it is not clear whether one can use the potentials in enviroments very different from where they were developed [6].

To make MD more reliable and more transferable one should start from some approximation of the Schrödinger equation. There are several good approaches such as the density functional theory (DFT). However, most of these approximations are computationally too cumbersome to be used in large-scale MD. Car and Parrinello (CP) [7] have introduced a MD method using DFT. One of the main ideas in the CP method is to update the electronic and ionic degrees of freedom in unison. However, the CP method is computationally very demanding and feasible for relatively small systems only, up to a few tens of atoms. In order to have "quantum mechanical" MD for larger systems we use a simpler approximation for the electronic problem, namely the tight-binding (or Hückel) approximation, and combine it with MD.

There have recently appeared other related MD methods using the tight-binding approximation in one form or another [8-10].

2. Molecular dynamics using the tight-binding approximation

In the tight-binding approximation a minimal basis set is usually applied (only those orbitals which are occupied or partially occupied in the free atom), and the matrix elements of the Hamiltonian are considered as adjustable parameters. These parameters (or matrix elements) are adjusted to fit e.g. known band structures.

Springer Proceedings in Physics, Vol. 48 — Many-Atom Interactions in Solids
Editors: R.M. Nieminen · M.J. Puska · M.J. Manninen

Let us first expand the electronic wave function of a collection of atoms as a linear combination of some suitably chosen localized basis functions ψ_i (Gaussian functions, atomic orbitals etc.)

$$\Phi^{\alpha}(\mathbf{r}) = \sum_{in} C_{in}^{\alpha} \psi_i(\mathbf{r} - \mathbf{R}_n) \tag{1}$$

where $\mathbf{R}_n$ denotes the position of the nth atom ($n = 1...N$), and α labels the electronic states. The Hamiltonian and overlap matrices are defined as

$$\begin{aligned} H_{ij,nm} &= \int \psi_i(\mathbf{r} - \mathbf{R}_n) H \psi_j(\mathbf{r} - \mathbf{R}_m) d^3r \\ S_{ij,nm} &= \int \psi_i(\mathbf{r} - \mathbf{R}_n) \psi_j(\mathbf{r} - \mathbf{R}_m) d^3r \ , \end{aligned} \tag{2}$$

where H is the single-particle Hamiltonian.
The matrix elements of $H_{ij,nm}$ have canonical dependence on the distance between the atoms and on the symmetry of the basis states [11]. We assume here that the basis functions are orthonormal ($S_{ij,nm} = \delta_{ij}\delta_{nm}$). The Schrödinger equation can be written as

$$\sum_{ijnm} H_{ij,nm} C_{in}^{\alpha} - E^{\alpha} C_{in}^{\alpha} = 0 \ . \tag{3}$$

Below, we change the numeration of $H_{ij,nm}$ to H_{nm} and replace the quantum number α with i.

The requirement that the wave functions Φ^i are orthonormal leads to the constraints for C^i

$$\sum_n C_n^i C_n^j = \delta^{ij} \ . \tag{4}$$

Now instead of solving Eq. (3) as an eigenvalue problem we search for a minimum of the energy expectation value $\langle \Phi | H | \Phi \rangle$. Thus one has to minimize the following functional

$$E(\{\mathbf{C}^i\}, \{\mathbf{R}_n\}) = \sum_{inm} C_n^i H_{nm} C_m^i = \sum_i \mathbf{C}^i \cdot H \cdot \mathbf{C}^i \ , \tag{5}$$

which also depends on the atomic positions $\mathbf{R}_n$ via H. One can now replace the original quantum mechanical eigenvalue problem by a classical potential problem, where the minimum of the potential $E(\{\mathbf{C}_i\}, \{\mathbf{R}_n\})$ corresponds to the solution of the eigenvalue problem (3). The dynamics that one derives from that classical system is not a true one, but if the mass parameter associated with the electronic parts much smaller than the true mass of the atoms, one can consider the dynamics of the atoms as a real one – this is the Born-Oppenheimer approximation [7]. In this limit one can do real molecular dynamics simulations, not only find static minima for collections of atoms.

We can write the Lagrangian for the tight-binding system as

$$\begin{aligned} \mathcal{L} = \frac{1}{2}\mu \sum_i \dot{\mathbf{C}}_i^2 &+ \frac{1}{2} m \sum_n \dot{\mathbf{R}}_n^2 - \sum_i \mathbf{C}^i \cdot H \cdot \mathbf{C}^i \\ &+ \sum_{ij} \lambda_{ij} (\mathbf{C}^i \cdot \mathbf{C}^j - \delta^{ij}) + U(\{\mathbf{R}_n\}) \ , \end{aligned} \tag{6}$$

where μ is the (fictitious) electronic mass parameter, m is the mass of the atoms and λ_{ij}

denotes the Lagrangi multipliers associated with the orthonormalization constraints. The 'external' potential $U(\{\mathbf{R}_n\})$ is introduced as a possible short-range correction potential for the tight-binding description [12].

The equations of motion for C_i's are easy to derive as

$$\mu\ddot{\mathbf{C}}^i = -H \cdot \mathbf{C}^i + \sum_j \lambda_{ij}\mathbf{C}^j \quad . \tag{7}$$

The equations of motion for the atoms are not as straightforward. One can use the Hellman-Feynman type of force

$$\frac{\delta E}{\delta \mathbf{R}_n} = \frac{\delta\langle\Phi|H|\Phi\rangle}{\delta \mathbf{R}_n} = \langle\Phi|\frac{\delta H}{\delta \mathbf{R}_n}|\Phi\rangle \quad . \tag{8}$$

We neglect the residual $\mathbf{R}_n$ dependence of λ_{ij} and obtain

$$m\ddot{\mathbf{R}}_n = -\sum_i \mathbf{C}^i \cdot \frac{\delta H}{\delta \mathbf{R}_n} \cdot \mathbf{C}^i + \frac{\delta U(\{\mathbf{R}_n\})}{\delta \mathbf{R}_n} \quad . \tag{9}$$

The integration of the equations of motion can be carried out by using some standard algorithm such as Beeman's [13], if the Lagrangi multipliers in Eq. (7) can be solved. One can then implement the ideas of Ryckaert *et al.* [14] to solve the Lagrangi multipliers for each time step demanding that the constraints Eq. (4) hold for every step.

3. Simulated annealing

Simulated annealing is very general minimization procedure for finding a global optimum. It has been introduced by Kirkpatrick *et al.* [15] and subsequently applied in various problems in physics [7],[16] as well as for several optimization problems in computer science [17]. Originally simulated annealing was used in the context of Monte-Carlo methods but the method we use is more suitable for dynamical systems. The method was first introduced by Car and Parrinello [7].

In simulated annealing one aims to minimize some energy functional (or in general a cost function) $E(C_i)$ which depends on the variables C_i $(i = 1...N)$. Usually there are some constraints σ_i, $(i = 1...M$ where M < N) for the variables C_i. A physical analogy is that E is a classical potential in space Ω and the variables C_i fix one point in that space. One can use normal Lagrangian formalism to calculate the equation of motion for a fictitious particle represented by coordinates $\mathbf{C} = C_i$ and an arbitrary mass.

The Lagrangian of the combined system is

$$\mathcal{L} = \frac{1}{2}\sum_{i=1}^{N} m_i\dot{C}_i^2 + E(C_i) + \sum_{i=1}^{M} \lambda_i\sigma_i \quad , \tag{10}$$

where λ_i is a Lagrangi multiplier associated with the constraint σ_i and the m_i's are the mass parameters for the particles.

By using a normal variational procedure one can derive equations of motion for variables C_i as

$$m_i \ddot{C}_i = \frac{\delta \mathcal{L}}{\delta C_i}. \tag{11}$$

To find the minimum of $E(C_i)$, one starts from some arbitrary point C_0 and velocity components $\dot{C}_i$ so that the kinetic energy corresponds to a given temperature T. By integrating the equations of motion one allows the system to evolve and slowly reduces the temperature (the kinetic energy of the system). If the cooling is slow enough, the system should finally reach the global minimum. How many time steps this will take depends crucially on the system itself but also on the cooling schedule. One simple way to cool the system is the so-called exponential cooling where the cooling speed depends linearly on T

$$\frac{\mathrm{d}T}{\mathrm{d}t} = -vT \ . \tag{12}$$

Above, t is iteration "time" and v is a constant. In this schedule the number of time steps which are needed to reach the minimum depends on the value of v. The exponential cooling is somewhat problematic because it has the tendency to cool too fast at the end of the calculation.

4. Application to silicon clusters

The tight-binding formalism presented here is based on the work by Tomanek and Schlüter [12]. Also Khan and Broughton [8] have used a method similar to ours for studying the Si_3 trimer and silicon surfaces, and Goodwin *et al.*[10] have used a related TB method for studying structures of small Si clusters.

In a tight-binding formulation for silicon we take four basis functions for each atom site - these basis functions are labeled (s, p_x, p_y, p_z).We choose the orthogonal tight-binding parametrisation first introduced by Chadi [18]

$$\begin{aligned} E_s &= -5.25\,\mathrm{eV},\ E_p = 1.20\,\mathrm{eV}, \\ V_{ss\sigma} &= -1.938\,\mathrm{eV},\ V_{sp\sigma} = 1.745\,\mathrm{eV},\ V_{pp\sigma} = 3.050\,\mathrm{eV},\ V_{pp\pi} = -1.075\,\mathrm{eV}. \end{aligned}$$

The V's are the Slater-Koster [19] parametrized nearest-neighbor hopping integrals, corresponding to a bulk interatomic distance 2.35 Å. These hopping integrals have the distance dependence $1/|\mathbf{R}_i - \mathbf{R}_j|^2$ [11]. The diagonal terms E are assumed to be independent of the distance between atoms.

The distance dependence used above for the V's does not give correct repulsion at short distances, and a short-range correction is needed. The correction potential is written as a sum of pair potentials

$$U(\mathbf{R}_n) = \sum_{n<m} E_{corr}(|\mathbf{R}_n - \mathbf{R}_m|) \ , \tag{13}$$

where E_{corr} can be calculated as the difference between the *ab initio* pair potential for Si_2 [12] and the corresponding TB potential

$$E_{corr}(r) = E_{Si_2}(r) - E_{TB}(r) \ . \tag{14}$$

Tomanek and Schlüter use two additional correction terms, one of which depends only on the number of occupied bonds in the cluster [12]

$$E_{bond} = -N(\phi_1 (n_b/N)^2 + \phi_2 (n_b/N) + \phi_3) \ . \tag{15}$$

The coefficients ϕ were chosen to reproduce the cohesive energies of both diamond and bcc structures of silicon:

$$\phi_1 = 0.225\,\text{eV} \quad \phi_2 = 1.945\,\text{eV} \quad \phi_3 = -1.03\,\text{eV}\ .$$

The other is a "Hubbard"–like term, which prevents large charge transfers between atoms :

$$E_{Hubbard} = U \sum_i (q_i - q_0)^2\ , \tag{16}$$

with the value of U around 1 eV.

In the simulated annealing we set the mass parameters to $\mu = 1.0$ and $m = 200-1000$. We have used the exponential cooling schedule and used values $v = 0.05-0.3$ for the cooling parameter. As for initial values for the atomic positions $\mathbf{R}_n$, we choose reasonables values by hand. For practical reasons we have used a few different initial configurations for each cluster, and for bigger clusters ($N = 8$ — 10) we have chosen the initial geometries near geometries proposed in other papers [20], [21], and used relatively high cooling speeds. Thus in the case of larger clusters we have not done full global minimum searches but find at least local mimima near the initial configurations.

The initial velocities for atoms $\dot{\mathbf{R}}_n$ and initial values for C_i's and $\dot{C}_i$'s have been taken randomly. After that the velocities have been scaled so that the kinetic energy corresponds to a given initial temperature.

We have studied here small clusters (2 – 10 atoms). The results are in general similar to those given by *ab initio* quantum chemistry (Hartree–Fock - based) calculations [20],[21], but in the case of Si_5 we find a different minimum. The results are presented in Tables I and II and Fig. 1.

Table I Binding energies of Si clusters.

N	E (eV)	E with U = 1 eV	E (ref [21])	E/N (eV)
2	3.07	3.07	3.12	1.54
3	7.33	6.96	7.61	2.44
4	12.63	12.45	12.68	3.16
5	15.58	15.38		3.12
5	15.29	14.58	16.48	
6	20.82	20.49	21.62	3.47
7	26.56	26.52	26.59	3.79
8	29.67	29.29	29.17	3.71
9	33.14	32.94	32.27	3.68
10	39.06	38.84	38.16	3.91

N is the number of atoms, and the second column gives the results for the energy minima. The third column is the energy of the same structure but including the Hubbard correction (see text). The fourth column contains the scaled energies obtained by Raghavachari and Rohlfing [21]. The last column lists the binding energy per atom.

Table II Bond lengths for Si clusters.

N	bond	d_{our} (Å)	$d_{ab\ initio}$ (Å)
2	1–2	2.27	2.23
3	1–2	2.26	2.17
angle	θ	85.3°	77.2°
4	1–2	2.39	2.30
	1–3	2.54	2.40
5	2–4	2.41	
	2–5	2.44	
	1–2	3.07	
	2–3	2.59	
	1–5	2.35	
5*	1–2	3.56	3.26
	1–4	2.48	2.39
6	1–2	2.77	2.86
	1–3	2.52	2.49
	1–5	2.62	2.69
	3–5	2.63	2.93
	5–6	2.36	2.38
7	3–4	2.55	2.48
	1–3	2.60	2.47
	1–2	2.88	2.58

N is the number of atoms, and the second column indicates the bonded atoms (see Fig. 1). The third column gives the bond lengths from our calculations. The fourth column contains the corresponding bond lengths from *ab initio* calculations[20],[21].
* In the case of Si_5 we have compared the bond lengths of the trigonal bipyramidal structure.

It is quite surprising that the present results are so close to more elaborate (*ab initio–* type) calculations, considering how simple a model we have used. There are few experimental results for small silicon clusters, so we basically have to compare our results with *ab initio* type calculations [12], [20-22]. The energies correspond quite well to the scaled energies by Raghavachari and Rohlfing (see Table I). The bond lengths are usually slightly longer (typically 0.1 Å , see Table II) than those from *ab initio* calculations. Also the results of TB calculations by Goodwin *et al.* [10] are very close to the results of Raghavachari and Rohlfing [21].

The fragmentation spectrum has been experimentally observed [23],[24] for small Si^+ clusters (N= 2–12). These experiments show that clusters with 6,7 and 10 atoms

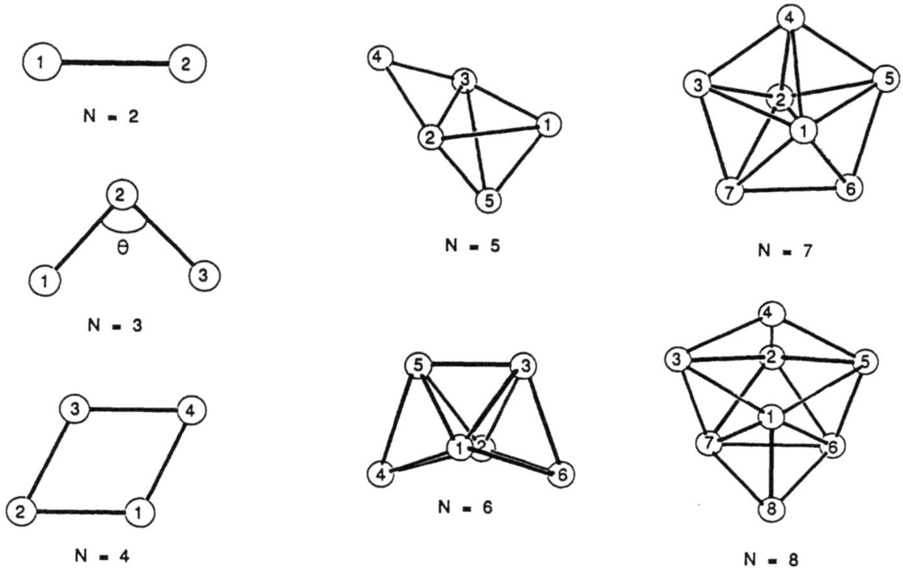

Fig 1. Structures of small silicon clusters (N = 2 – 8). In some clusters not all the bonds are drawn.

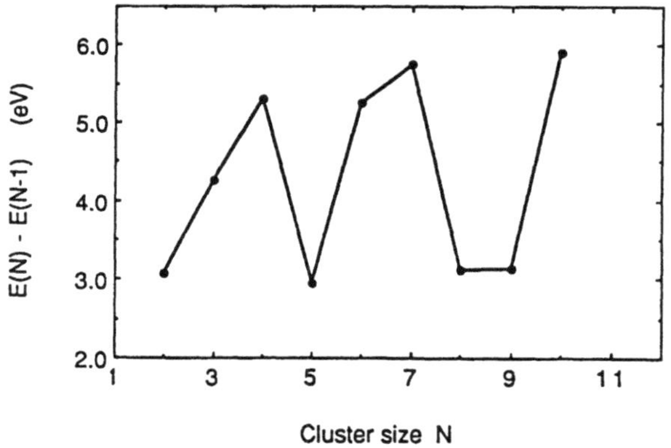

Fig 2. Fragmentation energy $E_N - E_{N-1}$ (in eV).

should be most stable. Our calculations gives a fragmentation spectrum (Fig. 2) which agrees with these results.

5. Discussion

The method presented here is a flexible way to extend classical molecular dynamics to include many-atom nonclassical interactions. It has the advantages of normal molecular dynamics, as one can study systems at nonzero temperatures, and one can also use the experience from existing MD simulation codes. The method takes the quantum mechanical effects into account at a more fundamental level than standard MD, and therefore we believe that it describes 'quantum mechanical' systems better and with fewer parameters than classical MD.

Comparing the method to *ab initio* methods, it is much faster and therefore useful also for larger systems. Our study of silicon clusters shows that a proper tight-binding parametrisation gives for small clusters results comparable to those of *ab initio* calculations. For larger systems, including defects and surfaces, the parametrisation should be even easier.

Acknowledgement

This work has been supported in part by a grant from the Academy of Finland.

References

[1] G.Ciccotti, D.Frenkel and I.R.McDonald (eds.),*Simulction of Liquid and Solids* (North-Holland, Netherlands 1987)

[2] F.Stillinger and T.Weber, Phys. Rev. **B31**, 5262 (1981)

[3] J.Tersoff, Phys. Rev. **B37**, 6991 (1988)

[4] A.D.Mistriotis, N.Flytzantis and S.C.Farantos, Phys. Rev. **B39**, 1212 (1989)

[5] E.Roger Cowley, Phys. Rev. Lett. **60**, 2379 (1988)

[6] B.P.Feuston, R.K.Kalia and P. Vashishta , Phys. Rev. **B35**, 6222 (1987)

[7] R.Car and M.Parrinello , Phys. Rev. Lett. **55**, 2471 (1985)

[8] F.S.Khan and J.Q.Broughton, Phys. Rev. **B39**, 3688 (1989)

[9] C.Z.Wang,C.T.Chan and K.M.Ho, Phys. Rev. **B39**, 8586 (1989)

[10] L.Goodwin, A.J.Skinner and D.G.Pettifor, to be published

[11] W.A.Harrison *Electronic Strcuture and the Properties of Solids* (Freeman, San Francisco, 1980)

[12] D.Tomanek and M.A.Schlüter , Phys. Rev. **B36**, 1208 (1987)

[13] D.Beeman, J.Comp.Phys. **20**, 130 (1976)

[14] J.-P.Ryckaert, G.Ciccotti and H.J.C.Berendesen, J.Comp.Phys. **23**, 327 (1977)

[15] S.Kirkpatrick, C.D.Gelatt Jr. and M.P.Vecchi, Science **220**, 671 (1983)

[16] M.Sprik and M.L.Klein, J. Chem. Phys. **89**, 1592 (1988)

[17] P.J.M van Laarhoven and E.H.L.Aarts, *Simulated Annealing:Theory and Applications* (D.Reidel Publishing Company, Holland) 1988

[18] D.J.Chadi , Phys. Rev. **B29**, 785 (1984)

[19] J.C.Slater and G.F.Koster , Phys. Rev. **94**, 1498 (1954)

[20] K.Raghavachari , J. Chem. Phys. **84**, 5672 (1986)

[21] K.Raghavachari and C.M.Rohlfing, J. Chem. Phys. **89**, 2219 (1988)

[22] P.Ballone, W.Andreoni, R.Car and M.Parrinello, Phys.Rev.Lett. **60**, 271 (1988)

[23] L.A.Bloomfield, R.R.Freeman and W.L.Brown, Phys. Rev. Lett. **54** 2246 (1985)

[24] Q-L.Zhang, Y.Liu, R.F.Curl, F.K.Tittel and R.E.Smalley, J.Chem.Phys. **88** 1671 (1988)

Part IV

Surface Phenomena

Linear Surface Model and the Environment-Dependent Effective Pair Potential

G. Tichy

Institute for Solid State Physics, Eötvös University,
H-1088 Budapest, Múzeum körút 6–8, Hungary

Abstract. The computer simulation of the many-atom potential gives several new results. This paper aims to understand what is the effect of the nonpair character of the potential on the surface. For this reason a simple model is presented.

1. Introduction

It is obvious that in metals the pair potential description is not suitable because the use of a pair potential gives false dependences between macroscopic quantities. For example the Cauchy relation referring to the elastic constants is valid and the vacancy formation energy is almost equal to the cohesive energy. These relations are far from true for real metals. To avoid these discrepancies, nonpair, i. e. many-atom, interactions were introduced on different theoretical backgrounds. In the last six years the many-body interatomic potential has aroused great interest. The embedded atom method (EAM) [1] and the Finnis-Sinclair (FS) potential [2] use the same functional form for the energy expression as follows:

$$E(\mathbf{r}_1 \ldots \mathbf{r}_N)=\frac{1}{2}\sum_{i,j} V(\mathbf{r}_{ij}) - \sum_i f(\rho_i) ,$$

where

$$\rho_i = \sum_j \Phi(\mathbf{r}_{ij}) \quad \text{and} \quad \mathbf{r}_{ij} = \mathbf{r}_i - \mathbf{r}_j . \tag{1}$$

The aim of the present paper is to understand the reason for the surface relaxation rather than to calculate it [3]. Therefore we investigate the effect of the nonpair character of the potential on the surface relaxation.

There have been several attempts to calculate the atom displacements on a metal surface [4]. The computer simulations showed, in simple cases only, that the inter-layer spacings change near the surface. If a non-oscillating pair potential was used, the first inter-layer spacing always increased, contradicting experimental observations, but if a many-body potential was used, the first distance decreased [5].

In this work we show a decomposition of the potential into two parts. It gives us an environment-dependent effective pair potential, and using it our analysis reveals the effect of the nonpair character on the inter-layer spacings.

Springer Proceedings in Physics, Vol. 48 Many-Atom Interactions in Solids
Editors: R.M. Nieminen · M.J. Puska · M.J. Manninen

2. Decomposition of the potential

The many-body potential has a pair and a nonpair term. The partitioning of the potential into these terms is not unique. If we add a linear term to the f(ρ) function, as F. Ercolessi et al.[6],

$$f_1(\rho) = f(\rho) + \beta\rho, \tag{2}$$

then we have to transform the pair term too:

$$V_1(\mathbf{r}_{ij}) = V(\mathbf{r}_{ij}) + 2\beta\,\Phi(\mathbf{r}_{ij}). \tag{3}$$

After the transformation a similar form was gained:

$$E(\mathbf{r}_1 \ldots \mathbf{r}_N) = \frac{1}{2}\sum_{i,j} V_1(\mathbf{r}_{ij}) - \sum_i f_1(\rho_i). \tag{4}$$

It is possible to decompose the energy function into two terms which are invariant to the above transformation:

$$\begin{aligned} E &= \frac{1}{2}\sum_{i,j}\left[V(\mathbf{r}_{ij}) - (f'(\rho_i) + f'(\rho_j))\Phi(\mathbf{r}_{ij})\right] - \\ &-\sum_i\left[f(\rho_i) - \rho_i f'(\rho_i)\right] = \frac{1}{2}\sum_{i,j} V_{eff}(\mathbf{r}_{ij},\rho_i,\rho_j) - \sum_i h(\rho_i). \end{aligned} \tag{5}$$

The first term can be called the effective pair potential, and the second one the individual energy of the atom. In fact the first term is not a real pair potential, because through ρ_ks it depends on the sites of other atoms as well.

Now the energy depends on two type of variables: $\mathbf{r}_i$ and ρ_i. They are not independent because ρ_i is a function of $\mathbf{r}_i$. As a matter of fact we can treat the two types of variables separately. The ρ_i variables are not very sensitive to the individual atomic displacements; they mainly depend on the environment of the special atoms only. If the coordination number changes, ρ_i changes drastically, as in the case of a vacancy or surface. If the atoms move but the topology of the lattice remains it does not change much.

Such separation of the variables is useful in several cases. The derivative of E with respect to ρ_i is zero and the force, the gradient of the potential, is the same whether the derivation is with respect to both types as a total derivative or only to one type of variable as a partial derivative:

$$\begin{aligned} &\frac{\partial E(\mathbf{r}_{ij},\rho_i)}{\partial \rho_k} = 0; \\ &\frac{dE(\mathbf{r}_{ij},\rho_i)}{d\mathbf{r}_{pq}} = \frac{\partial E(\mathbf{r}_{ij},\rho_i)}{\partial \mathbf{r}_{pq}} + \sum_k \frac{\partial E(\mathbf{r}_{ij},\rho_i)}{\partial \rho_k}\frac{d\rho_k}{d\mathbf{r}_{pq}} = \frac{\partial E(\mathbf{r}_{ij},\rho_i)}{\partial \mathbf{r}_{pq}}. \end{aligned} \tag{6}$$

The upper decomposition explains several items in the earlier calculations:

1./ If ρ_i does not change drastically, the pair potential description is useful. The condition is fulfilled when the density of the material has no large change. It is good for dislocation, stacking fault and grain boundary calculations, but it fails for vacancies, interstitials and surfaces.

2./ When the second derivative of the energy is also important, the use of the effective pair potential is not enough. This is the case of the Cauchy relation.

3./ Both the pair potential term and the individual energy term contribute to the cohesive energy. However, in the case of a vacancy, the individual term is not present because an atom is missing. Without relaxation and not regarding the environment dependence of the pair potential the vacancy formation energy is only the energy of the missing bonds, thus

$$E^F_{VAC} = -\frac{1}{2}\sum_j V_{eff}(r_{0j}) = -E_{COH} - h(\rho). \tag{7}$$

Numerical calculations have shown that the above expression for vacancy formation energy is sufficiently good. For example, in the case of a fcc crystal, relaxation gives only a 1-2% decrease. It is clear, because the coordination number does not fall too much, from 12 to 11.

3. The quadratic approximation of the effective potential

Using the effective pair potential it is simple to get the surface energies and the stresses on the surface. Neglecting the local environment dependence one can have relatively good results. However, these results are not as good as those for vacancy formation energy, because the contribution of the relaxation is much larger. At the surface the coordination number changes significantly (e. g. fcc (110) surface; from 12 to 7).

Our simple model is limited to the second nearest neighbour interactions. The effective potential for the first and second nearest neighbour and between them can be approximated by a quadratic functional form as

$$V_{eff}(r) = -A + \frac{B}{2}(r-b)^2, \tag{8}$$

where A, B and b are the parameters depending on the local environment. Naturally, this simple shaped potential does not result in a stable structure. From this potential for a given structure we can derive potentials between atomic layers with different Miller indices.

Between two layers ß and ß+i, the potential is $E^{ß,ß+i}$. $E^{ß,ß+1}$ and $E^{ß,ß+2}$ are the first and second nearest layer interaction energies. In our approximation

$$E^{ß,ß+i}(r) = -V_i + \frac{W_i}{2}(r-b_i)^2, \tag{9}$$

where r is the distance between the two layers being considered. Eq. (9) is obtained from eq. (8) as a Taylor expansion using the structure.

Our investigations were limited only to fcc and bcc lattices. The parameters for fcc(100), fcc(110) and bcc(100) planes are listed in Table 1. The parameters A and B were defined by

Table 1

plane	fcc(100)	fcc(110)	bcc(100)
V_1	-4A	-4A	-4A
W_1	2.5428B	1.814B	1.6788B
b_1	0.6605a	0.5579a	0.6013a
V_2	-A	$-3A+0.0667Ba^2/2$	-A
W_2	B	2.0648B	B
b_2	0.8284a	0.6499a	0.9282a

the atomic interactions. a is the lattice constant and b was calculated from the energy minimum.

Our approximation is not useful for other fcc and bcc planes. For example, in the case of fcc(111) and bcc(110) the third nearest neighbour interaction ought also to be considered and in the case of bcc(111), bcc(210) and bcc(310) the second nearest neighbour interaction between the atoms results in even third or more nearest layer interaction.

4. Surface relaxation

4.1 Using pair potential

Applying the values from Table 1, the inter-layer energies are determined by eq. (9). From the energies the force acting between two layers can be derived:

$$F^{\beta,\beta+i}=W_i(b_i-r). \qquad (10)$$

Let the surface be the zeroth layer and the spacings between layers be $d_0, d_1, \ldots d_\beta$, respectively, as shown in figure 1. So the equations of equilibrium are

$$\begin{aligned}
&0/\quad W_1(b_1-d_0)+W_2(b_2-d_0-d_1)=0\\
&1/\quad W_1(b_1-d_1)+W_2(b_2-d_1-d_2)=W_1(b_1-d_0)\\
&2/\quad W_1(b_1-d_2)+W_2(b_2-d_2-d_3)=\\
&\qquad\qquad W_1(b_1-d_1)+W_2(b_2-d_1-d_0)\\
&\beta/\quad W_1(b_1-d_\beta)+W_2(b_2-d_\beta-d_{\beta+1})=\\
&\qquad\qquad W_1(b_1-d_{\beta-1})+W_2(b_2-d_{\beta-1}-d_{\beta-2})
\end{aligned} \qquad (11)$$

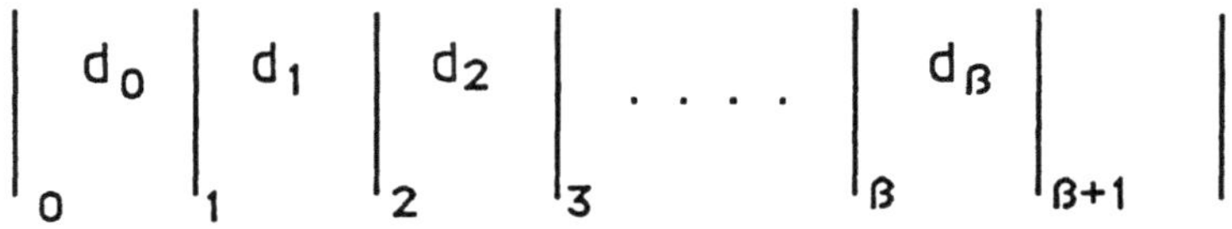

Fig.1. Numbering of the levels and distances at the surface.

The equilibrium equations for the first and second layers are specific, but from the third layer all of them are described by the general form. The general equation is a homogeneous equation for d_β:

$$W_2 d_{\beta+1} + (W_1+W_2) d_\beta - (W_1+W_2) d_{\beta-1} - W_2 d_{\beta-2} = 0 . \tag{12}$$

The solution of the equation has the form

$$d_\beta = c_0 q_0^\beta + c_1 q_1^\beta + c_2 q_2^\beta \tag{13}$$

where q_0, q_1 and q_2 are the roots of

$$W_2 q^3 + (W_1+W_2) q^2 - (W_1+W_2) q - W_2 = 0 , \tag{14}$$

the characteristic equation of third order. They are given as

$$q_0 = 1; \quad q_1 = -\frac{W_1 + 2W_2 - \sqrt{W_1^2 + 4W_1W_2}}{2W_2}; \quad q_2 = -\frac{W_1 + 2W_2 + \sqrt{W_1^2 + 4W_1W_2}}{2W_2}. \tag{15}$$

It is easy to show that $-1<q_1<0$ and $q_2<-1$. For the parameters given in Table 1 the roots of eq. (14) are listed in Table 2.

Table 2

plane	fcc(100)	fcc(110)	bcc(100)
q_0	1	1	1
q_1	-0.2320	-0.4041	-0.2956
q_2	-4.3108	-2.4744	-3.3832
c_0	0.5099a	0.3670a	0.5047a
c_1	0.0578a	0.0569a	0.0340a

The lattice is infinite in the positive direction and therefore c_2 must be zero. c_0 and c_1 are given from the first two equations of (11) and are also listed in Table 2:

$$c_0 = \frac{W_1 b_1 + 2W_2 b_2}{W_1 + 4W_2} \; ; \; c_1 = \frac{W_1 W_2 (2b_1 - b_2)}{(W_1 + 4W_2)(W_1 + W_2(1+q))} \qquad (16)$$

c_0 is the layer spacing of the bulk. The exact values of different planes are 0.5a, $1/(2\sqrt{2})a = 0.3536a$ and 0.5a, respectively. These values are not the same as what we obtained from our calculation for c_0. The second order approximation has led to small deviations. c_1 gives the deviation from the bulk value at the first layer spacing. It is positive, that is, the first spacing expands as expected from a calculation using pair interactions. Now let's see the effect of the local environment.

4.2 The effect of the local environment

The effective inter-layer potential depends on the environment of both layers concerned. Let's see the simplest assumption. At the front, at the zeroth layer, every atom has fewer neighbours than in the bulk. From the first layer the change of the environment is neglected. That means, the first-neighbour force at the surface is significantly stronger than in the bulk, as stated earlier by R. A. Johnson[7]. The parameters of the interactions which refer to the first layer are indexed with zero. In this way eqs. (11) have the form

$$\begin{aligned}
&0/ \quad W_1^0(b_1^0 - d_0) + W_2^0(b_2^0 - d_0 - d_1) = 0 \\
&1/ \quad W_1(b_1 - d_1) + W_2(b_2 - d_1 - d_2) = W_1^0(b_1^0 - d_0) \\
&2/ \quad W_1(b_1 - d_2) + W_2(b_2 - d_2 - d_3) = \\
&\qquad\qquad W_1(b_1 - d_1) + W_2^0(b_2^0 - d_1 - d_0) \qquad (17) \\
&\beta/ \quad W_1(b_1 - d_\beta) + W_2(b_2 - d_\beta - d_{\beta+1}) = \\
&\qquad\qquad W_1(b_1 - d_{\beta-1}) + W_2(b_2 - d_{\beta-1} - d_{\beta-2}).
\end{aligned}$$

After modification, the general form of eq. (12) is valid only from the third layer and an equation of three unknowns has to be solved. The unknown variables are d_0, c_0 and c_1.

The analytical solution of the equation is complicated. The main features can also be seen if the $W_1{}^0 = W_1$ and $W_2{}^0 = W_2$ simplifying conditions are applied.

The bulk spacing c_0 remains the same as it was in eq. (16), but c_1 and d_0 change. Let $\Delta_1 = b_1 - b_1{}^0$ and $\Delta_2 = b_2 - b_2{}^0$. Both are positive numbers because at the surface the repulsive term becomes weaker. We can even assume that $\Delta_1 > \Delta_2$, since the effect at the zeroth-first-layer interaction is larger than at the zeroth-second one. Thus

$$c_1 = \frac{W_1 W_2 (2b_1 - b_2)}{(W_1 + 4W_2)(W_1 + W_2(1+q))} - \frac{W_1(\Delta_1 - \Delta_2)}{W_1 + W_2(1+q)} \tag{18}$$

and $\quad d_0 = c_0 + c_1 - \Delta_2 \; .$

If Δ_2 is large enough the zero spacing, d_0, can be smaller than the bulk one, c_0. c_1 is also smaller than for pure pair potential. They can be positive or negative.

5. Conclusion

The decomposition of the many-atom potential can be a useful tool in the approximation of molecular dynamics or molecular statics. It has been shown why the pair potential calculations were so effective if the density did not change too much and why it was so ineffective in other cases. The effective potential and the individual atom terms help one to understand the relation of the cohesive energy to vacancy formation energy.

The above described linear surface model revealed the effect of the pair potential character in the case of a surface, and it was shown that the main effect is a smaller first spacing at the surface. Near the surface the atoms have fewer first neighbours, therefore the effective potential has a stronger attractive part and the equilibrium spacing has to decrease.

References

1. M. S. Daw, M. I. Baskes, Phys. Rev. B **29** (1984) 6443
2. M. W. Finnis, J. E. Sinclair, Phil. Mag. A **50** (1984) 45
3. G. J. Ackland, G. Tichy, V. Vitek, M. W. Finnis, Phil. Mag. A **56** (1987) 56
4. G. J. Ackland, M. W. Finnish, Phil. Mag. A **54** (1986) 301
   C. C. Matthai, D. J. Bacon, Phil. Mag. A **52** (1985) 1
   R. P. Gupta, Phys Rev. B **23** (1981) 6265
   S. M. Foiles, M. I. Baskes, M. S. Daw Phys. Rev. B **33** (1986) 7983
5. N. V. Doan, Phil. Mag. A**58** (1988) 179
6. F. Ercolessi, M. Parrinello, E. Tosatti, Phil. Mag. A**58** (1988) 213
7. R. A. Johnson, Surface Science, **151** (1985) 311

Simulation of Surface Premelting

P. Stoltze

Laboratory of Applied Physics, Technical University of Denmark,
DK-2800 Lyngby, Denmark

The premelting of Al(110) is studied by Molecular Dynamics simulations using the potential for Al derived from Effective Medium Theory. We find a progressive disordering of the surface region as the bulk melting point is approached. This disorder is mediated by the formation of adatom-vacancy pairs. The premelting appears anisotropic only for some choices of order parameter. This has implications for the detection of the disorder by a diffraction technique.

The understanding of the structure of the surface of a hot crystal is related to the problem of melting of metals by the hypothesis of Tammann (1910) that surfaces play an important role as sources of defects in the bulk melting process. While the melting process for bulk metals is a first order phase transition the presence of the surface reduces the symmetry and the order of the transition at the surface may differ from that of the bulk; the surfaces of metals may show *premelting*, defined as a progressive disordering of the surface of the metal as the melting point for the bulk is approached.

The premelting phenomenon has recently been studied intensively under ultra high vacuum conditions by ion scattering [1, 2, 3, 4, 5, 6], LEED [7, 8, 9] and x-ray diffraction [10].

The nature of the transition at surfaces when the bulk undergoes melting has been investigated in a continuum model [11, 12, 13, 14, 15] and in a model based on the interface thermodynamics [5]. Although these models are very successful in explaining some aspects of the premelting phenomenon the description of the disorder at the level of atomic structure is beyond continuum models as well as models based on macroscopic thermodynamics [16].

One reason for studying the properties of hot metal surfaces by computer simulation is that computer simulations bridge experiments and theory. While theories of atomic interactions may give detailed and accurate descriptions of the energy of the system, the properties at finite temperature are determined by the *free energy* rather than the *energy*. Computer simulations may be viewed as one way of determining the entropy of the potential.

Springer Proceedings in Physics, Vol. 48 — Many-Atom Interactions in Solids
Editors: R.M. Nieminen · M.J. Puska · M.J. Manninen

When the understanding of the atomic properties has been developed to the limit of agreement with experiments, interesting information may be derived by systematic calculations. These calculations may aid in the interpretation of experiments and may even supply data for physical quantities for which no experimental techniques exist.

Premelting of crystals has been studied by computer simulation using Molecular Dynamics and Lennard-Jones potentials. For fcc-crystals premelting has been found for the (100), (110) and (111)-surfaces [17, 18, 19]. The premelting is mediated by adatom-vacancy pairs [20]. The important role of surface phonons in the simulations and the differences between roughening and premelting have been studied [21].

Recently the behavior of Ni(110) at high temperature has been studied by computer simulation [22] using a potential derived by the Embedded Atom Method. These simulations show that this surface does not show premelting, in agreement with thermodynamic considerations [5].

For metals the delocalization of valence electrons causes the presence of a volume-dependent term in the total energy for the metal and a single pair potential will not be able to reproduce the cohesive energy, the bulk modulus and the surface energy for a metal. In this paper we report the results of one series of simulations of the premelting of Al(110) using a potential derived from the Effective Medium Theory [23, 24]. We have previously used this potential in simulations of the thermal expansion of bulk Al [25], in a comparison of the high temperature behavior of Al(110) and Al(111) [26] and in investigations of the properties of Al(110) [27, 28].

In this study of Al(110) we have used a setup of 512 atoms distributed in 16 layers, 32 atoms per layer. Periodic boundary conditions are used in X and Y and 4 static layers are placed at the underside of the system. This means that we are essentially simulating a semi-infinite crystal rather than a slab. The lattice constant for the static substrate equals the equilibrium lattice constant at the actual temperature. The equilibrium lattice constant has been determined by Monte Carlo simulations for the bulk crystal. The simulations are made by Molecular Dynamics. The integration of the equation of motion is performed by the Velocity-Verlet algorithm [29].

Typical snapshots of the system at 560, 745 and 930 K are shown in figure 1, 2 and 3. At the lower temperature the layer structure may easily be identified while at higher temperatures the premelting of the surface results in a partial dissolution of the layer structure. The presence of adatoms is noticeable. The snapshots as well as the density profiles and the calculated density of defects show that the disorder is mediated by the formation of adatom-vacancy pairs at the surface.

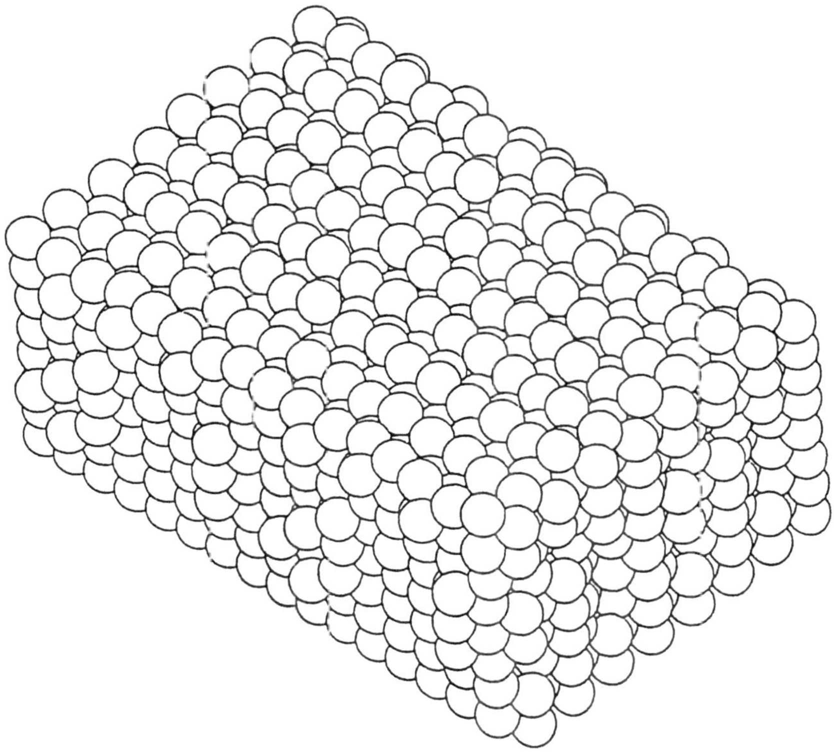

Figure 1: Snapshot of the Al(110) surface at 560 K. In the snapshot the system has been replicated in the X and Y directions in the production of the figure. In the projection the atoms have been labeled by layer number. At this temperature the surface is only slightly disordered and the presence of the close-packed rows is obvious.

As an example of a property which can be calculated but not measured we have determined the low index structure factors $|S|^2$ for individual layers

$$S_l = < \sum_i \exp(i\vec{g} \cdot \vec{r_i}) > \tag{1}$$

where the averaging is made over the position $\vec{r}$ of all atoms in layer number l. For all layers we find that the structure factor increases through the sequence

$$|S_{(2,0)}|^2 < |S_{(0,2)}|^2, |S_{(1,1)}|^2 < |S_{(1,0)}|^2, |S_{(0,1)}|^2. \tag{2}$$

At low temperatures the decrease in the structure factors are caused by essentially harmonic vibrations. This may be treated quantitatively in a Debye model [30]

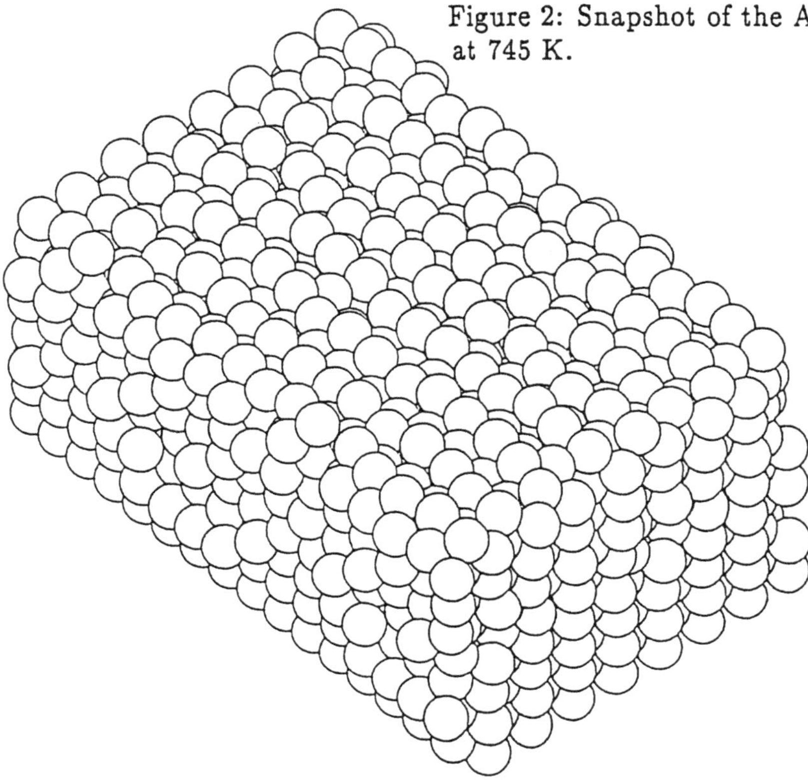

Figure 2: Snapshot of the Al(110) surface at 745 K.

$$\ln |S|^2 = \frac{3\hbar |\vec{g}|^2 T}{m k_B \theta_D^2}. \tag{3}$$

The dependence of $|S|^2$ on $|\vec{g}|^2$ in the Debye model explains the sequence of structure factors, equation (2)

The Debye temperature for the surface layer as determined from the simulations is $\theta_D \sim 250$ K compared to $\theta_D \sim 320$ K found for the deeper layers. The thickness of the surface region determined from the Debye temperature is 3-4 layers. The variations in Debye temperature may be interpreted as a softening of the lattice at the surface and a hardening of the lattice near the static substrate.

From the structure factors for the layers we may calculate an average structure factor using a penetration depth, λ,

$$S = \frac{< S_l \exp(\frac{z_l}{\lambda}) >}{< \exp(\frac{z_l}{\lambda}) >}. \tag{4}$$

For radiation with an effective penetration depth λ undergoing kinematic diffraction the averaged structure factor will equal the normalized intensity

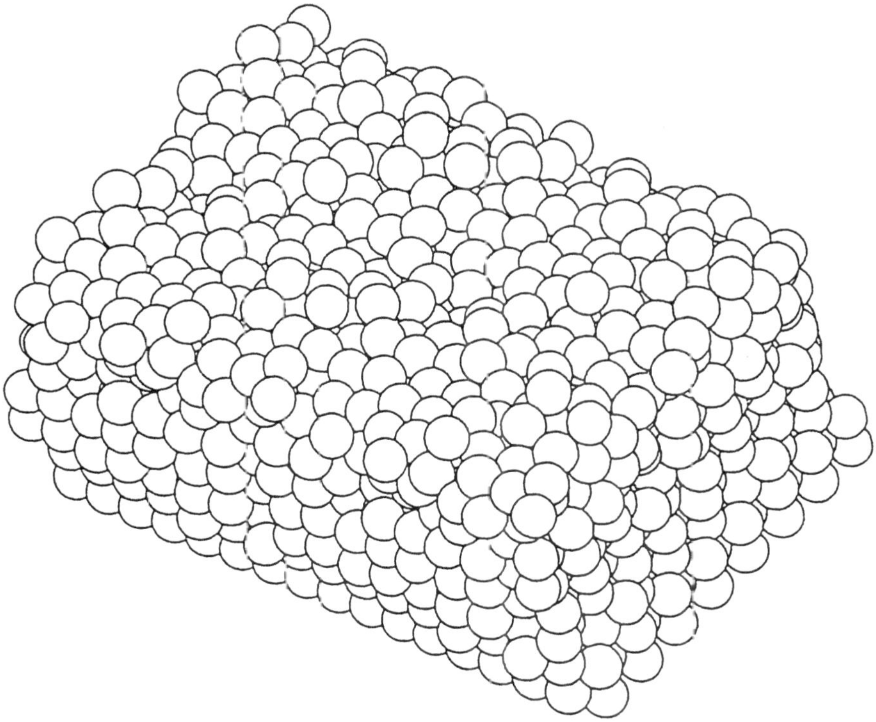

Figure 3: Snapshot of the Al(110) surface at 930 K. At this temperature the surface is more disordered and several adatom-vacancy pairs can be found.

of scattered radiation. In this interpretation $\vec{g}$ is the momentum transfer vector and the intensity should be normalized by the intensity at zero temperature. By application of the kinematic scattering approximation, we will not get the dependence of the spot intensity on $|\vec{g}|^2$ right. However, in the kinematic scattering approximation we will obtain a good approximation to the dependence of the intensity on temperature [31].

The mean free path for electrons is approx 7 Å at 50 - 100 eV [32], corresponding to a penetration of 3.5 Å or 2.5 interlayer spacings. In the calculations of the averaged structure factors, figure 4, we have used a penetration of 2.5 interlayer spacings.

If one were to use the structure factor $|S|^2$ as an order parameter, this order parameter would have the awkward feature that purely harmonic vibration would cause $|S|^2$ to drop below unity. For this choice of order parameter, purely harmonic vibration would have to be considered as a part of the disorder. We may use the calculated Debye temperature to correct this problem by defining another order parameter, $|S'|^2$,

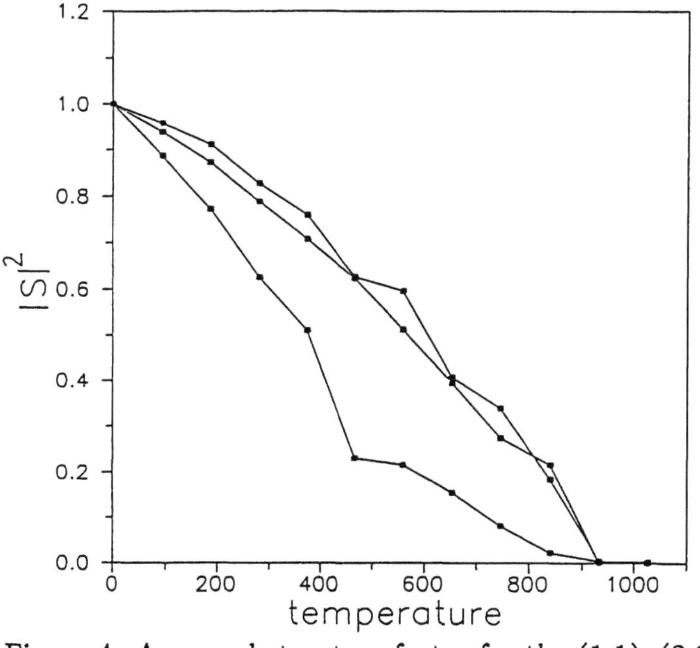

Figure 4: Averaged structure factor for the (1,1), (2,0) and (0,2) peaks as a function of temperature (in K). While $|S_{(1,1)}|^2$ and $|S_{(0,2)}|^2$ have comparable values, $|S_{(2,0)}|^2$ is lower.

$$\ln|S|^2 = \frac{3\hbar g^2 T}{m k_B \theta_D^2} + \ln|S'|^2. \tag{5}$$

For a perfect lattice and for a lattice undergoing harmonic vibrations $|S'|^2$ is unity. The calculated $|S'|^2$ is shown in figure 5. It appears that this choice of order parameter leads us to conclude that the loss of order in the lattice is anisotropic as $|S'_{(2,0)}|^2$ drops significantly faster than $|S'_{(1,1)}|^2$ or $|S'_{(0,2)}|^2$. The same conclusion is reached if we consider $|S'|^2$ calculated for individual layers.

An apparent anisotropy of the onset of the melting has been found from LEED intensity measurements for Pb(110) [8, 9]. If the conclusion based on the order parameter $|S'|^2$ is taken at face value, we are led to the conclusion that the disorder close to the melting point is strongest *along* the close-packed rows of the (110)-surface.

By comparing equation (5) to the results for the Debye-Waller factor for anharmonic lattices [33] it is found that anharmonic vibrations will cause $|S'|^2$ to drop below unity, and further, that this drop depends on $|\vec{g}|^2$. This suggests that one may analyze the order parameter $|S"|^2$ defined by

$$\ln|S|^2 = g^2 \ln|S"|^2. \tag{6}$$

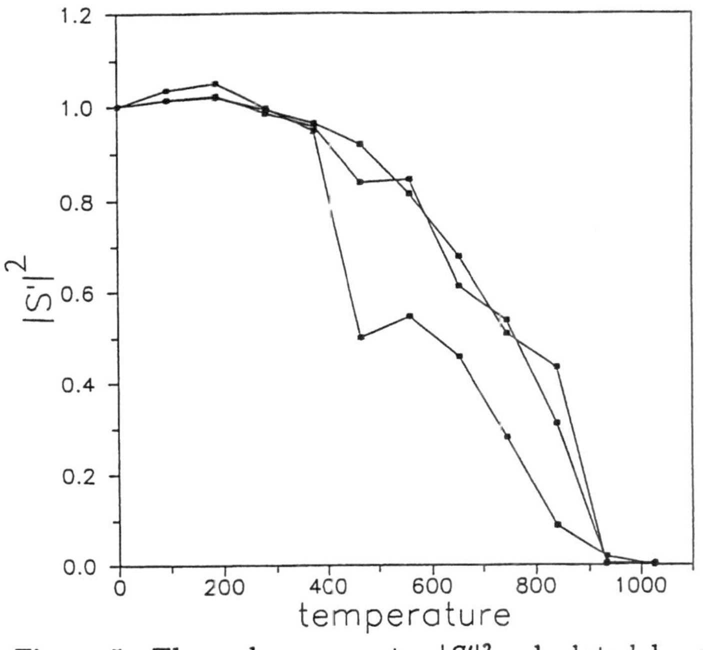

Figure 5: The order parameter $|S'|^2$ calculated by correcting the exponentially averaged structure factor in figure 4 using the Debye temperature, equation (5). As in figure 4 we find that while $|S'_{(1,1)}|^2$ and $|S'_{(0,2)}|^2$ have comparable values, $|S'_{(2,0)}|^2$ is lower.

Vibrations will cause this order parameter to drop below unity and thus contribute to the disorder as measured by $|S"|^2$. However, the leading terms in $|S"|^2$ are independent from $|\vec{g}|^2$. In other words $|S"|^2$ is independent from the momentum transfer $\Delta\vec{p} = \hbar\vec{g}$ made in the *observation* of the disorder by a diffraction technique.

Figure 6 shows the order parameter $|S"|^2$ calculated from the averaged structure factor $|S|^2$ in figure 4. The coincidence of $|S"_{(1,1)}|^2$, $|S"_{(2,0)}|^2$ and $|S"_{(0,2)}|^2$ strongly suggest that the anisotropy observed by using $|S|^2$ (figure 4) or $|S'|^2$ (figure 5) as order parameters may be accounted for by the dependence of $|S|^2$ and $|S'|^2$ on $|\vec{g}|^2$. In other words the anisotropy is not in the atomic structure itself but enters through the anisotropy in the observation. As the unit cell of a FCC (110) surface is rectangular, the Bragg condition implies different values for $|\vec{g}|^2$ along x and y.

In conclusion, we have studied the premelting of Al(110) by Molecular Dynamics simulations. The potential used in the simulations is the potential for Al derived from Effective Medium Theory. From the derivation and from the previous applications of this potential we expect that is should give a

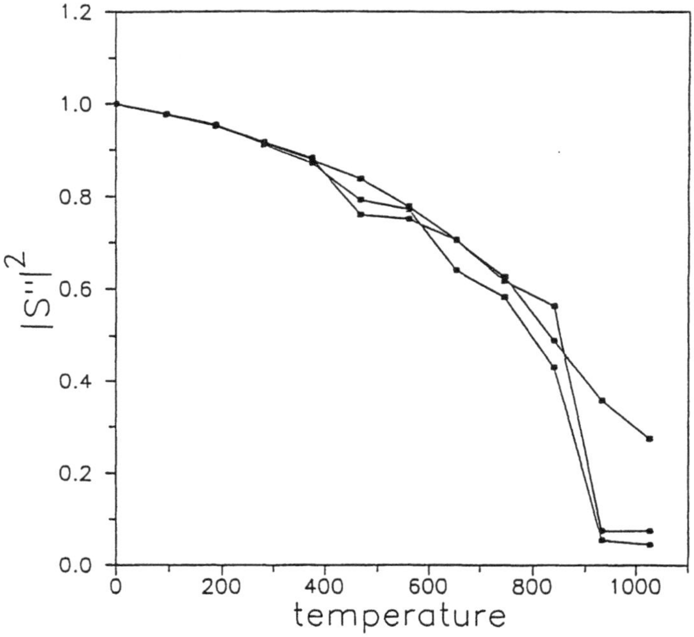

Figure 6: The order parameter $|S"|^2$ calculated by correcting the averaged structure factor in figure 4 by the length of the wave vector $\vec{g}$, equation (6).

good description of metallic systems, while the computational effort remains moderate.

At high temperatures we find evidence for premelting in the form of a progressive disorder of the surface region. This disorder is mediated by the formation of adatom-vacancy pairs. The correlation of the onset of the disorder with the formation of adatom-vacancy pairs is not inconsistent with Tammann's hypothesis on the importance of surfaces for bulk melting.

Some structure factors have been calculated for the layers as well as averaged data using an effective penetration depth relevant for comparison with LEED experiments. From the averaged structure factors we have calculated two other order parameters. The order parameter $|S'|^2$ is $|S|^2$ corrected for harmonic vibrations, equation (5). If $|S|^2$ or $|S'|^2$ are used as measures of the order of the surface region, the onset of the disorder appears to be anisotropic. Inspired by the theory of the Debye-Waller factor for anharmonic lattices we have calculated the order parameter $|S"|^2$ by correcting $|S|^2$ for the dependence on the length of the wave vector, equation (6). If $|S"|^2$ is used as a measure of the order of the surface region, the onset of the disordering appears isotropic. This strongly suggests that the apparent anisotropy in the disordering of the surface region is caused by the *detection* of the disorder, experimentally by a diffraction, numerically by Fourier transformation.

The discussions I have had with Jens K. Nørskov, Uzi Landman and Karsten W. Jacobsen on topics related to this work are gratefully acknowledged.

References

[1] J. W. M. Frenken and J. F. van der Veen, Phys. Rev. Lett. **54**, 134 (1985).

[2] J. W. M. Frenken, P. M. J. Marée and J. F. van der Veen, Phys. Rev. B **34**, 7506 (1986)

[3] B. Puis, A. W. Denier van der Gon, J. W. M. Frenken and J. F. van der Veen, Phys. Rev. Lett. **59**, 2678 (1987).

[4] B. Puis, J. W. M. Frenken and J. F. van der Veen, Phys. Scr. **T19**, 382 (1987)

[5] J. F. van der Veen, B. Puis and A. W. Denier van der Gon, in "Chemistry and Physics of Solid Surfaces" 7 (1988)

[6] J. W. M. Frenken, J. P. Toennies and Ch. Wöll, Phys. Rev. Lett. **60**, 1727 (1988).

[7] P. von Blankenhagen, W. Schommers and V. Vogele, J. Vac. Sci. Technol. A **5**, 649 (1987)

[8] K. C. Prince, U. Breuer and H. P. Bonzel, Phys. Rev. Lett. **60**, 1146 (1988)

[9] R. Lipowsky, U. Breuer, K. C. Prince and H. P. Bonzel, Phys. Rev. Lett. **62**, 913 (1989)

[10] P. H. Fuoss, L. J. Norton and S. Brennan, Phys. Rev. Lett. **60**, 2046 (1988)

[11] R. Lipowsky, Phys. Rev. Lett. **49**, 1575 (1982)

[12] H. Nakanishi and M. E. Fisher, Phys. Rev. Lett. **49**, 1565 (1982)

[13] R. Lipowsky and W. Speth, Phys. Rev. B **28**, 2983 (1983)

[14] R. Lipowsky, Z. Phys. B **55**, 345 (1984)

[15] R. Lipowsky, Ferroelectrics **73**, 69 (1987).

[16] J. M. Ziman, Models of Disorder. Cambridge University Press (1979).

[17] J. Q. Broughton and G. H. Gilmer, J. Chem. Phys. **79**, 5095 (1983)

[18] J. Q. Broughton and G. H. Gilmer, J. Chem. Phys. **79**, 5105 (1983)

[19] J. Q. Broughton and G. H. Gilmer, J. Chem. Phys. **79**, 5119 (1983)

[20] V. Rosato, G. Ciccotti and V. Pontikis, Phys. Rev. B **33**, 1860 (1986)

[21] V. Pontikis and P. Sindzingre, Phys. Scr. **T19**, 375 (1987).

[22] U. Landman *et al.* (unpublished).

[23] K. W. Jacobsen, J. K. Nørskov and M. J. Puska: Phys. Rev. B **35**, 7423 (1987)

[24] K. W. Jacobsen, Comments Condens. Matter Phys. **14**, 129 (1988)

[25] P. Stoltze, K. W. Jacobsen and J. K. Nørskov: Phys. Rev. B **36**, 5035 (1987)

[26] P. Stoltze, J. K. Nørskov and U. Landman, Phys. Rev. Lett. **61**, 440 (1988)

[27] P. Stoltze, J. K. Nørskov and U. Landman: Surf. Sci. (in press).

[28] P. Stoltze: Simulations of the premelting of Al(110) (submitted).

[29] D. W. Heerman, Computer Simulation Methods in Theoretical Physics. Springer (1986).

[30] J. M. Ziman, The principles of the theory of solids. Cambridge University Press (1965)

[31] D. P. Woodruff and T. A. Delchar, Modern Surface Science Techniques. Cambridge University Press (1986).

[32] A. Zangwill, Physics at Surfaces. Cambridge University Press (1988)

[33] A. A. Maradudin and P. A. Flinn, Phys. Rev. **129**, 2529 (1963).

Hydrogen Diffusion on Metal Surfaces – Nonadiabatic Effects

G. Wahnström

Institute of Theoretical Physics, Chalmers University of Technology,
S-41296 Göteborg, Sweden

Abstract. Recent progress in deriving adiabatic potential energy surfaces for complex many-atom systems has greatly improved our understanding of their structure and ground-state properties. The same information can be used as realistic input for the interatomic forces in molecular dynamic simulations. However, the application to metals may pose additional problems due to the presence of conduction electrons. The coupling between the motion of a light particle, as e.g. hydrogen, and the lattice vibrations in a metal is weak and electronic excitations among the conduction electrons can become important in providing an additional mechanism for energy transfer. Here, it is found from numerical simulations that for H diffusing on the Ni (001)-surface at T=250 K it is important to include electron-hole pair excitations in order to characterize correctly the diffusive motion.

1. Introduction

Several technologically important processes involve interaction between hydrogen and metals, as e.g. heterogeneous catalysis [1] and hydrogen storage in bulk [2]. A thorough understanding of these processes must include a careful description of both static and dynamic aspects of the interaction. Here, we concentrate on the dynamics of the hydrogen motion and in particular the possibility of nonadiabatic effects. Hydrogen is light and therefore the coupling to the lattice vibrations in the metal is weak. Electronic excitations can become important in providing an additional and maybe effective mechanism for energy dissipation [3]. This implies nonadiabatic changes in the electronic structure and one has to go beyond the adiabatic approximation in dynamic simulations [4].

1.1 The adiabatic approximation

The calculation of adiabatic potential energy surfaces is based on the Born-Oppenheimer approximation for separating nuclear and electronic degrees of freedom. The nuclei, with coordinates $\{\mathbf{R}_i\}$, are assumed to move in the potential $V_0(\{\mathbf{R}_i\})$ set up by the instantaneous ground-state of the electron system. If we assume classical mechanics for the nuclear motion the dynamics is obtained by solving Newton's equation of motion,

$$m_i \ddot{\mathbf{R}}_i = -\nabla_i V_0(\{\mathbf{R}_i\}), \tag{1}$$

Springer Proceedings in Physics, Vol. 48 — Many-Atom Interactions in Solids
Editors: R.M. Nieminen · M.J. Puska · M.J. Manninen

where m_i is the mass of particle (nucleus) i. The multi-dimensional function $V_0(\{\mathbf{R}_i\})$ is the ground-state adiabatic potential energy surface. The use of a single adiabatic potential energy surface is based on the assumption that the energy difference,

$$\Delta E_n(\{\mathbf{R}_i\}) = E_n(\{\mathbf{R}_i\}) - E_0(\{\mathbf{R}_i\}), \tag{2}$$

between an electronically excited state and the ground-state for a given set of nuclear coordinates is large compared with typical energies characterizing the nuclear motion. This is usually true for small systems and for insulators. For metals, however, there is a quasi-continuum of excited states, electron-hole pair excitations, ranging from zero to several electron volts. The energy difference in Eq. (2) can become infinitesimal and the use of a single adiabatic potential energy surface can not be justified.

1.2 Nonadiabatic effects

The nonadiabatic effect that we are going to consider is the possibility of direct excitation or de-excitation of electron-hole pairs in the metal by the moving hydrogen atom. We treat systems in thermal equilibrium around room temperature. The time-scale for the electronic motion is very fast compared with the time-scale for the hydrogen motion. The electrons will adjust themselves very rapidly, but not strictly instantaneously, to the motion of the hydrogen atom. In this limit of near adiabaticity the inclusion of electron-hole pair excitations into dynamic simulations is straightforward, provided the hydrogen motion is treated classically. One simply has to add a friction term and a stochastic force with white noise spectrum to the equation of motion. Eq. (1) for the hydrogen atom is modified to [5]

$$m_H \ddot{\mathbf{R}}_H = -\nabla_H V_0(\{\mathbf{R}_i\}) - m_H \overleftrightarrow{\eta}(\{\mathbf{R}_i\}) \cdot \dot{\mathbf{R}}_H + \mathbf{F}^{st}(t)\,, \tag{3}$$

where the mass of the hydrogen atom is denoted by m_H and its position by $\mathbf{R}_H$. The positions of all atoms, including the hydrogen atom, are denoted collectively by $\{\mathbf{R}_i\}$. The friction tensor $\eta(\{\mathbf{R}_i\})$ is an electronic property with its only dependence on the atoms, the nuclei, through their instantaneous configuration $\{\mathbf{R}_i\}$. It is related to the stochastic force $\mathbf{F}^{st}(t)$ through the fluctuation-dissipation theorem [5],

$$< \mathbf{F}^{st}(t) \cdot \mathbf{F}^{st}(t') > = 2\, m_H \overleftrightarrow{\eta}(\{\mathbf{R}_i\})\, k_B T\, \delta(t-t')\,, \tag{4}$$

where T is the temperature, k_B Boltzmann's constant and <...> denotes a thermal average. Nonadiabatic effects also influence the damping of the lattice vibrations but we will neglect that effect and assume that the equations of motion for the different metal atoms are unchanged and given by Eq. (1).

Eq. (3) can be derived formally [6,7], starting from the basic equation of motion, and a microscopic expression for the friction tensor can be given. Two main assumptions are involved. First is the use of classical mechanics for motion of the atoms. This is justified for heavy particles, but hydrogen may exhibit important quantum effects even at room temperature. Some comments on possible quantum effects will be made in connection to the numerical calculations in Sec. 2.3.

The second approximation is the use of a simple friction description for the influence of the electronic degrees of freedom. This is based on the assumption that the electrons relax to a local equilibrium, the adiabatic configuration, on a time scale much shorter than that for significant hydrogen motion. Trouble could possibly arise if two locally distinct configurations of the electrons become degenerate in energy. It is then not clear if the electrons, if disturbed, will relax rapidly to a well defined local equilibrium, since the system can easily change between two distinct electronic configurations.

We have also assumed that the force field experienced by the hydrogen atom is unchanged when including the electronic excitations. This is very likely to be true for the long wave-length fluctuations introduced through the friction description. However, if locally distinct electronic configurations are relevant this will not be true, since the force on the hydrogen atom can be quite different depending on the electronic configuration.

1.3 The friction coefficient

Several authors have derived microscopic expressions for the friction coefficient [6,8-12]. We are interested in situations where there is a substantial overlap between the electronic wavefunctions of the hydrogen atom and the metal atoms, as e.g. in chemisorption. This implies that the electron more or less localized around the proton and the conduction electrons should be treated on the same footing. In this case the friction coefficient can be expressed in terms of the density-density correlation function for the electronic motion [8b]

$$\overset{\leftrightarrow}{\eta}(\{R_i\}) = \frac{1}{m_H k_B T} \times \mathrm{Re}\left\{ \int dr \int dr' \, \nabla_H V(R_H, r) \cdot \int_0^\infty dt < \delta n(r,0)\, \delta n(r',t) > \cdot \nabla_H V(R_H, r') \right\}, \qquad (5)$$

where $\delta n(r,t)$ is the fluctuating part of the density operator for the electrons relative to the adiabatic configuration, $V(R_H,r)$ is the Coulomb interaction between the proton and the electrons, and Re{...} denotes the real part. The density-density correlation function should be evaluated in the presence of a fixed configuration of nuclei with the proton located at $\mathbf{R}_H$. An estimate of the order of magnitude of this expression leads to [8b]

$$\hbar\eta \sim \frac{m_e}{m_H} \varepsilon_F, \qquad (6)$$

where m_e is the electron mass and ε_F the Fermi energy.

A proper evaluation of the correlation function in Eq. (5) is very difficult to do from the first principle. It should include all important many-body effects among the electrons in the presence of a fixed configuration of nuclei. To simplify one could first try to evaluate the friction coefficient for a proton in a homogeneous electron-gas. From symmetry considerations follow that the friction coefficient in this case is diagonal in its Cartesian coordinates and all components are equal. It is also independent of the location of the proton.

The density functional theory provides a one-electron scheme for calculating ground-state properties. It is not formally justified to use it in calculating the friction coefficient, which is of dynamic origin. However, arguments in favour of such an extension have been presented [13]. In an one-electron scheme the friction coefficient can be expressed in terms of the phase-shifts $\delta_l(\varepsilon_F)$ of the l:th partial wave at the Fermi-surface [8b];

$$\hbar\eta = \frac{m_e}{m_H}\,\varepsilon_F\,\frac{8}{3\pi}\sum_{l=0}^{\infty}(l+1)\sin^2[\delta_l(\varepsilon_F)-\delta_{l+1}(\varepsilon_F)]\,. \qquad (7)$$

Puska and Nieminen [14] have calculated these phase-shifts using the density functional theory within the local density approximation. In Table I we give some values for $\hbar\eta$ of electron-densities appropriate for metals, taken from ref. [14], together with the estimate obtained from Eq. (6). These values could be used when simulating hydrogen motion in bulk metals.

Table I. The friction coefficient η for a hydrogen atom moving in a homogeneous electron gas, taken from ref. [14] together with the estimate obtained from Eq. (6). The electron density is given by $\rho=3/4\pi r_s^3$ and a_0 is the Bohr radius.

r_s/a_0	ε_F (eV)	$\frac{m_e}{m_H}\varepsilon_F$ (meV)	$\hbar\eta$ (meV)
1.5	22.3	12.1	4.59
2.0	12.5	6.8	3.99
3.0	5.6	3.0	2.42
4.0	3.1	1.7	1.45
5.0	2.0	1.1	0.87

At the surface the electron-density is reduced compared with the bulk value and it is inhomogeneous. That implies that the friction coefficient has spatial as well as directional dependences, but these will be disregarded in the numerical simulations presented in Sec. 2.2. Similar calculations as in ref. [14] have been reported for the inhomogeneous case using free-electron like surfaces [10,15]. These calculations show that a reasonable value for the friction coefficient is obtained by using the results from the homogeneous case but with the electron-density equal to the local density at the location of the proton. The chemisorbed hydrogen atom is moving rather closely to the surface and for nickel (r_s/a_0=2.0) we will use the value $\hbar\eta$=2 meV.

1.4 Numerical simulation

Based on Eq. (3), together with Eq. (1) for the metal atoms, the dynamics of hydrogen interacting with metals can be simulated, including low-lying electron-hole pair excitations. The adiabatic potential energy surface $V_0(\{R_i\})$ and the friction tensor $\eta(\{\mathbf{R}_i\})$ are the two quantities that are needed as input. Standard numerical algorithms [16] can be used for the time-evolution and the stochastic force can be obtained from a random number generator [16]. The stochastic force has a Gaussian distribution [16], with its mean equal to zero and the variance given by Eq. (4). In order to obtain proper statistics one has to repeat the calculation with different set of stochastic forces.

2. Numerical example - hydrogen diffusion on a metal surface

Recent measurements using laser induced desorption (LID) have allowed the extraction of hydrogen self-diffusion coefficients on clean metal surfaces [17]. In the case of hydrogen on the Ni(001)-surface the experimental results for temperatures between 200 K and 300 K are explained in terms of uncorrelated jumps between adjacent lattice sites with the activation energy and the prefactor being equal to $\Delta E = 0.17$ eV [18] ($\Delta E = 0.15$ eV [19]) and $D_0 = 4.5 \cdot 10^{-3}$ cm^2/s [18] ($D_0 = 2.5 \cdot 10^{-3}$ cm^2/s [19]), respectively. The LID-experiments do not reveal the microscopic details of the diffusive motion. Since the spot heated by the laser is of the order 0.25 mm^2 and the time delay being 1-10 minutes, it is very difficult to determine conclusively the basic mechanism for the diffusive motion. Computer simulations on realistic model systems can therefore be of great value; e.g. the relative role of phonons and electron-hole pair excitations for energy transfer can be studied.

2.1 Dynamic correction factor

The high potential barriers for diffusion act as bottlenecks for the motion along the surface. Most of the time the hydrogen atom is located and vibrating at a particular surface lattice site and only very rarely does it make a jump and change site. Direct molecular dynamics simulation of rare events is time consuming and in practice often unfeasible. For these kind of systems an elegant alternative approach has been developed [20-23]. In cases with high potential barriers for diffusion the slow diffusive motion can be expressed in terms of rate constants for jumps between different sites [23a]. Denote the rate by which the hydrogen atom jumps from site i to site j with $k_{i \to j}$. These sites need not to be adjacent in configuration space. For a system with cubic symmetry the self-diffusion coefficient D is related to the rate constants through [23a]

$$D = \frac{1}{2d} \sum_{j(\neq i)} k_{i \to j} \, l_{ij}^2 \,, \tag{8}$$

where d is the dimensionality associated with the diffusion process and l_{ij} is the distance between site i and j.

The key observation is that for a system in thermal equilibrium the rate constants $k_{i\to j}$ can be factorized rigorously into two parts [20,23a];

$$k_{i\to j} = k^{TST}_{i\to} \, f_d(i\to j)\,. \tag{9}$$

One equilibrium factor $k^{TST}_{i\to}$, defined as the flux out through all surfaces bordering site i, the dividing surfaces, and one dynamic factor $f_d(i\to j)$, which accounts both for re-crossings of the dividing surface and for correlated multiple jumps.

The factorization of the rate into an equilibrium factor and a dynamic factor is not unique and it depends on the choice of the dividing surface. Only under the condition that the dividing surface is located in a region where the equilibrium mean density for the hydrogen atom is low, the factorization is likely to be useful. The rare event nature of the process is then included in the first factor, the rate constant within the transition state theory (TST) [24], and it does not require any knowledge of the dynamics. It can be calculated with chosen accuracy for an arbitary potential energy surface using the Monte Carlo technique [25] or by other related methods [21,26].

The second factor, the dynamic correction factor, is computed from trajectories initiated at the dividing surfaces and followed only for relatively short times. It is the fact that these trajectories have to be followed only of the order of a pico-second that makes it possible to determine the self-diffusion coefficient using the molecular dynamics technique. This time should be compared with the mean residence time at a site which is found experimentally to be of the order 100 ps for H on the Ni (001)-surface [18,19].

2.2 Numerical results for H on the Ni (001)-surface [27]

We focus the attention on the character of the diffusive motion and only results relevant for the dynamic correction factor will be presented. In all calculations the temperature is 250 K. The dividing surfaces for one surface lattice site on the Ni (001)-surface are defined by connecting the four nearest substrate atoms with planes which extend to infinity in the direction perpendicular to the substrate surface. These planes are defined with respect to the undistorted equilibrium positions of the substrate atoms which implies that only the adatom coordinates determine the site the system is located in.

The key quantity in determining the dynamic correction factor is the time-dependence of the population in different sites resulting from trajectories initiated from the same dividing surface and in the same direction. We denote this function by $R(I\to j)$, where I indicates the dividing surface (see Fig. 1) and j denotes the different lattice sites (see Fig. 1).

If transition state theory (TST) for the escape rate together with the assumption of uncorrelated jumps is correct, then all trajectories will thermalize in the sites towards which they are initially traveling. $R(I\to j)$ is then equal to 1 for j=1 and equal to zero for all other sites, for times much less than the mean residence time. The self-diffusion coefficient can then be written as,

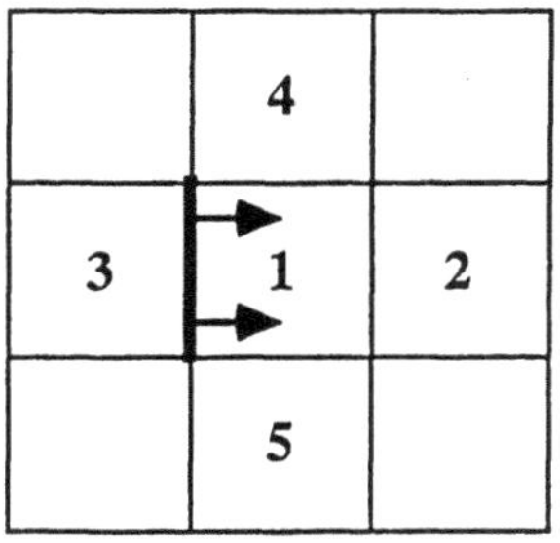

Fig. 1. The two arrows show the initial direction for the trajectories used in calculating $R(I \rightarrow j)$. The location of the different surface lattice sites j=1, 2, 3, 4, and 5 is also shown.

$$D^{TST} = \frac{1}{2d} k_{i\rightarrow}^{TST} l^2, \tag{10}$$

where l is the distance between adjacent surface lattice sites.

First, we study the effect of energy transfer to the lattice vibrations, the phonons, by solving Eq. (1). In Fig. 2 the result from 1000 different trajectories is shown. The line marked by $m_a=m_H$ shows the result for a hydrogen adatom. The mass ratio between the hydrogen atom and the nickel atoms is small, $m_H/m_{Ni}=0.017$, and the energy transfer is very inefficient. This is clearly seen in Fig. 2. After 1 ps only 10 % of the adatoms are still in lattice site 1. In the same figure the results with heavier adatoms are also shown, but with identical potential energy surfaces. When the mass of the adatom is equal to the mass of

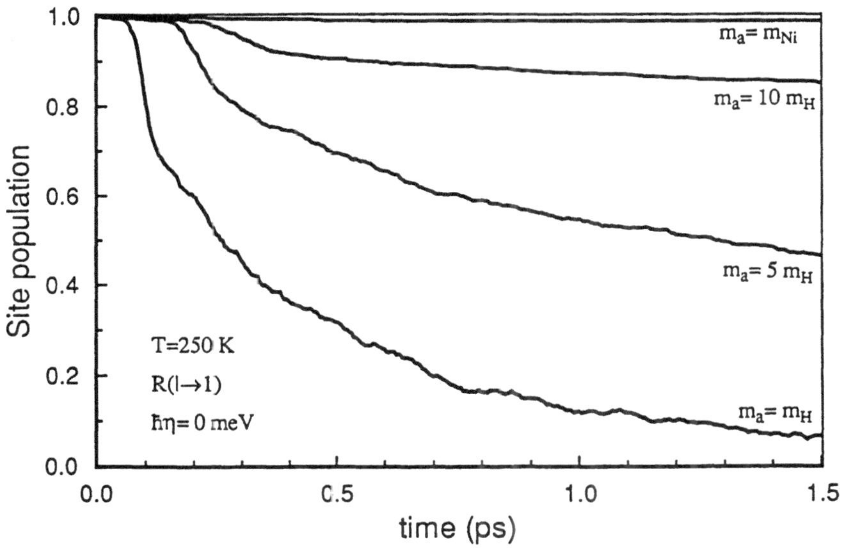

Fig. 2. The time-dependence of the population in site 1 $R(I \rightarrow 1)$, resulting from 1000 trajectories, with coupling only to the lattice vibrations. All trajectories are initiated at the dividing surface shown as a bold line in Fig. 1 and in the direction given by the arrows in the same figure.

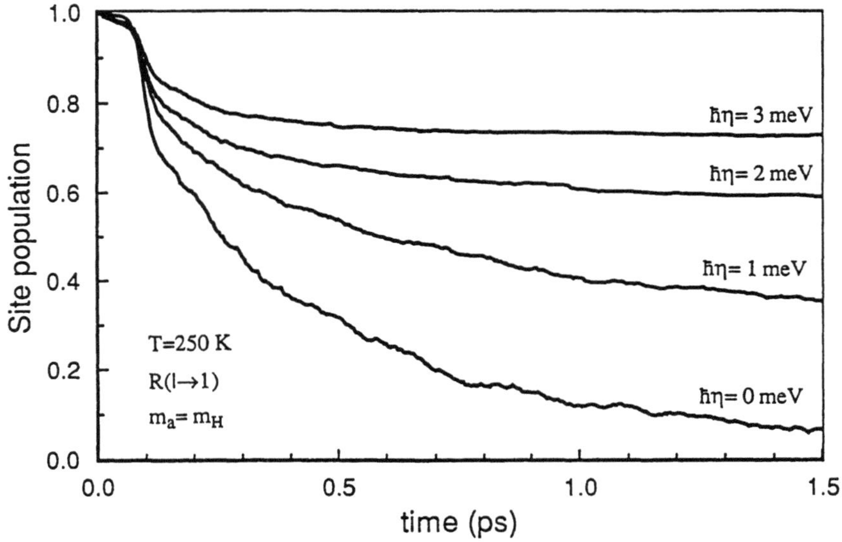

Fig. 3. The same as in Fig. 2, including coupling to the electron-hole pair excitations.

the substrate atoms, $m_a = m_{Ni}$, the coupling to the lattice vibrations is sufficiently strong to inhibit effectively correlated jumps. This finding is in accordance with both FIM experiments [29,30] and with previous computer simulations [23a]. The key result here is that if the mass ratio between the adatom and the substrate atoms is small the adatoms which initially are located at the dividing surface are moving rather freely, though somewhat erratically. Energy transfer to the vibrational degrees of freedom of the lattice is very inefficient and there is no tendency to thermalization within the time $t<1.5$ ps.

Next, we add coupling to electron-hole pair excitations by adding a friction term and a stochasic force to the equation of motion for the hydrogen atom, according to Eq. (3). In Fig. 3 the results with three different values for the friction coefficient are shown. The value $\hbar\eta=2$ meV is supposed to be most appropriate for hydrogen on a nickel surface, and the values $\hbar\eta=1$ meV and $\hbar\eta=3$ meV are reasonable lower and upper limits, respectively [28]. Using $\hbar\eta=2$ meV about 60 % of the adatoms get stuck within 1 ps in lattice site 1, the site towards which they initially are traveling. Based on this result we claim that energy dissipation to electron-hole pairs is more important than to phonons in the case of hydrogen diffusion on the Ni (001)-surface. It also shows that for this system one has to go beyond the Born-Oppenheimer approximation and include the nonadiabatic coupling to electron-hole pairs in order to characterize correctly the diffusive motion.

In the calculation the populations in thirty different sites are followed, five are shown in Fig. 4. They all reach constant values and the thermal rate constants $k_{i \to j}$ are well-defined. The TST value for the escape rate is in error due to recrossings of the dividing surface (see R(I→3) in Fig. 4). The main reason for this

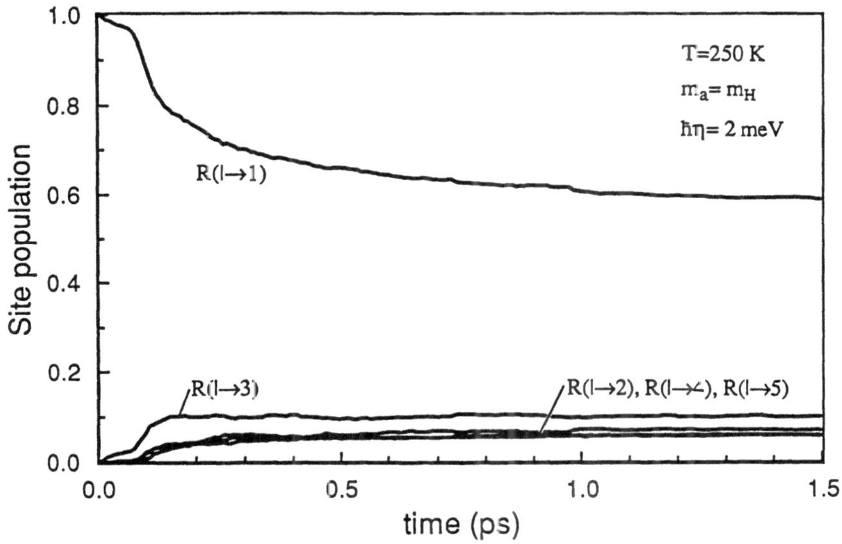

Fig. 4. The same as in Fig. 3, showing the population in five different sites, defined in Fig. 1.

is not the presence of fluctuating forces but merely the strong coupling between the perpendicular and parallel degrees of freedom for the motion of the hydrogen atom, introduced by the shape of the potential energy function [31,32]. Recrossing events diminish the value of the escape rate with a factor of 2 compared with the TST-value. Other trajectories (see R(I→2), R(I→4) and R(I→5) in Fig. 4) contribute to correlated jumps in the forward direction, which increase the value of the self-diffusion coefficient. By taking these two effects together a small deviation from the TST-result D^{TST}, defined in Eq. (10), is obtained, $D/D^{TST} = 0.8$. These findings for the dynamic correction factor are consistent with available experimental results. The lack of accurate potential energy surfaces and the insufficient experimental information on the short time behaviour make a more detailed comparison somewhat premature.

The need of an accurate description of the potential energy surface becomes particularly urgent in a proper evaluation of D^{TST}, which is a separate issue. Here, quantum effects are important, which is stressed in the next subsection.

2.3 Quantum effects

A basic assumption is that the motion of the atoms, and in particular the hydrogen atom, is treated classically. The complexity of a full time-dependent quantum simulation is immense. The much simpler classical trajectory approach can be of great value in cases where it is reasonable to assume that it provides an appropriate description of the dynamics. Hydrogen is light, however, and even at room temperature quantum effects can become important.

The most important quantum effect is the one associated with the discrete nature of the energy levels, including the zero-point motion, which influences static features. The vibrational frequency for hydrogen adsorbed at the center position on the Ni (001)-surface is experimentally found to be $\hbar\omega_0$=66 meV [33]. That corresponds to the temperature $\hbar\omega_0/k_B$=766 K and quantum effects will be important at T=250 K. In diffusion one can correct for this approximately by calculating the equilibrium factor in Eq. (9) quantum mechanically [24]. For the present model potential this results in an increase of the diffusion rate by about a factor 5 at T=250 K [28b].

Another quantum effect, tunneling, is important for the dynamics. There is no unique definition of tunneling in a multi-dimensional system. At sufficiently high temperatures the diffusion rate will be dominated by adatoms having energies around or above the barrier top. It is then appropriate to expand the potential energy function around the barrier top, the bridge position for the Ni (001)-surface. The contribution from tunneling and non-classical reflexion can be estimated by determining the conventional imaginary frequency $i\omega^{\#}$, given by the unbound degree of freedom at the saddle point [24]. In our case it is equal $\hbar\omega^{\#}$=28 meV ($\hbar\omega^{\#}/k_B$=321K) and by using Wigner's expression [24] for the tunneling correction factor we find an increase of the diffusion rate by only 7 % at T=250 K [28a].

There are more subtle quantum effects, such as interference. At finite temperatures it is most likely that these effects are washed out by the thermal motion of the metal atoms [28b]. In conclusion, we argue that classical mechanics is a reasonable approximation for the hydrogen motion when evaluating the dynamic correction factor at T=250 K and with the present model potential.

3. Discussion

Molecular dynamics is a powerful tool in the investigation of different dynamic phenomena in complex many-atom systems. As input interatomic forces are needed and in recent years considerable interest has been directed to developments of realistic adiabatic potential energy surfaces for metallic systems [34]. The electrons are crucial for the functional form of these potential energy surfaces but in dynamical simulations they do not explicitly appear. The electrons are assumed to follow adiabatically the slow motion of the nuclei. However, the presence of low-energetic electron-hole pair excitations among the conduction electrons in a metal introduces a nonadiabatic effect that may be important even when the nuclei are moving very slowly. There is no energy gap in the excitation spectrum and the use of a single adiabatic potential energy surface can not be justified.

In the limit of near adiabaticity this effect can be included into classical dynamical simulations by simply adding a friction term and a stochastic force with white noise spectrum to Newton's equation of motion for the nuclei. A key issue is the strength of the nonadiabatic coupling which promotes transitions between nearby potential energy surfaces. Here, we rely on that the calculations for free-electron like systems based on the density-functional scheme give reasonable values.

We have discussed one situation, hydrogen diffusion on a metal surface, where we claim that it is important to include electron-hole pair excitations in order to characterize correctly the diffusive motion. A key fact is the large difference in mass between hydrogen and e.g. nickel. This makes the coupling between the hydrogen motion and the lattice vibrations in the metal very weak. It is also important that hydrogen is chemisorbed on nickel. There is a substantial overlap between the electronic wavefunctions of the hydrogen atom and the substrate atoms which makes the coupling to electron-hole pairs much stronger than in physisorption.

Electronic excitations can become important also for hydrogen diffusing in bulk. Recently a careful molecular dynamic study has been performed for hydrogen in palladium, based on simple pair-potentials [35]. Most of the experimental results are reproduced quite well except for one point; the simulation data show a too large correlation effect between successive jumps of the hydrogen atom. This is found by comparing the simulation results with neutron scattering data for the wave-vector dependence of the quasi-elastic peak in the incoherent scattering function, which is a sensitive measure of the character of the diffusive motion. The most likely explanation for this discrepancy is the lack of coupling to electron-hole pair excitations in the simulation [35].

Acknowledgments. The author would like to thank Mike Gillan, Bo Hellsing, Bengt Lundqvist and Alf Sjölander for valuable discussions.

References

1. *Hydrogen Effects in Catalysis,* edited by Z. Paal and P. G. Menon (Dekker, New York, 1988).
2. *Hydrogen in Metals II,* edited by G. Alefeld and J. Völkl (Springer, Berlin, 1978).
3. K. Schönhammer and O. Gunnarsson, in *Many-Body Phenomena at Surfaces,* edited by D. Langreth and H. Suhl (Academic Press, London, 1984) p. 421; B. I. Lundqvist, *ibid.* p. 453.
4. J. C. Tully, Ann. Rev. Phys. Chem. **31**, 319 (1980).
5. D. Forster, *Hydrodynamic Fluctuations, Broken Symmetry, and Correlation Functions,* (W. A. Benjamin, Reading, Massachusetts, 1975) p. 121.
6. W. L. Schaich, J. Chem. Phys. **60,** 1087 (1974); see also ref. [5] for a pedagogical review of Brownian motion theory.
7. In ref. [6] the Fokker-Planck equation for the distribution function is derived. The corresponding equation of motion is Eq. (3), the Langevin's equation. For more details see ref. [5].
8. (a) E. G. d'Agliano, W. L. Schaich, P. Kumar, and H. Suhl, in *Collective properties of physical systems,* edited by Bengt Lundqvist and Stig Lundqvist (Academic Press, New York, 1973) p. 200; (b) E. G. d'Agliano, P. Kumar, W. Schaich and H. Suhl, Phys. Rev. **B11**, 2122 (1975).
9. A. Blandin, D. Hone and A. Nourtier, J. Physique Lett. **36**, L-109 (1975); A. Blandin, A. Nourtier and D. W. Hone, J. Physique **37**, 369 (1976).
10. B. Hellsing and M. Persson, Physica Scripta **29**, 360 (1984).
11. P. Minnhagen, J. Phys. C**15**, 2293 (1982).
12. F. Sols, F. Flores, and N. García, Surf. Sci. **137**, 167 (1984).
13. V. Peuckert, J. Phys. C11, 4945 (1978); see also the appendix in ref. [10].

14. M. J. Puska and R. M. Nieminen, Phys. Rev. B27, 6121 (1983).
15. M. Persson and B. Hellsing, Phys. Rev. Lett. **49**, 662 (1982); B. Hellsing, M. Persson and B. I. Lundqvist, Surf. Sci. **126**, 147 (1983).
16. D. W. Heermann, *Computer Simulation Methods in Theoretical Physics*, (Springer, Berlin, 1986).
17. S. M. George, in *Diffusion at Interfaces*, edited by M. Grunze, H.-J. Kreuzer, and J. J. Weimer (Springer, Berlin, 1988); E. G. Seebauer, A. C. F. Kong and, L. D. Schmidt, J. Chem. Phys. **88**, 6597 (1988), and references therein.
18. S. M. George, A. M. DeSantolo, and R. B. Hall, Surf. Sci. **159**, L425 (1985).
19. D. R. Mullins, B. Roop, S. A. Costello, and J. M. White, Surf. Sci. **186**, 67 (1987).
20. J. C. Keck, Adv. Chem. Phys. **13**, 85 (1967).
21. C. H. Bennett, in *Diffusion in Solids: Recent Developments*, edited by A. S. Nowick and J. J. Burton (Academic Press, New York, 1975). C. H. Bennett, in *Algorithms for Chemical Computations*, edited by R. E. Christoffersen (American Chemical Society, Washington D. C., 1977).
22. D. Chandler, J. Chem Phys. **68**, 2959 (1978).
23. (a) A. F. Voter and J. D. Doll, J. Chem. Phys. **82**, 80 (1985); J. D. Doll and A. F. Voter, Ann. Rev. Phys. Chem. **38**, 413 (1987); (b) A. F. Voter, J. D. Doll, and J. M. Cohen, J. Chem. Phys. **90**, 2045 (1989); A. F. Voter, Phys. Rev. Lett. **63**, 167 (1989).
24. P. Pechukas, in *Dynamics of Molecular Collisions, Part B*, edited by W. H. Miller (Plenum Press, New York, 1976) p. 269.
25. A. F. Voter, J. Chem. Phys. **82**, 1890 (1985).
26. M. J. Gillan, J. H. Harding, and R.-J. Tarento, J. Phys. C20, 2331 (1987).
27. For more details see ref. [28].
28. (a) G. Wahnström, Chem. Phys. Lett. **163**, 401 (1989); (b) G. Wahnström, unpublished.
29. T. T. Tsong, Prog. Surf. Sci. **10**, 165 (1980).
30. S. C. Wang and G. Ehrlich, Surf. Sci. **206**, 451 (1988).
31. G. Wahnström, K. Haug, and H. Metiu, Chem. Phys. Lett. **148**, 158 (1988); G. Wahnström, J. Chem. Phys. **89**, 6996 (1988); K. Haug, G. Wahnström, and H. Metiu, J. Chem. Phys. **90**, 540 (1989).
32. V. P. Zhdanov, Surf. Sci. **214**, 289 (1989).
33. P.-Å. Karlsson, A.-S. Mårtensson, S. Andersson, and P. Nordlander, Surf. Sci. **175**, L759 (1986).
34. K. W. Jacobsen, proceedings of this workshop; M. Daw, *ibid.*
35. M. J. Gillan, J. Phys. C19, 6169 (1986).

Molecular Dynamics Simulations for Molecular Beam Epitaxy

K. Terakura [1], *K. Hara* [2], *M. Ikeda* [2], *M. Nomura* [2], *M. Mikami* [3], *and T. Oguchi* [4]

[1] Institute for Solid State Physics, University of Tokyo, 7-22-1 Roppongi, Minato-ku, Tokyo 106, Japan
[2] Fujitsu Laboratories Ltd., 10-1, Morinosato-Wakamiya, Atsugi-shi, Kanagawa 243-01, Japan
[3] Fujitsu Ltd., Scientific System Department, System Engineering Group, 1-17-25 Shinkamata, Ohta-ku, Tokyo 144, Japan
[4] National Research Institute for Metals, 2-3-12 Nakameguro, Meguro-ku, Tokyo 153, Japan

Abstract. Molecular dynamics simulations were performed in order to simulate the process of the molecular-beam epitaxy for two-component Lennard-Jones systems. Two different cases are studied. In the first case, the atomic size of the incident beam is varied from that of the substrate and a systematic change in the overlayer growth pattern is studied. In the second case, the interaction parameter between the incident beam atom and the substrate atom is varied to study the two different crystal growth modes, i.e., the Stranski-Krastanov type and the Volmer-Weber type. A movie was made for the former case and efforts will be made to convey the dynamical aspects of the crystal growth, the very important information of the movie.

1. Introduction

The molecular-beam epitaxy (MBE) is regarded as a promising technique for synthesizing new artificial materials and has been applied extensively to semiconductor superlattices and metallic superlattices [1,2]. At the same time, MBE is a useful technique for studying the fundamental process of crystal growth. Several diffraction techniques are available for analyzing the dynamics of MBE and the structure of synthesized materials. Nevertheless, it is still fairly difficult to obtain reliable information of the local atomic arrangement and the dynamical aspect of the crystal growth in the atomistic scale. On the other hand, it is nowadays possible to perform large-scale computer simulations by using supercomputers in order to supplement the experimental situation. Some attempts[3-9] have already been made to simulate the MBE process by means of molecular dynamics (MD) technique, which is particularly suitable for studying the dynamical aspects.

In our previous paper[9], we presented our simulation results for two-component Lennard-Jones systems where the radius of the incident beam atom σ_b is different from that of the substrate atom σ_s. We demonstrated that the overlayer growth pattern changes systematically as a function of the atomic size mismatch defined by $\xi = (\sigma_b - \sigma_s)/\sigma_s$. We studied the cases ranging from -0.4 to +0.4 as ξ. Some interesting effects of the substrate temperature T_s were also observed. The details were discussed in ref. 9

In the present paper, we first discuss specifically some additional results for ξ = 0.0, 0.05 and 0.10 with T_s = 10 K, which are contained in our movie. We present several figures of atomic configurations in the intermediate process of MBE in order to convey the information from the movie. We then present some preliminary results for the case where the interaction between the incident beam atom and the substrate atom ε_{bs} is different from those between incident beam atoms ε_{bb} and those between substrate atoms ε_{ss}. This calculation was done to simulate the different modes of crystal growth known as the Stranski-Krastanov (SK) type and the Volmer-Weber (VW) type[10]. In this paper, we study only the case with $\varepsilon = \varepsilon_{bb} = \varepsilon_{ss}$. As expected, we observed the VW type crystal growth for $\varepsilon_{bs}/\varepsilon < 1$ and the SK type one for $\varepsilon_{bs}/\varepsilon > 1$. For sufficiently large $\varepsilon_{bs}/\varepsilon$ and relatively high substrate temperature T_s, the interface alloying effect was also observed.

2. Model and Methodology

The substrate is composed of three parts. The bottom part for fixing the whole system consists of fixed atoms, on which two or three atomic layers form the middle part, and is used to adjust the substrate temperature T_s by a periodic adjustment of the velocity of atoms there. The top-most layer is free to move and is exposed to incident beam atoms whose velocity is controlled by the Maxwell distribution corresponding to the beam temperature T_b. The beam atoms are injected to the system in a random-time sequence but on average every 100 time steps. (One time step is 10^{-14} sec.)

The atoms interact through the Lennard-Jones potential given as

$$V_{ij}(r) = 4\varepsilon_{ij}[(\sigma_{ij}/r)^{12} - (\sigma_{ij}/r)^{6}],$$

where the subscript i(j) takes s or b representing substrate or beam. Two different cases are studied. In the first case, we assume a common ε for all ε_{ij} and consider a variation of σ_{ij}. The atomic size mismatch parameter ξ is defined as $(\sigma_b - \sigma_s)/\sigma_s$ and the relation σ_{sb} $(= \sigma_{bs})$ $= (\sigma_b + \sigma_s)/2$ is assumed. In these equations, the double suffixes bb and ss are abbreviated as b and s, respectively. In the second case on the other hand, we assume a common σ $(=\sigma_s)$ for all σ_{ij} and consider a variation of ε_{ij}. In this paper, we consider the case where ε_{sb} $(=\varepsilon_{bs})$ is varied with a constraint $\varepsilon = \varepsilon_s = \varepsilon_b$. In all cases, the cutoff distance of the potential is set to be $2.06\sigma_s$. We set the parameters σ_s, ε, and the temperature to be suitable for solid argon: $\sigma_s = 3.40$Å, $\varepsilon = 0.0104$ eV, T_s = 10 or 30 K, and T_b = 100 K. Note that the melting temperature of argon is 83.95 K. However, as all the equations can be expressed in a nondimensional form, the results can be converted to other situations by scaling.

We take the x and y axes in the (001) plane and the z axis in the [001] direction. In the process of MD simulations, a periodic boundary condition is assumed in the x and y directions but the MD cell is open in the z direction.

3. Results and Discussions

3.1. Case 1: Atomic Size Mismatch

Details of the results for ξ ranging from -0.4 to 0.4 were presented in ref.9. Here we show some additional results for ξ =0.0, 0.05 and 0.10 with T_s = 10 K from our movie. In these cases, only the central $(n-4)\times(n-4)$ part of the $n\times n$ substrate surface is exposed to the incident beam. n = 20 for ξ = 0.0 and n = 16 for ξ = 0.05 and 0.10. In order to have a crude idea about the overlayer growth pattern, we first study the relative stability of the three patterns shown in Fig.1 as a function of ξ. The open circles denote the substrate atoms in the top-most surface and the closed ones the overlayer atoms. The adsorption energy per one overlayer atom in units of 4ε is shown in Fig.2. As the hollow site of the substrate is the most stable position for a single adsorbed atom, the growth of the square lattice

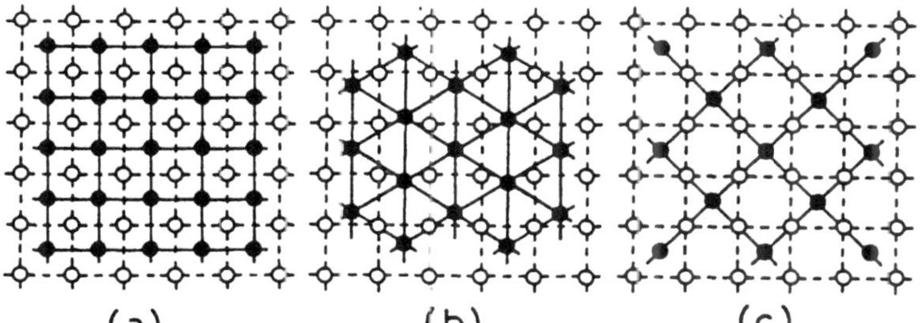

Fig.1 Three patterns are considered as the possible overlayer atomic arrangements for positive ξ. The open (closed) circles denote substrate (overlayer) atoms.

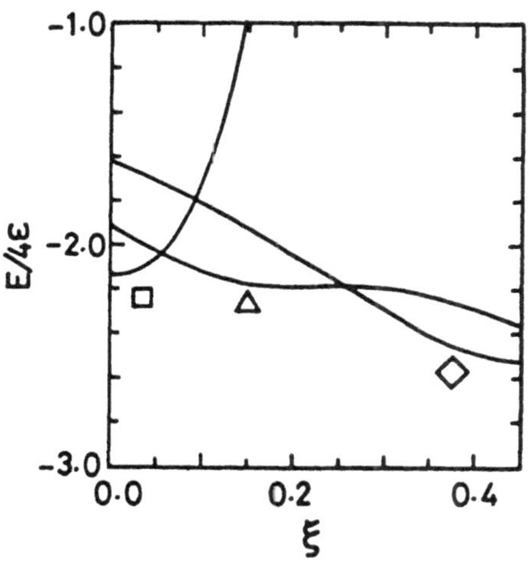

Fig.2 The ξ dependence of the adsorption energy for the three overlayer patterns shown in Fig.1. □ is for the simple square lattice, Δ for a triangular lattice and ◇ for $\sqrt{2}\times\sqrt{2}$ -R45°. The energy is for one overlayer atom in units of 4ε.

identical to the substrate one is a natural consequence for $\xi=0.0$. As ξ increases beyond about 0.06, a triangular lattice becomes more stable and then a super structure of $\sqrt{2}\times\sqrt{2}-$ R 45° is stabilized for $\xi>0.25$. Note however that the possible patterns are strictly limited in the above analysis and the realized one in our MD simulation becomes rather complicated for ξ near the phase boundary in Fig.2.

ξ = 0.0

The incident beam atom is identical to the substrate atom and a simple layer-by-layer growth is realized. Figure 3 shows the patterns of overlayer atoms in the process of MBE for n=20 and T_s=10 K. At this low substrate temperature, the incident beam atom can move over only a couple of lattice constants once it reaches the substrate surface and therefore each overlayer contains a significant number of vacancies.

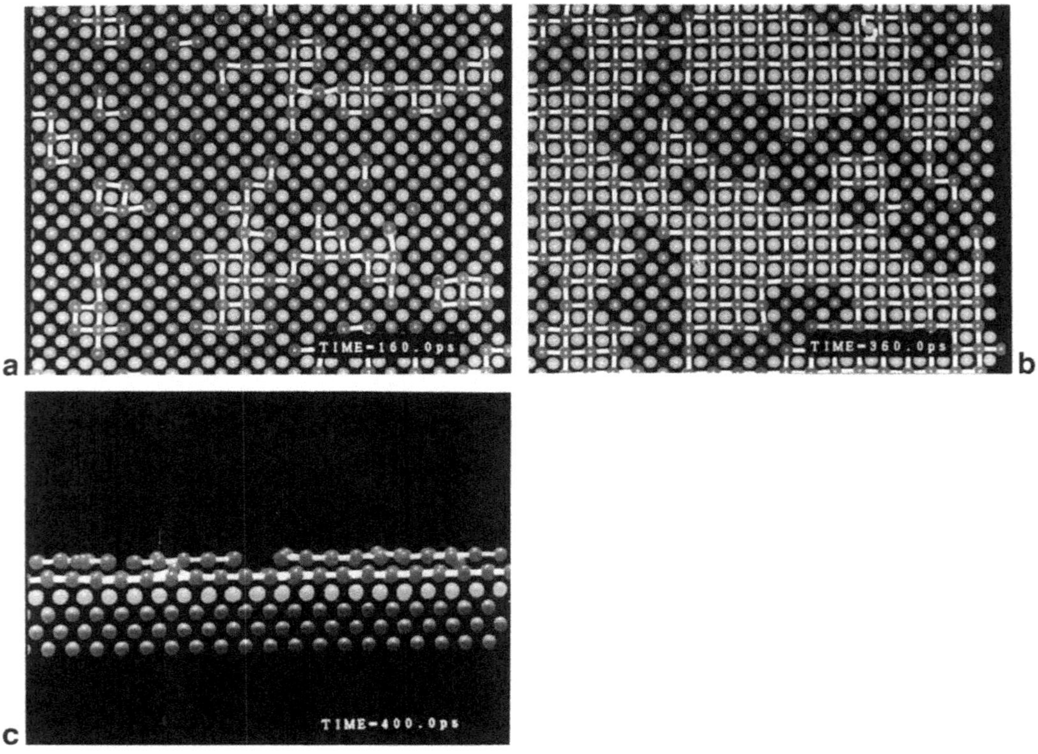

Fig.3 The simulation results for ξ =0.0 and Ts=10 K. The deposited atoms are connected by white bars when they are in the first-neighbor distance within a layer. There are some isolated deposited atoms, but they are hard to see. (a) and (b) are the top views of the first overlayer at 16,000th and 36,000th time step, respectively, and (c) a side view at 40,000th time step.

ξ = 0.05

As this value of ξ is close to the boundary of a square lattice and a triangular lattice as can be seen in Fig.2, the realized structure may contain the mixed aspects of the two phases. From the pattern at the final stage (Fig.4f), we see that the overlayer growth is essentially into the square lattice of the substrate surface. However, the atomic size mismatch causes a significant strain energy for the perfect commensurate structure and composite discommensuration lines (DCL) are introduced to reduce the overall strain energy. (The atomic arrangements associated with the composite as well as elemental DCL's are shown in Fig.5.)

The dynamical process of the DCL's formation can be clearly seen by our movie. To convey such information, we show in Fig.4 the time sequence of the growth pattern in the first overlayer for every 8000 time steps. Several short DCL's are formed in an early stage of MD simulation. As the layer becomes complete, these short DCL's are gradually organized into a smaller number of longer DCL's to end up with a single DCL running from one edge of the overlayer to its other

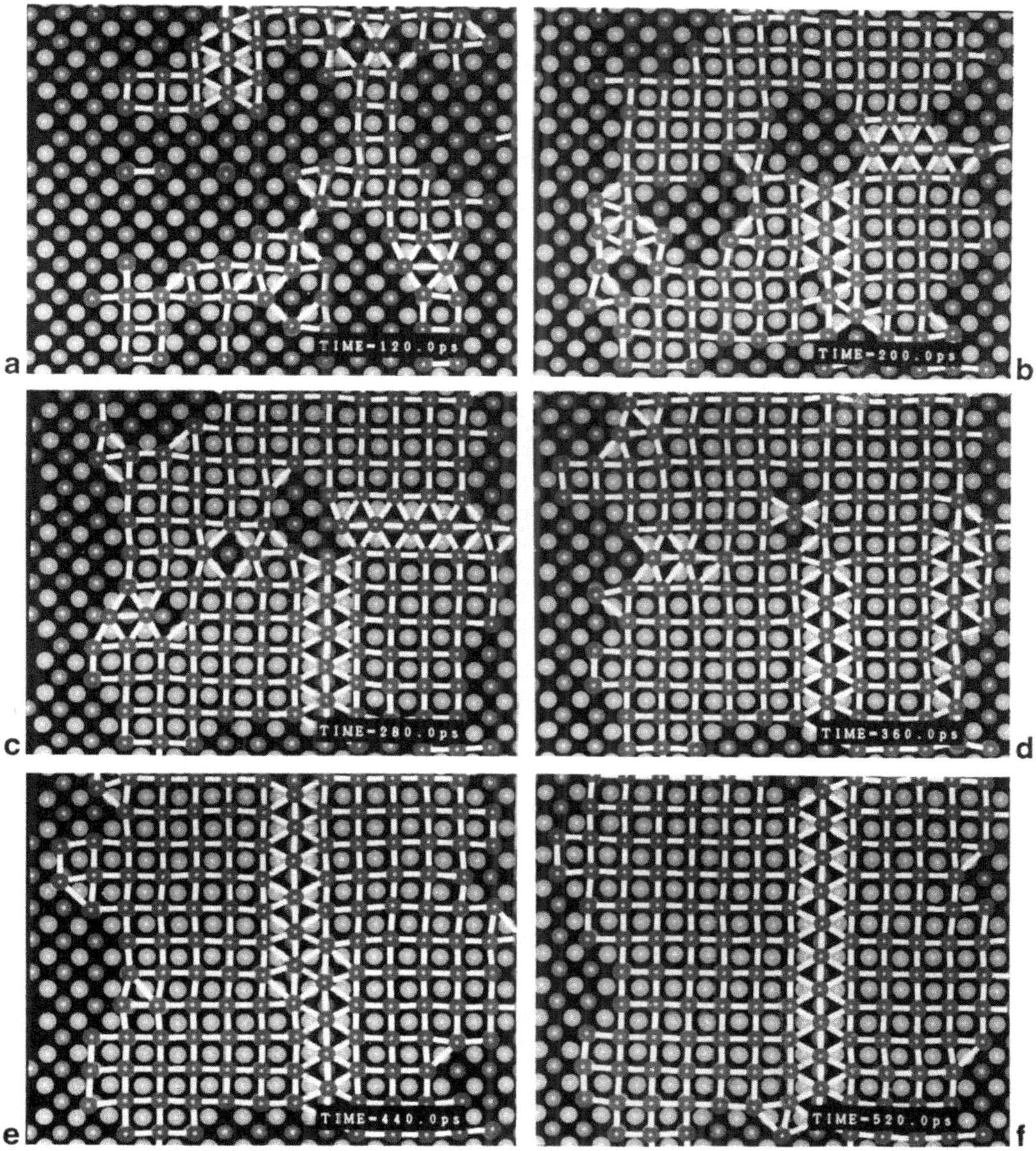

Fig.4 A time sequence of the growth pattern for $\xi=0.05$ in the first overlayer. Starting from the 12,000th time step (a) to 53,000th time step (f) with an interval of 8,000 time steps. $T_s=10$ K.

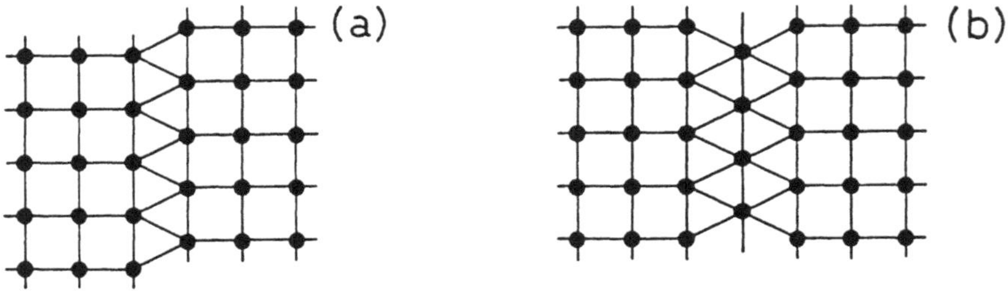

Fig.5 Schematic illustrations of the atomic arrangements associated with (a) an elemental discommensuration line and (b) a composite discommensuration line.

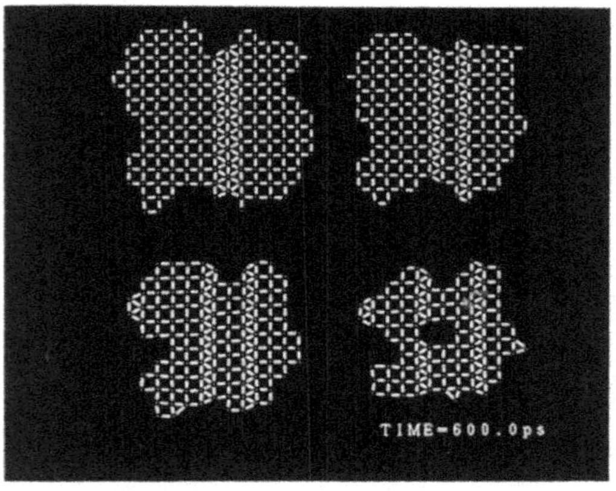

Fig.6 The simulation results for ξ =0.05 and T_s=10 K at the 6× 10^4th time step. Top views of the overlayers: the first layer at upper left, second at upper right, third at lower left and fourth at lower right. The white bars connect the neighboring atoms within each layer.

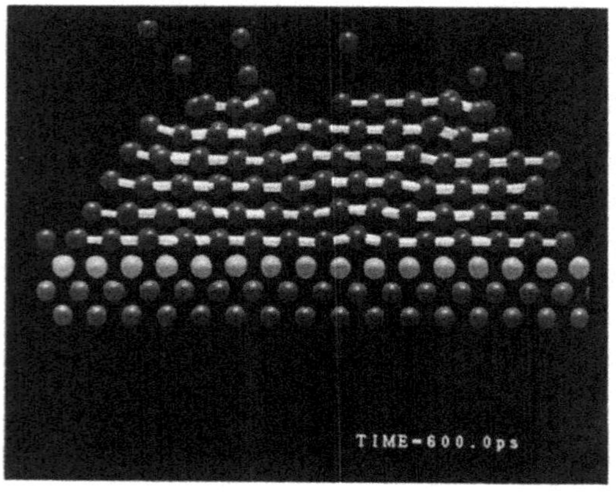

Fig.7 A side view of the results shown in Fig.6 from the (110) direction. The three layers from the bottom are the substrate, on which the deposited atoms form a well-ordered lattice. The seventh atom from the right in the first overlayer is located along the center of the composite discommensuration line and is shifted slightly upward.

edge as seen in Fig.4f. The atomic arrangements from the 1st to the 4th overlayer are summarized in Fig.6. It is interesting to see that the number of rows of square lattice separating the elemental DCL's increases for upper overlayers. A side view of the stacking pattern seen in the direction of DCL is shown in Fig.7. The atoms in the DCL and in the square lattice sandwiched by the elemental DCL's are located slightly higher in the z direction compared with other atoms.

The results shown in Figs.4-7 are for $n=16$. We performed MD simulations with $n=20$ and found that one DCL has a finite length with one atomic vacancy along it[9].

ξ = 0.10

As expected from Fig.2, the overlayer forms a triangular lattice. Figure 8 shows a time sequence of the growth pattern in the first overlayer for every 4,000 time steps. At a very early stage, the incident beam atoms are mostly trapped at the hollow site of the substrate surface. However, when only three or four atoms aggregate, they already form triangles locally. It is interesting to see in Fig.8 how a multi-domain structure in the intermediate coverage state is organized into a single-domain structure as the coverage saturates.

It was experimentally observed that the overlayer of Ag on Cu(001) surface forms a triangular lattice[11]. As the lattice constant of Ag metal is about 13% larger than that of Cu metal, this system may be an example to which the present simulation can be applied.

3.2. Case 2: Change in the Beam-Substrate Interaction

So far we discussed mostly the ordering pattern and the ordering process of overlayer atoms for different ξ, the atomic size mismatch parameter. Now we present some results of preliminary calculations concerning the two different crystal growth modes known as the Stranski-Krastanov (SK) type and the Volmer-Weber (VW) type[10]. To measure the change in the beam-substrate interaction, we introduce a parameter η defined as $\varepsilon_{bs}/\varepsilon$ with $\varepsilon = \varepsilon_{bb} = \varepsilon_{ss}$. Ten different cases, T_s = 10 K and 30 K for each of η = 0.4, 0.5, 2.0, 3.0 and 4.0, are studied, among which the results for η = 0.4, 0.5 and 3.0 are presented below. The system size is 12× 12 and the whole surface area is exposed to the incident beam. The calculations are performed up to the 4×10^4th time step, at which the 4th overlayer is partially formed.

η = 0.4

Figure 9a shows the atomic arrangement in the first overlayer next to the substrate surface for T_s = 10 K. A strong tendency of island formation produces a significant size of void near the center of the overlayer. This is a natural consequence of the weaker beam-substrate attractive interaction than the beam-beam interaction. As the substrate temperature T_s increases, the diffusion distance of the beam atoms on the substrate surface increases so that the system can approach its equilibrium state more easily. In the present case, the tendency of island formation becomes more obvious as seen in Fig.9b. Clearly the crystal growth of the VW type is realized for η = 0.4.

η = 0.5

The crystal growth is of the VW type in this case also. However, the atomic arrangement for η =0.5 (Fig.10) is significantly different from that for η = 0.4 (Fig.9). This may be interpreted in the following way. As η increases, the tendency of forming three-dimensional islands is reduced. This in turn enhances the size of two-dimensional islands on top of the substrate. For T_s = 30 K, the pattern of Fig.10b may be obtained from Fig.9b simply by enlarging the

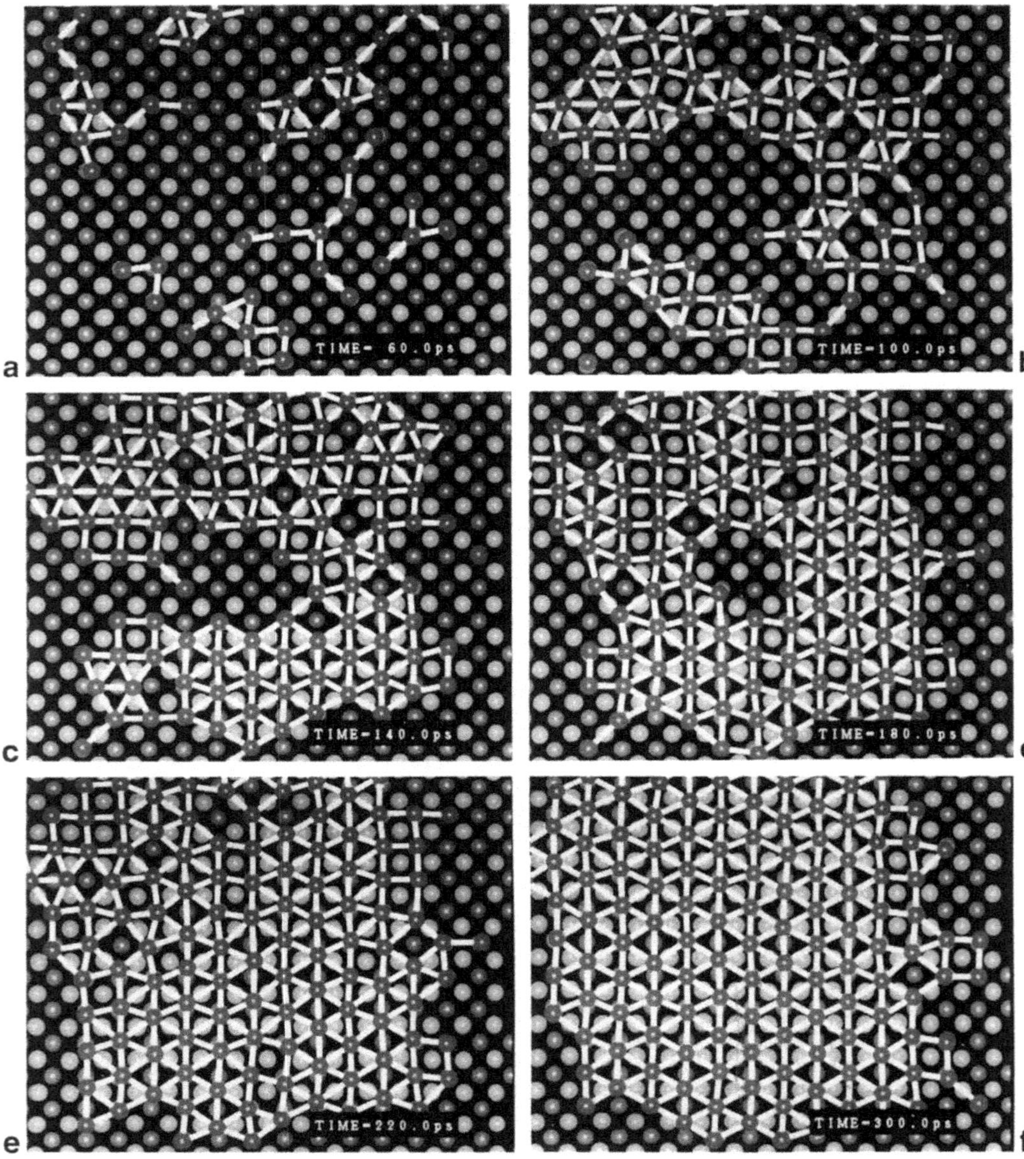

Fig.8 A time sequence of the growth pattern for ξ =0.10 in the first overlayer. Starting from the 6,000th time step (a) to 22,000th time step (e) with an interval of 4,000 time steps. (f) for the 30,000th time step. T_{w}=10 K.

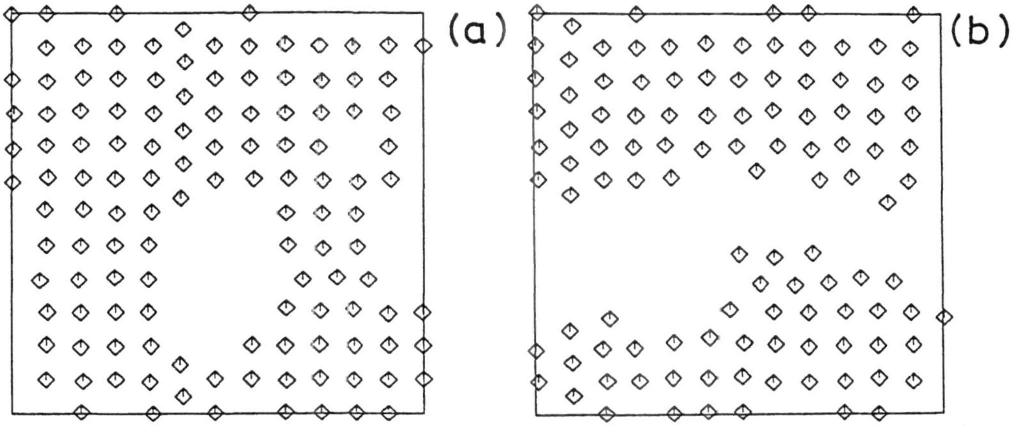

Fig.9 The simulation results for η =0.4 at the 4×10^4th time step. The first overlayer for (a) T_s=10 K and (b) T_s=30 K.

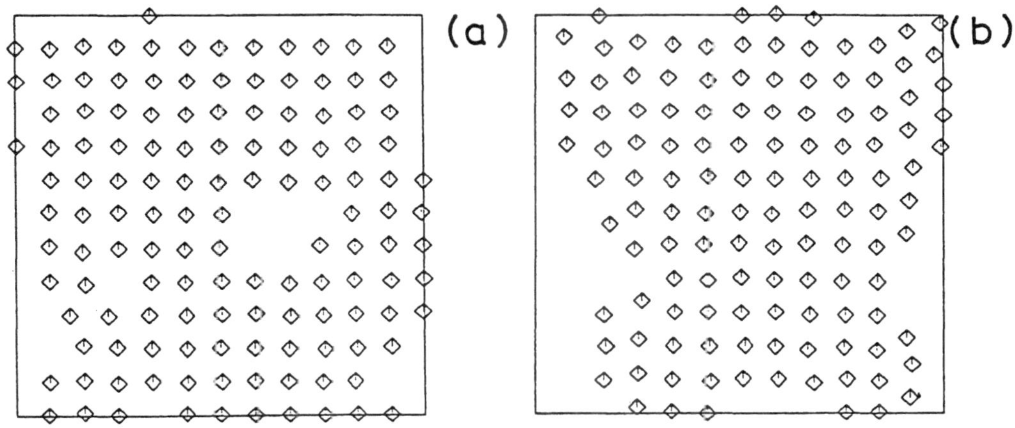

Fig.10 The simulation results for η =0.5 at the 4×10^4th time step. The first overlayer for (a) T_s=10 K and (b) T_s=30 K.

size of islands so that their tails in Fig. 9b overlap slightly. The fact that the pattern of Fig.10b is constricted in the middle may support this interpretation. The use of a larger system size is desirable.

η = 3.0

As the beam-substrate interaction is more attractive than the beam-beam interaction, the incident beam, after reaching the substrate surface, prefers to sit next to the substrate atoms. Therefore the overlayer covers the substrate surface uniformly. This is indeed the case as can be seen from Figs.11b and d and the crystal growth is of the SK type. It is also obvious that the interfacial alloying will further reduce the free energy of the system through both internal energy and entropy. For η =3.0 and T_s =30 K, the ratio of the substrate atoms mixing into the first overlayer is only 3.5%. This

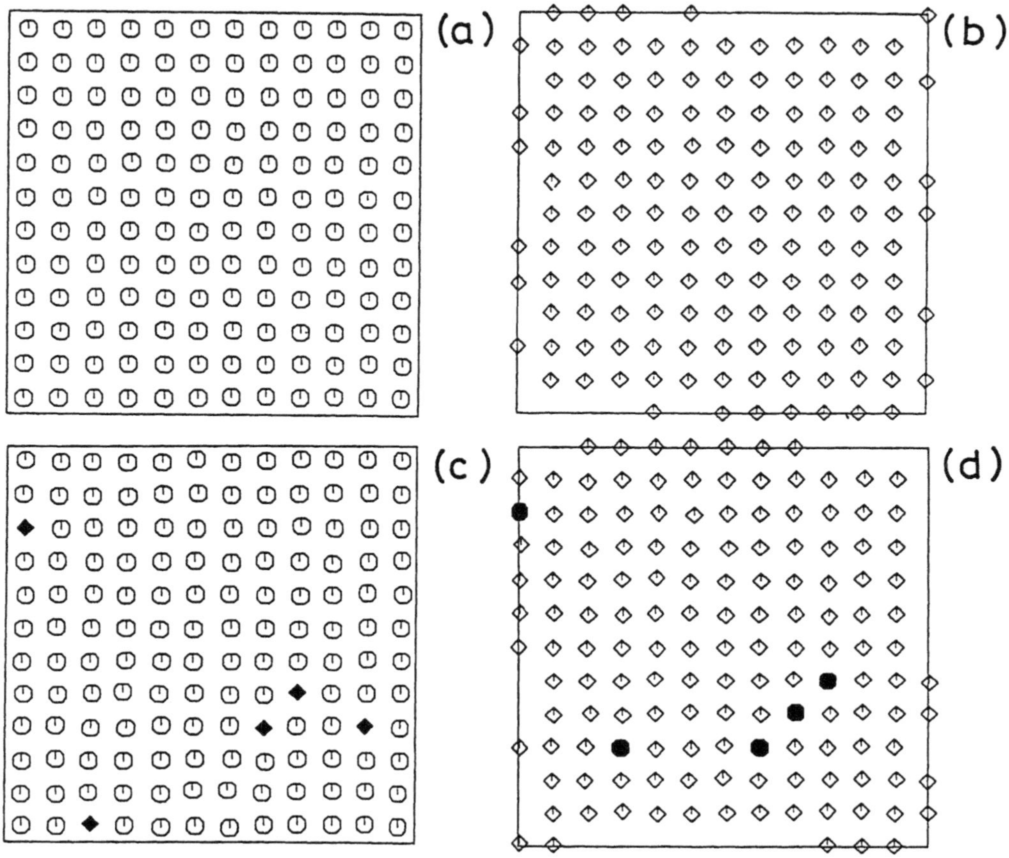

Fig.11 The simulation results for η =3.0 at the 4×10^4th time step. (a) the substrate surface and (b) the first overlayer for T_s=10 K. (c) the substrate surface and (d) the first overlayer for T_s=30 K. In (c), ◆ denotes the beam atoms mixing into the substrate and ● in (d) denotes the substrate atoms mixing into the first overlayer.

ratio increases up to 33% for η = 4.0 and T_s = 30 K. However, this can happen only at rather high T_s due to kinematical reasons (Figs.11c and d). Even for η =4.0, the interfacial alloying was not observed at T_s=10 K.

4. Concluding Remarks

We have performed molecular-dynamics simulations for the molecular-beam epitaxy for the two-component Lennard-Jones systems. Two different cases were considered. In the first one, the effect of the atomic size mismatch between the substrate atom and the incident beam atom was studied on the overlayer growth pattern. We had made a movie for some typical cases and tried in this paper to convey the information of the movie. In particular, dynamical aspects of forming the discommensuration lines for ξ =0.05 and the triangular lattice for ξ =0.10 were discussed in detail. In the second case, the interaction parameter between the substrate atom and the incident beam atom was artificially varied and the two crystal growth modes, the Stranski-Krastanov type and the Volmer-Weber type, were studied.

More detailed calculations for the second case are now in progress and the results will be published elsewhere[12]

References

1. T. Shinjo, Metallic Superlattices, edited by T. Shinjo and T. Takada (Elsevier, Amsterdam, 1987), p.1.
2. For example, some articles in Thin Film Growth Technique for Low-Dimensional Structures, edited by R. F. C. Farrow, S. S. P. Parkin, P. J. Dobson, J. H. Neave and A. S. Arrott (Plenum, New York, 1987).
3. M. Schneider, A. Rahman and I. K. Schuller, Phys. Rev. Lett. **55**, 604 (1985).
4. M. Schneider, A. Rahman and I. K. Schuller, Phys. Rev. B **34**, 1802 (1986).
5. M. Schneider, I. K. Schuller and A. Rahman, Phys. Rev. B **36**, 1340 (1987).
6. S. D. Sarma, S. M. Paik, K. E. Khor and A. Kobayashi, J. Vac. Sci. Technol. B **5**, 1179 (1987).
7. E. T. Gaulinski and J. D. Gunton, Phys. Rev. B **36**, 4774 (1987).
8. J. Lampinen, R. M. Nieminen and K. Kaski, Surf. Sci. **203**, 201 (1988)
9. K. Hara, M. Ikeda, O. Ohtsuki, K. Terakura, M. Mikami, Y. Tago and T. Oguchi, Phys. Rev. B **39**, 9476 (1989).
10. M.H. Grabow and G.H. Gilmer, Semiconductor-Based Heterostructures: Interfacial Structure and Stability, edited by M. L. Green, J. E. E. Baglin, G. Y. Chin, H. W. Deckman, W. Mayo and D. Narasinham (Metallurgical Society, Pensilvania, 1986), p.3.
11. J. G. Tobin, S. W. Robey, L. E. Klebanoff and D. A. Shirley, Phys. Rev. B **28**, 6169 (1988).
12. K. Hara, M. Ikeda, M. Nomura and K. Terakura, to be published.

Initial Stages of Oxidation of Al(111)

U. Yxklinten, M. Limbäck, and B.I. Lundqvist

Institute of Theoretical Physics, Chalmers University of Technology, S-41296 Göteborg, Sweden

Abstract. The initial stages of oxidation of aluminium have been elucidated by calculations within the effective medium theory: (i) An ordered chemisorbed oxygen overlayer is unstable with respect to states with oxygen occupying subsurface sites; (ii) There is no island formation of oxygen on the Al(111) surface; (iii) For subsurface oxygen it is equivalent whether the oxygen atoms form small islands or not.

1. Introduction

The initial stages of oxidation of aluminium involve a large number of configurations of Al and O atoms. Many sensible questions can be raised, for instance the following ones:

Which sites are energetically most stable, the overlayer chemisorption sites or the subsurface ones? Common belief, supported by both experiments and theory, is that the preferred chemisorption site is the fcc threefold hollow site [1-5]. So far [6], however, it has been hard to distinguish whether this fcc site is energetically favoured compared with the hcp threefold hollow site.

How big is the penetration barrier for an oxygen atom going from the chemisorption to the subsurface site? Available theoretical estimates of the barrier place it in the range of 0.6-3.0 eV [7-9].

Is a stable monolayer of oxygen formed on the surface before subsurface sites are occupied? The question is far from settled [3,4,10].

In this brief note [11] attempts to answer these questions within the effective medium theory [12] are presented. The calculations for both a single oxygen and an oxygen overlayer on the surface show the hcp threefold hollow site to be as stable as the fcc one. The penetration barrier is found to lie in the lower region of the mentioned range, since relaxation among the surface Al atoms is significant. An ordered oxygen overlayer is shown to be unstable towards subsurface site occupation.

2. The effective medium approach

In the version of the effective medium theory, EMT, with interatomic forces [12], there are two types of terms contributing to the total energy: (i) the cohesive energy, $E_c(n)$, which depends upon the electron density that each atom senses, and

Springer Proceedings in Physics, Vol. 48 — Many-Atom Interactions in Solids
Editors: R.M. Nieminen · M.J. Puska · M.J. Manninen

(ii) the atomic sphere correction energy, E_{as}, which is an electrostatic correction that comes from the overlap of the neutral spheres. The total energy is then given by

$$E_{tot} = \Sigma E_{c,Al}(n_i) + \Sigma E_{c,O}(n_j) + \Sigma E_{as,Al-Al} + \Sigma E_{as,Al-O} \,,$$

where $E_{c,Al}(n_i)$ is the cohesive energy for an Al atom and $E_{c,O}(n_j)$ that for an O atom. The E_{as}-terms correspond to different atomic sphere corrections.

3. Chemisorbed contra subsurface oxygen

3.1 Chemisorption

The binding energy per O atom is almost constant for small ordered coverages of oxygen. This can be explained in terms of compensation of the loss of bonds among the surface Al atoms. When a (111) surface is created by cutting a bulk piece the surface atoms lose three nearest neighbors, and the outer layer relaxes inwards to compensate for the lost bonds. As O atoms chemisorb on the surface the nearest neighbor surface Al atoms move out towards the O atom, i.e. they weaken their bonds with the second layer due to the bonds formed with the O atoms in the EMT, the outward relaxation is found to be larger than the contraction for the clean surface, as illustrated in fig. 1. This can be explained by the fact that the Al-O bond is stronger than the Al-Al bond. At a certain oxygen coverage the lost Al bonds are compensated, and adding more oxygen does not lower the energy of the surface Al atoms any further, fig. 2.

As the oxygen content increases on the surface, the binding energy per chemisorbed oxygen atom decreases. There is an effective **repulsion** between

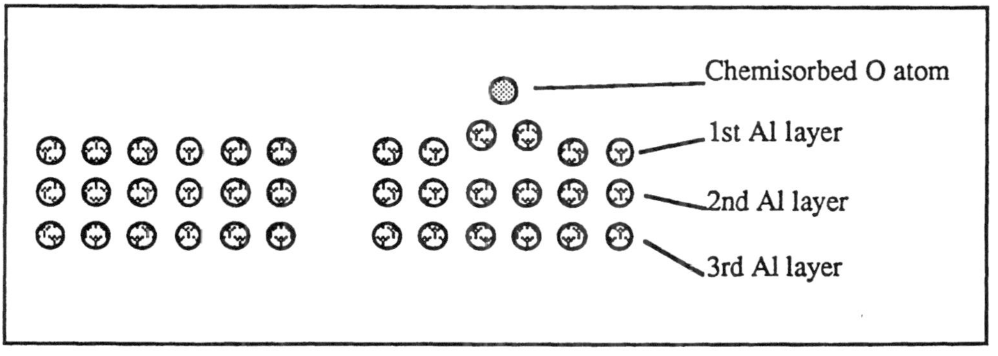

Figure 1. Schematic picture showing the clean surface and the corrugation of the Al surface as an O atom is chemisorbed. The inward relaxation for the clean surface is 1% of the interplanar distance. The outward relaxation of the Al atoms nearest the O atom is ~0.2 bohr, which correspond to 5% of the interplanar distance.

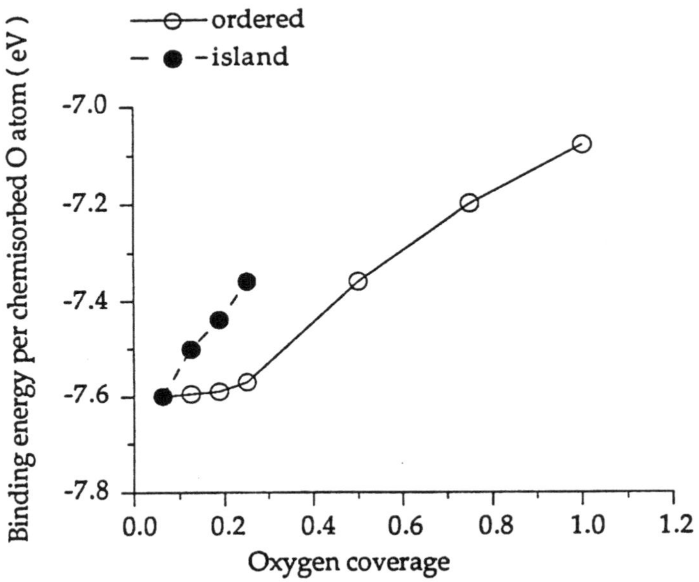

Figure 2. The binding energy per chemisorbed O atom on the Al(111) surface, as a function of oxygen coverage, θ. The graph contains two curves, one for island formation and one where the oxygen coverage is uniformly spread out in an ordered way. The O atoms are chemisorbed in the fcc threefold hollow sites. Note that binding energy calculations based on the local-density approximation in the density-functional theory systematically give too high binding energy values for oxygen [13].

oxygen atoms on the surface, as shown by calculating the difference in energy between configurations with the O atoms uniformly spread out on the surface contra small oxygen islands formation, fig. 2.

3.2 Penetration barrier

The barrier for oxygen penetration through the surface Al layer increases with increasing oxygen coverage on the surface. Oxygen experiences, when penetrating through the fcc threefold hollow site, a barrier in the interval 0.7-1.0 eV. A reported value of 0.63 eV [14] is in good agreement with our value. An experimental barrier height is 0.90 eV [15].

The increase in barrier height with increasing oxygen coverage can be explained in simple terms. For low ordered coverages the surface Al triangle rather easily opens up to let the penetrating O atom through. As the coverage increases the other chemisorbed O atoms stiffen up the surface so that the Al triangle will find it harder to expand. Therefore the barrier increases with increasing oxygen coverage.

3.3 Instability of a single monolayer of oxygen on the Al(111) surface

To test if a stable monolayer of oxygen might form on the Al(111) surface a series of calculations have been performed of total energies for different configurations; the oxygen chemisorption site has been kept as the fcc threefold hollow site; the subsurface site is the modified octahedral site.

The gain in energy for an O atom going from the chemisorption to the subsurface site is always found to be positive, going from ~0.1 to ~0.5 eV as the surface coverage, θ, goes from 0 to 1.

4. Subsurface island formation

Having shown that subsurface oxygen is energetically favoured, it is natural to ask whether the subsurface oxygen atoms cluster into islands or are evenly spread out.

Let us consider subsurface island formation. The energetically most favourable subsurface site, according to the EMT, is the modified octahedral site. This is in good agreement with findings of other groups [8,16,17]. We will compare energies of configurations where on one hand the subsurface oxygen is spread out in an ordered way, e.g. an ordered (2x2) subsurface coverage, and on the other where the subsurface oxygen have formed small islands. In fig. 3 the calculated binding energy per subsurface O atom is shown as a function of island size.

The subsurface binding energy, defined as the subsurface binding energy divided by the total number of subsurface oxygen atoms, decreases with increasing oxygen subsurface coverage. The decrease is relatively weak so neither island formation nor an ordered structure for subsurface O atoms is favoured.

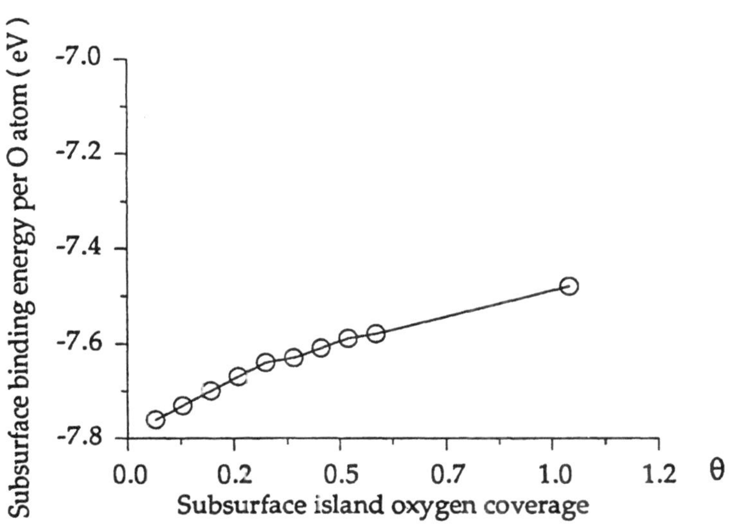

Figure 3. Binding energy per subsurface O atom, in the modified octahedral site, as a function of island size.

5. Discussion

There is a coexistence of chemisorbed and subsurface oxygen on the Al(111) surface. This conclusion is based on our finding that subsurface oxygen, independently of the chemisorbed oxygen coverage, is energetically more stable than chemisorbed oxygen. In a similar study of the O/Cu(111) system [18] an effective repulsion between chemisorbed oxygen atoms also has been found. There is, however, a clear difference between these two systems. For O/Cu(111) an effective attraction among the subsurface atoms is found [18], i.e. the binding energy per subsurface oxygen atom increases with increasing subsurface coverage. This is not the case for the O/Al(111) system, therefore no tendency among the subsurface oxygen atoms to form islands is found. Island formation is found to be as likely as an ordered homogeneous distribution among the subsurface oxygen atoms.

Acknowledgment

Valuable support from the National Swedish Board for Technical Development (STU) is gratefully acknowledged. Many thanks to Karsten Jacobsen for interesting discussions.

References

1 C.W.B. Martinson and S.A. Flodström, Sol. State Comm. **30**, 671 (1979)
2 W. Eberhardt and F.J. Himpsel, Phys.Rev. Lett. **42**, 1375 (1979)
3 R.Z. Bachrach, G.V. Hansson and R.S. Bauer, Surf. Sci. **109**, L560 (1981)
4 V. Martinez, F. Soria, M.C. Muñoz and L.C. Sacedon, Surf. Sci. **128**, 424 (1983)
5 R.L. Strong, B. Firey, F.W. deWette and J.L. Erskine, Phys. Rev. **B26**, 3483 (1982)
6 D.R. Salahub, L.C. Niem and M. Roche, Surf. Sci. **100**, 199 (1980)
7 K.Y. Yu, J.N. Miller, P. Chye, W.E. Spicer, N.D. Lang and A.R. Williams, Phys. Rev. **B14**, 1446 (1976)
8 B.N. Cox and C.W. Bauschlicher, Surf. Sci. **115**, 15 (1982)
9 J.O.A. Idiodi and B.I. Lundqvist, Institute Report, CTH
10 I.P. Batra and O. Bisi, Surf.Sci. **123**, 283 (1982)
11 A more complete report will be published elsewhere, (U. Yxklinten, to be published)
12 K.W. Jacobsen, J.K. Nørskov and M.J. Puska, Phys. Rev. **B35**, 7423 (1987)
13 R.O. Jones, J. Chem. Phys. **82**, 325 (1985)
14 Y. Wang, P. Nordlander, S.Wei and M. Ronay, Defect and Diffusion Forum Vol.61, 1 (1988)
15 F.P. Fehlner, J. Electrochem. Soc. **129**, 2369 (1982)
16 R.L. Strong and J.L. Erskine, J. Vac. Sci. Tech. **A3**, 1428 (1985)
17 D.M. Bylander and L. Kleinman, Phys. Rev. **B28**, 523 (1983)/**B30**, 2997 (1984)
18 K.W. Jacobsen, "Bonding in metallic systems, an effective medium approach", Nordita 1987, Copenhagen, Denmark

Index of Contributors

Ackland, G. 249

Barnett, R.N. 103
von Boehm, J. 191

Carlsson, A.E. 257

Daw, M.S. 48

Goodwin, L. 242

Hafner, J. 12,144
Häkkinen, H. 168
Hara, K. 303
Heine, V. 12
van Himbergen, J.E. 232
Hohl, D. 214

Ikeda, M. 303

Jacobsen, K.W. 34
Johnson, R.A. 85
Jones, R.O. 214

Laasonen, K. 264
Landman, U. 103
Limbäck, M. 314
Luedtke, W.D. 103
Lundqvist, B.I. 314

Manninen, M. 2,199
Mikami, M. 303
Moriarty, J.A. 158
Myllyrinne, S. 199

Nieminen, R.M. 2,191,204,264
Nomura, M. 303
Nørskov, J.K. 199

Oguchi, T. 303

Pettifor, D.G. 64
Puska, M.J. 2,134

Skinner, A.J. 242
Stoltze, P. 281

Terakura, K. 303
Thetford, R. 176
Tichy, G. 274

Virkkunen, R. 204

Wahnström, G. 291

Yxklinten, U. 314

Zwartkruis, A. 232

K. Binder, University of Mainz; **D.W. Heermann**, University of Wuppertal

Monte Carlo Simulation in Statistical Physics

An Introduction

1988. VIII, 127 pp. 34 figs. (Springer Series in Solid-State Sciences, Vol. 80) Hardcover DM 49,- ISBN 3-540-19107-0

Contents: Introduction: Purpose and Scope of This Volume, and Some General Comments.- Theoretical Foundations of the Monte Carlo Method and Its Applications in Statistical Physics.- Guide to Practical Work with the Monte Carlo Method.- Appendix.- References.- Subject Index.

D.W. Heermann, University of Heidelberg

Computer Simulation Methods

in Theoretical Physics

2nd ed. 1990. XI, 145 pp. 30 figs. Softcover DM 39,- ISBN 3-540-52210-7

Contents: Introductory Examples.- Computer Simulation Methods.- Deterministic Methods.- Stochastic Methods.- Appendices.- References.- Subject Index.

www.ingramcontent.com/pod-product-compliance
Ingram Content Group UK Ltd.
Pitfield, Milton Keynes, MK11 3LW, UK
UKHW051449290926
15471UKWH00008B/1106
* 9 7 8 3 6 4 2 7 5 7 5 5 6 *